Welding - Modern Topics

Edited by Sadek Crisóstomo Absi Alfaro, Wojciech Borek and Błażej Tomiczek

Published in London, United Kingdom

IntechOpen

Supporting open minds since 2005

Welding - Modern Topics
http://dx.doi.org/10.5772/intechopen.83204
Edited by Sadek Crisóstomo Absi Alfaro, Wojciech Borek and Błażej Tomiczek

Contributors
Gaurav Bhargava, Guillermo Alvarez Bestard, Rogfel Thompson Martinez, Sadek Crisostomo Absi C. Absi Alfaro, Elina Mylen Montero Puñales, Nageswara Rao Muktinutalapati, Geetha Manivasagam, Sudhagara Rajan S, Jithin Vishnu, Carlos Capdevila, David San Martin, Javier Vivas, Francisca G. Caballero, Karel Dušek, David Bušek, Petr Veselý, Nicky Kisku, Svitlana Maksymova, Eduardo De Magalhães Braga, Lino A. S. Rodrigues, Ednelson Da Silva Costa, Tarcio Dos Santos Cabral, Stefan Lucian Toma, Daniela Lucia Chicet, Radu Armand Haraga, Viorel Paleu, Costica Bejinariu

First published in London, United Kingdom, 2021 by IntechOpen
IntechOpen is the global imprint of INTECHOPEN LIMITED, registered in England and Wales, registration number: 11086078, 5 Princes Gate Court, London, SW7 2QJ, United Kingdom
Printed in Croatia

British Library Cataloguing-in-Publication Data
A catalogue record for this book is available from the British Library

Additional hard and PDF copies can be obtained from orders@intechopen.com

Welding - Modern Topics
Edited by Sadek Crisóstomo Absi Alfaro, Wojciech Borek and Błażej Tomiczek
p. cm.
Print ISBN 978-1-83881-895-1
Online ISBN 978-1-83881-896-8
eBook (PDF) ISBN 978-1-83881-897-5

For EU product safety concerns:
IN TECH d.o.o., Prolaz Marije Krucifikse Kozulić 3, 51000 Rijeka, Croatia, info@intechopen.com or visit our website at intechopen.com.

Meet the editors

Dr. Sadek Crisóstomo Absi Alfaro obtained a degree in Mechanical Engineering from the Federal University of Rio Grande do Sul (UFRGS), Brazil, in 1979; a master's degree in Metallurgical Engineering from the Federal University of Minas Gerai (UFMG), Brazil, in 1983; and a Ph.D. and postdoctoral degree in Welding Engineering from SIMS Cranfield University, UK, in 1989 and 1990, respectively. Since 1983, Dr. Sadek Crisostomo Absi Alfaro has been a lecturer at Brasilia University. From 2003 to 2017, he was Full Professor in the Mechanical/Mechatronics Engineering Department, where he developed and coordinated research and development projects for undergraduate and postgraduate studies. Currently, he is a senior researcher at Brasilia University. He is a member of several national and international professional associations and the founder of the Automation and Control Group in Manufacturing Processes (GRACO) and the Mechatronics Engineering course at the University of Brasília. Dr. Sadek Crisostomo Absi Alfaro has experience in the areas of mechanical and mechatronics engineering, with emphasis on advanced manufacturing processes including welding, automation, and sensing and experimental statistics.

Dr. Wojciech Borek, Ph.D. and MSc in Engineering, is an assistant professor in the Department of Engineering Materials and Biomaterials at the Silesian University of Technology in Gliwice, Poland, and a member of the association of Polish Cluster of Innovative Forging Technologies "HEFAJSTOS." His scientific interests include materials science, heat treatment, thermomechanical treatment, plastic deformations, and Gleeble simulations, especially Gleeble welding simulation. He is also a specialist in steels, stainless steel, high-manganese austenitic steels, and light metal alloys. He is an author and coauthor of around 100 scientific publications worldwide including ten chapters in books and more than thirty publications in the Web of Science database. He has won ten national and international awards and honors. Dr. Borek has served, or is currently serving, as a contractor for more than eight research and didactic projects in Poland and abroad, a reviewer of numerous scientific publications, and co-promoter of two doctoral dissertations.

Dr. Błażej Tomiczek, Ph.D. and MSc in Engineering, is an assistant professor in the Laboratory of Nanotechnology and Materials Technologies at the Faculty of Mechanical Engineering of the Silesian University of Technology in Gliwice, Poland. His scientific interests include materials science, metal matrix composites, powder metallurgy, heat treatment, laser sintering and joining methods, mechanical alloying, and pressure infiltration techniques. He is an author and co-author of around forty scientific publications worldwide, including two book chapters and more than twenty papers in the Web of Science database. He has won more than fifteen national and international awards and honors. He has experience as a manager and contractor for more than ten research projects and a co-supervisor of two doctoral dissertations.

Contents

Preface

The book is divided into three parts and twelve chapters. Section One addresses automation and control in welding. Chapter 1 presents a bibliographical review of the scientific literature related to qualitative and quantitative indexes to evaluate the stability of the Gas Metal Arc Welding (GMAW) process. It also examines the factors that affect stability and stability indexes. Chapter 2 introduces the reader to how data mining processes, machine learning, deep learning, and reinforcement learning techniques have had good results in the analysis and control of systems as complex as the welding process. Chapter 3 shows how different sensing, modelling, estimation, and control techniques are used to overcome the challenge of welding bead geometry on-line control. Chapter 4 discusses selected sensing techniques and estimation algorithms used on-line for monitoring and collecting values on the welding parameter process. Special attention is given to sensor fusion techniques. Some real applications and innovative research results are also discussed.

Section Two illustrates some modern welding processes. Chapter 5 looks at the results of metallographic and micro X-ray spectral analysis investigations of dissimilar brazed joints of molybdenum–stainless steel and shows the features of formation of brazed seams at the application of brazing filler metals of a Cu-Mn-Ni(Me) system. Chapter 6 lists some examples of errors that can occur in soldering, as well as describes selected defects such as non-wettability of the solder pads, dewetting, wrong solder mask design, warpage, head-in-pillow, cracks in the joints, pad cratering, black pad, solder beading, tombstoning, dendrites, voids, flux spattering from the solder paste, popcorning, and whiskers. Chapter 7 presents the technology for obtaining ultra-hard layers based on tungsten carbide (WC) and titanium carbide (TiC) by the arc spraying process, using a classic spray device equipped with a conical nozzle system and tubular wire additional material containing ultra-hard compounds (WC, TiC).

Section Three addresses some important thermal-mechanical treatments related to welding quality. Chapter 8 describes how to creep strength at high temperature could be improved by a microstructural optimization through nano-precipitation, guided by computational thermodynamics and thermomechanical control process optimization. Chapter 9 presents a heat treatment route as an important route for the development of high-strength alloy steel. Many heat treatment processes are applied depending on alloy compositions and desired mechanical properties. Chapter 10 categorizes heat treatment of metastable beta titanium alloys into two steps: solution treatment in beta or alpha+beta phase field and ageing at appropriate lower temperatures. The chapter pays special attention to heat treatment of beta titanium alloys for biomedical applications because of the growing interest in this class of alloys. Chapter 11 examines the important role that grain boundary design plays in achieving required end mechanical properties in the final product form: hot rolled or cold rolled coil. Grain refinement in steel design is a particularly attractive strengthening mechanism, as it benefits both fracture toughness and mechanical behaviour at lower temperatures, particularly in the case of high strength and line pipe steels. Chapter 12 presents a contextualized approach to destructive and non-destructive techniques used to measure residual stresses (RS) generated by arc welding. It also discusses the influence of distortion and stress on welded structures and presents possible control techniques.

Sadek Crisóstomo Absi Alfaro
Mechanical Engineering Department,
University of Brasilia,
Brasilia, Brazil

Wojciech Borek
Department of Engineering Materials and Biomaterials,
Faculty of Mechanical Engineering,
Silesian University of Technology,
Gliwice, Poland

Błażej Tomiczek
Laboratory of Nanotechnology and Materials Technologies,
Faculty of Mechanical Engineering,
Silesian University of Technology,
Gliwice, Poland

Section 1

Automation and Control in Welding

Chapter 1

Stability on the GMAW Process

Elina Mylen Montero Puñales
and Sadek Crisóstomo Absi Alfaro

Abstract

The gas metal arc welding (GMAW) process is highly used in industrial production; therefore great efforts are made to select the appropriate procedure to ensure the highest quality. An area of study directly correlated to the quality of GMAW and widely studied is the control of process stability. The objective of this chapter is to present a bibliographical review of the scientific literature related to qualitative and quantitative indexes to evaluate the stability of the GMAW process. The documents present a compilation of the factors that affect stability, stability indexes, and, finally, a synthesis of the study. With a review of the literature, it was concluded that the highest percentage of investigation was aimed at the study of metal transfer stability, specifically with the short-circuit transfer mode. It is also evident that the main processing techniques to develop the indexes were the mathematical formulation; the statistical analysis; image processing; and monitoring of acoustic signals. In this text, the discussion surrounds the papers, the thesis, and other documents found on the theme.

Keywords: welding, GMAW, quality index, GMAW, process stability

1. Introduction

GMAW as a welding process presents a high degree of production, reliability, and automation capacity. With the appropriate parameter configuration, it allows welding in almost all positions and with almost all existing metal alloys. One peculiarity of the GMAW process is that, depending on the intensity of the current and voltage, different types of metal transfer can be observed. The metal transfer mode characterizes the way molten metal is deposited. The three main modes of metal transfer are short circuit, globular, and spray. The most relevant parameters involved in the process can be mentioned: amperage, voltage, welding speed, and stick out. Another peculiarity is that the process can be defined as chaotic; involves the interaction of several nonlinear welding variables; and presents a stochastic behavior. Therefore, great efforts are made to select the appropriate procedure to ensure the highest quality.

Quality can be defined as the union of a client's requirements with respect to a product. In the particular case of welding, the main objective is to get a weld bead as close as possible to the requirements. The welding quality can be monitored in two moments: online while the process is running and offline after the welding bead is obtained.

The offline evaluation considered geometric factors such as proper penetration, reinforcement, and the length of the pieces. Destructive tests can be carried out and consist of taking samples of weldments to evaluate the metallic continuity, mechanical strength, and other determining factors for the correct performance in service. Sometimes these tests lead to the destruction of the body tested. On the other hand, Wu et al. [1] affirm that online quality control allows the saving of financial resources through the reduction of defects in the production line. For this purpose, sensors for visual imaging, sound acquisition, infrared cameras, and ultrasonic sensing methods have been implemented.

One concept that is strongly correlated to the online quality is the control of the process stability. According to Ponomarev [2], the stability of the GMAW process is evaluated online by three factors: metallic transfer regularity, arc stability, and the operational behavior of the welding process. Meneses [3] also ensures that the higher the transfer stability, the higher the penetration and the lesser the amount of spatter.

The objective of this work is to present a bibliographical review of the scientific literature related to weld quality evaluation, focused mainly on those studies that present qualitative and quantitative indexes to evaluate the stability of the GMAW process. The chapter is structured as follows: Section 2 discusses Stability Control in the GMAW process; Section 2.1 discusses GMAW process operation; Section 2.2 discusses factors that affect stability; Section 2.3 presents a Summary of Stability index; and finally, Section 3 reveals a synthesis of the study and future research directions.

2. Stability control in the GMAW process

2.1 GMAW process operation

GMAW process is characterized by producing an arc between a consumable electrode that is constantly fed, a protective gas, and the piece to be welded, as represented in **Figure 1**.

Conductor tube: It is a welding torch component device and fulfills the function of guiding the gas flow in the welding process.

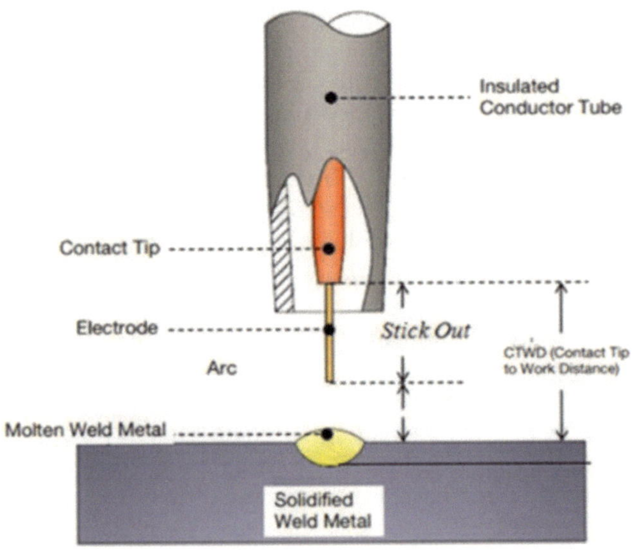

Figure 1.
Basic diagram of the MIG/MAG process (modified from [4]).

Contact tip: It is a torch device that has the function of guiding and supplying voltage to the wire.

Electrode: It is the consumable copper-coated steel electrode that melts with the electric arc and transfers to the melting pool.

Workpiece: composed of the metal bodies to be joined by the weld.

CTWD (contact tip to work distance): It is often confused with the distance between the contact tip and the work piece, which coincides when the nozzle front cut is also the same as the contact tip front cut.

Stick out: It is the length of free wire after it has passed through the contact tip.

The gas composition aims to stabilize the arc and protect the welding material from atmospheric gases such as nitrogen and oxygen, which can cause fusion defects, porosity, and weld metal embrittlement if they come in contact with the electrode, the arc, or the welding metal. Depending whether the gas is inert (Ar or He) or active (CO_2, or mixtures including N_2 or O_2), it can sbe classified as metal active gas (MAG) or metal inert gas (MIG).

The weld bead geometry depends directly to the parameters that govern the process. **Figure 2** outlines these geometric parameters in the cross section of a weld bead. The most important parameters affecting penetration and geometry in the GMAW process are welding current, arc voltage, torch travel speed or welding speed, stick out, torch tilt, and the diameter of the electrode.

According to [6] the process parameters of GMAW can be divided into five basic groups (as shown in **Figure 3**):

- Fixed, that cannot be modified by the operator and it is defined in the process design.
- Adjustable online, that can be modified during the process.
- Adjustable offline, that can be modified only before starting the process.
- Quantifiable online, that is measurable during the process.
- Quantifiable offline, that is measurable only after the process ended.

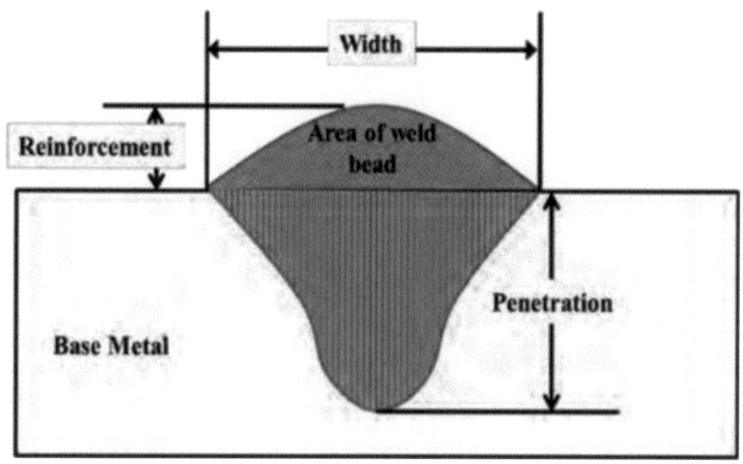

Figure 2.
Weld bead geometric characteristics [5].

Fixed

- Shielding gas composition.
- Position of the piece or pipe.
- Electrode characteristics (Ø, composition material, among others)
- Thickness of the base metal.
- Physic properties of the base metal.
- Polarity of the weld electric current.

Adjustable online

- Welding voltage.
- Inductance of the welding power source.
- Wire-feeding speed, wire speed or wire feed rate.
- Welding speed or torch speed
- Orbital angle (torch angle relative to pipe center)
- Gun angle or attack angle (torch angle side view relative to base metal surface)
- Work angle (torch angle end view relative to base metal surface)
- Contact tip to work distance (CTWD)
- Electrical stick out (ESO) or electrode extension.
- Nozzle to plate distance.
- Shielding gas flow rate. *2

Adjustable offline

- Inductance of the welding power source.
- Shielding gas flow rate.

Quantifiable online

- Welding voltage.
- Welding electric current intensity.
- Width of weld bead or bead width.
- Weld buildup or reinforcement of the weld bead.

Quantifiable offline

- Weld bead depth, depth of penetration or weld depth of fusion.
- Width of weld bead or bead width.
- Weld buildup, reinforcement of weld bead or bead reinforcement.

Figure 3.
Classifications of GMAW parameters [6].

2.2 Factors that affect the stability

The operational behavior has a big influence on stability. If the gas is not supplied accurately, the arc may not initialize, there would be no stable or continuous plasma ionization, and the protective effect will be affected; nitrogen, oxygen, and water vapor enter the welding region and directly contact with the arc and melting metals, reducing the arc stability and forming a variety of welding defects. In the same way presence of grease, paint, dust, humidity, and extreme temperature produce a variation on the welding voltage.

But arc stability is directly influenced by the parameters of the process. It is possible to mention that a relationship exists between the arc length and process stability. Increasing the length of the arc (due to the increase of the contact nozzle to work the piece length) will lead to a destabilization of the process, producing variations in the intensity of the welding current and the arc voltage. At the same time, when the voltage is too small, the arc length is short, so the droplet does not fully grow and then contacts with the molten pool.

The parameter's wire feed speed also has an influence. By increasing the feed rate of the wire, the diameter of the drop decreases; very high or lower values coincide with the most unstable conditions. But the degree of this influence depends on the shielding gas used and the welding voltage.

Furthermore, the variation of the current affects the metallic transfer regularity, and furthermore the transfer regularity reflects the stability of the process. Then it can be said that these factors are going to be influenced by the dynamic behavior of the GMAW welding process, particularly by the physical variations during the different transfer modes. Consequently, to understand how these factors influenced the stability, it is necessary to delve into the characteristics of the metal transfer.

The metal transfer has a direct influence on the stability of the arc and final geometry of the weld bead. The metal transfer is controlled by several parameters such as current, voltage, electrode diameter, and shielding gas composition. It directly influences the way that metal droplets are transferred; the uniformity and the volume of the drop; and variations in arc length.

The three first transfer modes are short circuit, globular spray, and pulsed GMAW. In addition to these modes of transfer, there are others classified as free-flight transfer modes which happen when the arc voltage is high and includes repelled globular, projected spray, streaming spray, and rotating spray. The present study focuses on the three first natural modes of transference.

Spray transfer is characterized by small, uniform drops with diameters close to the size of the electrode. This transfer is obtained with high intensities and high voltages; its current intensities are from 150 to 500 A and its voltages from 24 to 40 v. Inert shielding gas favors this type of transfer. The process is presented with high arc stability, with high currents and deep penetration in the workpiece, and a high frequency of detachment. It allows high penetration to be achieved. Voltage and welding current oscillograms do not differ significantly, as shown in **Figure 4**.

In the globular transfer, the drop grows until exceeding the size of the electrode, and the detachment occurs by the action of the gravitational force. Typical parameters in globular transfer are voltage 20–36 v, current intensity 70–255 A. It has been unwanted in the industry for its instability and high grade of spatter. During this transfer mode, the output currents are kept oscillating depending on the detachment of the drop, as shown in **Figure 5**.

Pulsed transfer is considered a particular case of spray transfer but is characterized by great stability that is achieved by controlling the process variables, in

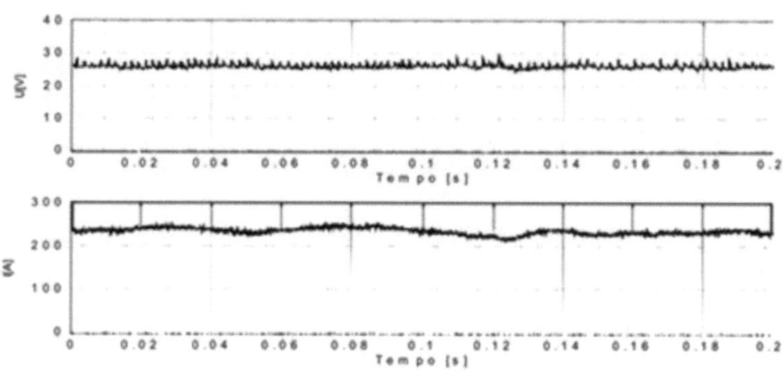

Figure 4.
Waveform factors spray transfer mode [7].

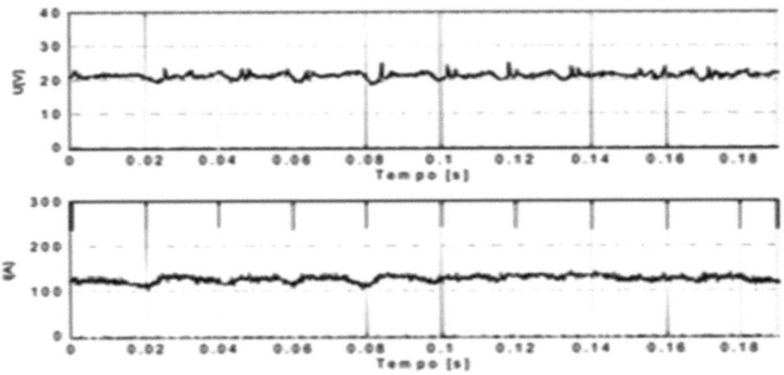

Figure 5.
Waveform factors of globular transfer mode [7].

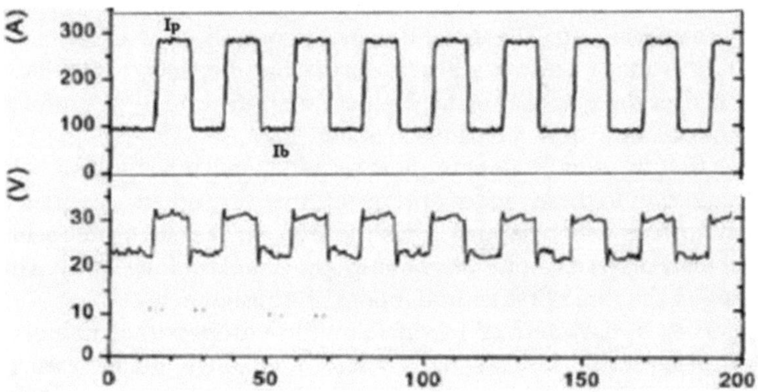

Figure 6.
Waveform factors (modified from [8]).

particular the current. The welding equipment generates two levels of current. In the first, the base current (Ib) is kept low so that there is no transfer, but only the onset of wire fusion; in the second, the peak current (Ip) is higher than the globular transition current causing the transfer, under optimal operating conditions, of a single drop. Typical parameters in pulsed transfer are voltage 20–30 v and current intensity 100–300 A, as shown in **Figure 6**.

Another parameter that influences the stability of the process is the transition current, which changes the frequency and diameter of the transferred drops.

In case of a given current of short-circuiting transition, the droplet transfer exists in the form of short-circuiting, and the welding is stable. When the welding current increases, the droplet transition changes from the short-circuiting mode to the mixed mode, so the welding process and electric signal become unstable.

On the other hand, the globular-spray transition current also presents instability; a big number of spatters but the arc is no longer extinguished. Studies show that with the increase of CO_2 in the gas mixture, an increase of the transition current is produced.

Finally, a peculiarity of the short-circuit transfer mode is the existence of regular contact between the electrode and the workpiece. Typical short-circuit parameters are voltage 16–22 v and current intensity 50–150 A. When the short circuit occurs, the arc is extinguished establishing two characteristic phases: the arcing period and the short-circuit period. Droplet growth occurs in the arcing period, whereas during the contact period, the metal is transferred. Also, the voltage and current oscillate to high and low at the same frequency of the metal transfer (**Figure 7**).

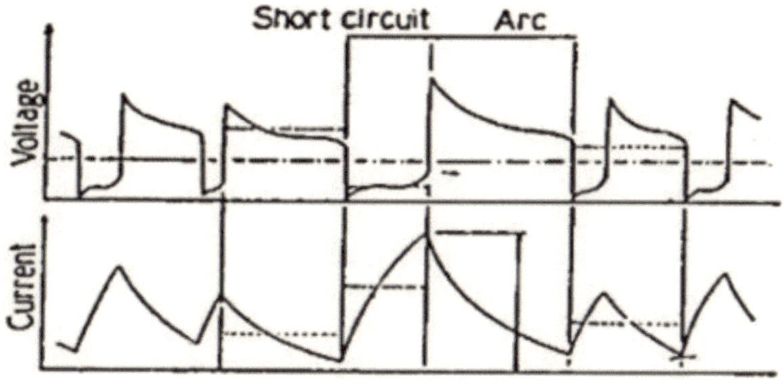

Figure 7.
Waveform factors (modified from [9]).

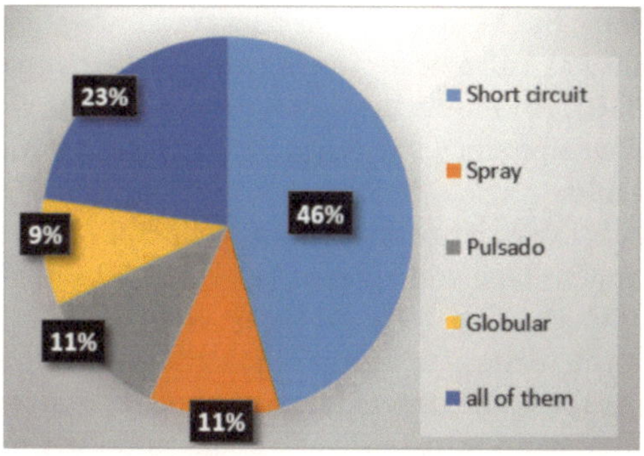

Figure 8.
Papers classified by transfer modes.

Furthermore, a relationship between the waveform factor of the short circuit and the arc stability exists. Some parameters (relating to time and current) used to quantify stability are easy to calculate from the waveform factor, as the short-circuit time, the arcing time, the transfer period, and the short-circuit frequency. Mita et al. [9] also affirm that the correlation between those parameters and the stability becomes weaker with increasing current.

2.3 Summary of stability indexes

Using the abovementioned concepts, several indexes have been proposed to infer the stability and quality of the welding process. They were calculated using image processing techniques, acoustic monitoring, and analysis of the electrical signals. **Figure 8** shows the percentage of papers classified by transfer modes, and it was found that the highest percentage of indexes focused on the short-circuit transfer mode.

2.3.1 Statistical analysis to identify disturbances

Knowing that a signal behaves according to a stochastic process, it is possible to determine a probabilistic model and apply some algorithms to process this signal. Hence, several works have focused on the study of the electrical signals at the moment of disturbance, using a statistical treatment.

Adolfsson and Bahrami [10] calculate the variance of weld voltage (every 1024 signals). The study validates the hypothesis that the instability of the process (caused by disturbances) correlates with a decrease in the variance of the weld voltage; in a similar manner, the short-circuit transfer rate decreased; conversely, no decrease occurs in the estimated variance of the weld current. The results obtained were used in the development of an online fault detection algorithm. This work shows a promising stability index but is only oriented to short-circuit transfer mode and was not extended to other transference modes. Note that the moments of disturbance were caused by making cuts in the workpiece and not varying input variables of the process such as wire feed speed, welding speed, and contact tip to work distance (CTWD), which also influences the stability.

Luksa [11] calculates the mean value of short circuit; the variance of welding current; the time of arc burning; and the short-circuit frequency values (every 2200 signals samples). He identifies two types of disturbances those caused by external factors such as grease and paint that affect the gas shield of the welding arc and a second group caused by variations in the wire extension. As was mentioned in the previous work, the author indicates that the variance of weld voltage decreased in the disturbance moment. But he also affirms that the short-circuit rates increase and optimal process stability can also occur during step disturbance, which contradicts the results found by [10]. An interesting contribution of this work is the study of the correct data window size since very large or small data window size can lead to erroneous stability results.

Finally, Wu et al. [12] used statistical process control (SPC), creating a sequential chart of the welding voltage and current (every 2000 signals). Coinciding with the index previously presented, a decrease in the estimated variance of the welding voltage occurs during the disturbance step. They also understand as a result an increase in the kurtosis for both the welding voltage and current. The results were generalized for the three main transferences modes and used in the construction of an SPC.

2.3.2 Arc stability

In 1988, the authors Mita et al. [9] enunciated the correlation between the stability of the arc and standard deviation of the arcing time; the standard deviation of short current; and the average value of short-circuit frequency. They used linear regression to prove this correlation and to create a new stability index. They showed that short-circuit frequency is influenced by several welding parameters mainly the wire feed rate and the arc voltage. Also, affirm that the stability of the process grows when the standard deviation of the short-circuit frequency decreases. However, the proposed index was tested in all current ranges, and the authors conclude that god arc stability can be obtained in all transfer modes.

Hermans and Ouden [13] propose a criterion for arc stability (Eq. 1, **Table 1**), based on the short-circuit frequency using the relationship between the arc time and the short-circuit time. To do this, they analyzed the behavior of the weld pool taking images with a high-speed camera. The authors concluded that the moment in which the oscillation frequency of the welding pool and the short-circuit frequency are synchronized, the greatest stability is reached.

Ogunbiyi and Norris [14] perform a summary of several criteria presented by other authors and propose three indexes to calculate the stability of the metal transfer. These indexes are Transfer index (Eq. 2, **Table 1**), transfer stability index (Eq. 3, **Table 1**) and dip consistency index (Eq. 4, **Table 1**), which are based on the correlation between metal transfer modes, arc stability and current waveform. The study confronts the three main modes of metal transfer, an advantage in relation to other studies. They calculate the indexes based on the relationship between minimum, mean, and maximum welding current. The indexes and the mathematical

Objective	gas	Transfer mode	Equation	Measured parameters and Variables
Calculate Short Circuit frequency	CO2	Short Circuit	short circuit frequency $Fs = \frac{1}{(ta+tc)}$ (1)	When *ta* is arc time and *tc* Short circuit time
Identification of the metal transfer mode and arc stability	Ar+Co2 Ar+He+CO2	Globular Spray Short Circuit	Transfer index $TI = 1 - \frac{I_{min}}{I_{mean}}$ (2) Transfer stability index $TSI = 1 - \frac{I_{max}}{I_{mean}}$ (3) Dip consistency index DCI $DCI = 1 - \frac{V_{bk}}{V_{mean}}$ (4) Power ratio $PR = \frac{I_{bk}.V_{bk}}{I_{mean}*V_{mean}}$ (5)	I mean =average of the welding current I max =maximum value of the current Imin= lowest value of the current. V mean =average of the voltage Vbk =average of all the voltage Ibk =average of all the current
Metal Transfer Regularity	Ar+CO2	Short Circuit	Regularity index $IVcc = \frac{\sigma tcc}{tcc} + \frac{\sigma tab}{tab}$ (6) Cutting frequency index $Fcc = \frac{Valim*(d)^2}{(k*d)^3}$ (7)	σtcc = standard deviation of the short-circuit time; σtab = standard deviation of open arc time, tcc = average of the short-circuit time; tab = average open arc time. d = wire-electrode diameter; Valim = wire feed rate; k = constant to estimate the drop diameter

Table 1.
Summary of arc stability indexes.

formulation are presented in **Table 1**. They also use the voltage waveform to predict the mode of metal transfer because more variations are observed in the voltage moving from spray to short-circuit transfer. They perform a generalization and propose a new index power ratio (PR) (Eq. 5, **Table 1**), used for identification of the metal transfer mode and arc stability. Finally, an online monitoring system was created capable of predicting the status of the process.

Simpson [15] presents a stability index using an image processing method known as signature images. This index is calculated successively from the comparison of two images of dimensional histograms of the voltage and current data, allowing the detection of faults for the three main modes of metal transfer. Although it is a method of image processing which does not require high-speed cameras, instead, it is necessary for a good data acquisition system to work in real-time. Therefore, it can be considered as a cheap and feasible method to implement in the industry.

Finally, the group Laprosolda of the Federal University of Uberlândia, Brazil [12, 13, 16, 17], in a similar approximation, based on numerical and statistical techniques, propose two indexes for the short-circuit transfer mode: the regularity index (IVcc) (Eq. 6, **Table 1**) criteria for quantifying the short-circuit transfer stability in the MIG/MAG welding process, taking into account the constancy of the short-circuit and open-arc times, and cutting frequency index (ΔFcc) (Eq. 7, **Table 1**) criteria to determinate the voltage regulation range that guarantees greater stability of metal transfer in GMAW short circuit. Using the parameters wire-electrode diameter, wire feed rate and drop diameter as a function of the wire diameter, they address metal transfer behavior (especially regarding the correlation between the stability of transfer mode and the welding defects). The use of these indices allowed the authors to test the correlation between the inductance; the regularity of the metallic transfer; and the influence of the variation in the contact tip to work distance (CTWD), with three different types of gases. In addition, the proposed indices have been widely used in other studies; some of them are discussed below.

Souza [18] presents a work related to mapping the droplet transfer modes to help welders in the choice of the best welding setting parameters needed. The maps were proposed for spray and short-circuit transfer modes. They used the IVcc and ΔFcc parameters to allow focusing voltage range and to obtain transfer regions with proper operating characteristics for the short-circuit mode. The study demonstrates that the index has the characteristic of decreasing and then again increasing its value with increasing welding voltage. As smaller index values indicate better stability, it appears that the process has poor stability at very low and very high voltages.

Meneses [3] presents an implementation of a model that represents the GMAW process in orbital welding. She also developed a study of the metal transfer control, with the objective of achieving a high level of quality of welded joint in different conditions. The mentioned indices were used to make evaluating the hypothesis possible so that more short circuits had greater stability in the process. That allows users to choose a correct parameter setting depending on their needs, in order to obtain a stable transfer with appropriate welding conditions.

Costa [19] performed the validation of the stability on the welding process for the short-circuit transfer mode. The regularity index (IVcc) and cutting frequency (Fcc) index were used, and this was able to identify the tension levels that result in greater transfer regularity, lower level of spatter, higher deposition efficiency, and better surface quality of the weld bead. In the next step, they used the deposition performance and allowed to estimate the amount of material lost by slag and fumes, along with the amount of generated spatter. It was also able to evaluate the effects of the feed rate and the influence of the type of protection gas on the behavior of short circuits. Finally, he developed a thermal efficiency analysis where he concludes that there is no relationship between the values of thermal efficiency and the regularity of transfer.

In conclusion, those indices are powerful tools to determinate the stability in the GMAW process and can be monitored in real time. The short-circuit frequency is one of the most suitable parameters to determine stability in the short-circuit transference mode, either by correlating it with the oscillation frequency of the weld pool or by calculating its standard deviation. The so-called Vilarinho index developed by the group Laprosolda has been widely adopted in Brazil, and it is the index of stability for short-circuit transfer of which the largest number of references was found.

2.3.3 Analysis of current and voltages waveforms

The analysis of current and voltage waveforms is used in the same way as an indicator of stability. Power spectral density and time-frequency analysis methods were used and allowed the decomposition in time and frequency of the waveforms.

Adolfsson and Bahrami [10] used spectral domain analysis of measurement data to detect differences in the power spectral densities of the weld voltage and current in disturbance moments. It made the creation of an algorithm that detects changes in the frequencies and that enables the detection of faults possible. They also affirm that a decrease in the variance was reflected in a decrease in the area in the power spectral density. This work was discussed previously in Section 2.3.1.

Also, Huang et al. [20] used time-frequency entropy techniques to estimate the stability of short-circuiting gas metal arc welding, demonstrating that when the welding is more stable, the time-frequency entropy increases. To obtain the results, the authors made variations in the input variables such as current, voltage, and welding speed, demonstrating that it is possible to use this technique to define the parameters that provide more stability. Finally, the results can be used to perform the process classification in a stable and non-stable arc. It would be interesting in future works to get an integration of these techniques with supervised machine learning algorithms to perform stability classification.

Chu et al. [21] perform an analysis of power spectral density of the current and voltage signals also for processes with short-circuit transfer mode using Fourier transformation to do that. To determine if the testing processes were stable, a correlation was made between the weld bead geometry and the voltage and current values. They affirm that the welding process with a unique frequency corresponds to uniform welds and good weld surface quality, enabling the detection of stable ranges and areas with defects.

Cayo and Alfaro [22] make a comparison between time domain and frequency domain to define which is most appropriate to calculate the stability of the S-GMAW welding process. Applying the two methods to the welding arc sound, the time domain was found to be the most appropriate technique. They also demonstrate that the acoustical ignitions frequency and short-circuit frequency decrease in regions of instability. The results obtained can be used for the development of an online system to identify regions of disturbances.

Macías et al. [23] use image processing to analyze the image generated by the time-frequency diagram obtained from acoustic monitoring. Proving that the minimum standard deviation of the metal transfer weld indicates that the process is stable, as previously mentioned. The authors did not implement online monitoring but highlight the existing flexibility in terms of image processing and online signal processing. It should be noted that in future works, the authors integrate their results into a neural network with artificial intelligence to predict stability in the process.

Then, it can be concluded that power spectral density is a powerful method for the quantification of stability and allows to identify faults in the process through the detection of changes in the waveform frequency. Then, it can be concluded that power spectral density is a powerful method for the quantification of stability and allows to identify faults in the process through the detection of changes in the waveform frequency, being possible to correlate with the quality of the geometry of the weld bead.

The current and voltage signals have also been used to create cyclograms that show the welding voltage as a function of welding current to obtain a process stability indicator. Cyclograms are a novel method for stability analysis in the welding process. They constitute a visual representation by graphs of the voltage values as a function of the current (**Figure 9**). It has been widely used as a stability indicator for the short-circuit transfer mode.

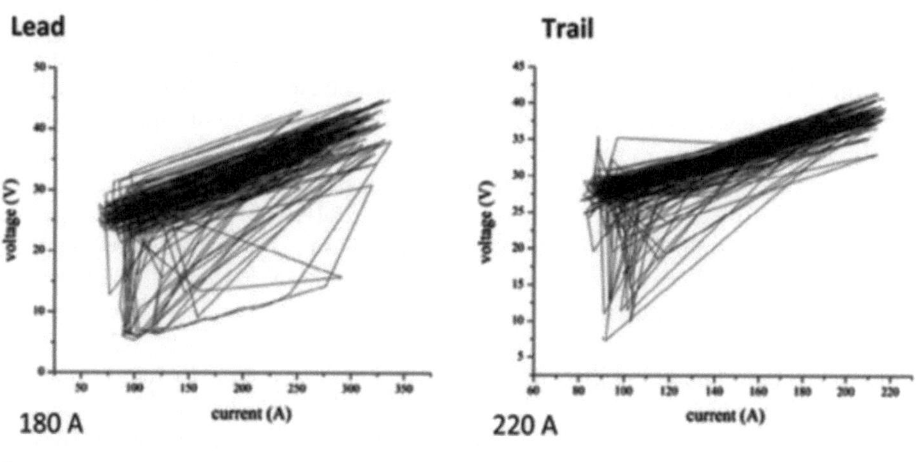

Figure 9.
Representation of the cyclograms (modified from [24]).

According to Moinuddin and Sharma [24], using the cyclograms it is possible to represent characteristics of droplet detachment and arc burning stage. The authors also carried out an analysis of probability density distribution of arc voltage, weld bead, and microstructure analysis for various welding conditions, allowing to extend the stability study to spray transfer mode. The study showed that there is a strong correlation between the microstructure and the stability of the arc. Besides, the different types of electrodes and their electrical conductivity capacity also has influence on the resulting microstructure in a welded bead. A stable arc produces greater penetration and improves melting efficiency. The authors mention that the study can be expanded taking into account other parameters such as electrode type, electrode extension, shield protection gas, welding speed, and other current modes such as pulsed.

Cayo [25] uses the cyclograms to detect defects in the weld reflected in the arc and current–voltage signals. The cyclograms allowed to identify three types of disturbances, a variation of the stand of, presence of grease and absence of protection gas. Each type of defect showed changes in the cyclograms, allowing to analyze the changes in voltage and current. One of the advantages of the cyclograms is that it provides a visual result that allows a quick analysis of the values obtained in the process. Again a powerful stability indicator is shown, but it has been oriented only to the analysis of the short-circuit transfer mode.

Suban [26] uses this index to determine a more stable short-circuit material transfer. As a result, open arc, short-circuit, and spray transfer moments are identified depending on the type of gas used. In addition, the author performs an analysis of the probability distribution of voltage and current using Fourier analysis. Among the conclusions, the authors emphasize that with pure CO_2, more stability is achieved. This method is simple and can be implemented in real time.

2.3.4 Control of droplet size

The control of droplet size ensures transfer stability. For measuring this variable, image processing, laser shadowing, and sound processing techniques are generally used. The appropriate control ensures proper transfer mode; increases the quality of welding, and decreases the number of defects. Large drops do not represent a suitable condition.

The transfer of the drop is dependent on welding current and arc voltage waveforms influenced by gravity force, electromagnetic force, plasma drag force, and surface tension. Suban [26] ensures that to maximize stability, the time between the transfers of two subsequent drops should always be the same.

Mousavi and Kulkarni [27, 28] demonstrate that a relationship between droplet detachment and statistical parameters of current exists, assuring that lesser standard deviation and coefficient to variation was considered to be of uniform droplet detachment and arc length uniformity.

Soderstrom and Mendez [29] use high-speed laser shadowgraphs and fast Fourier transform (FFT) of the voltage signal for droplet diameter and detachment frequency measurement. It has been found that a relationship between average droplet diameter and current for the different diameter electrodes exists. In addition, it states that the increase in CO_2 above normal standards causes an erratic detachment.

Then it can be concluded that there is a correlation between the waveform of the current and the detachment of the drop. A lower coefficient of variation in the mean of the welding current represents uniformity in the detachment frequency. Additionally, for variable transfer time, the welding arc tends to be unstable and the current signals exhibit irregular behavior. In the case of short-circuit transfer mode, it is recommended to detach one dropper short. Equally Pal et al. [30] affirm that in the pulsed welding processes, the detachment of the drop should occur during the pulses and the diameter of the drop should be similar to the diameter of the electrode. Finally, adequate control and study of the metallic transfer allow guaranteeing the quality in the geometry of the welded bead.

2.3.5 Spattering index

The amount of spatters generated during the welding process has been another indicator widely used; the spatters are a product of instability in the arc and should be minimized. The largest amount of study is developed in the short-circuit area. The moment when the short circuit occurs and the arc is reset is when the largest number of spatters is produced. Also, if the mean of the short-circuit time is irregular, more spatters will be generated.

Silva et al. [31] propose a criterion for the spattering index correlating spattering rate (S—Eq. 1, **Table 2**) and the deposition rate (D—Eq. 2, **Table 1**). The purpose was to demonstrate that the correct control of these indicators allows to choose appropriate parameters for any specific welding application.

On the other hand, Kang and Rhee [32] develop statistical regression models to predict the amount of spatter in the short-circuit transfer for GMAW. It is shown, in the same way, that voltage and welding current waveforms can be satisfactorily used to predict the presence of spatters. Kang et al. [33] in a similar work use four different linear and nonlinear regression models composed of the waveform factors to develop the spatter prediction model. Proving that the amount of spatter depends on the number of arc extinctions, arc extinctions occur when the welding voltage is

Objective	gas	Transfer mode	Equation	Measured parameters and Variables
Calculate Spattering Index	CO2	Globular Spray Short Circuit	spattering index $SI = \frac{S}{-} * 100$ (%) (1) spattering rate S = (Felect or Fwire) – D. (2) Deposition rate D = 3.6 x (Mfcp – Micp) / tarc (3) Covered electrode fusion rate Felect = 3.6 x (Miel – Mfel) / tarc . (4) Penetration index PI = (p / t) x 100 [%]. (5) The convexity index CI = (r / w) x 100 [%]. (6)	The penetration index (PI) was defined by relating the depth of the weld bead (P) to the sheet thickness CI was defined as a relationship between the bead reinforcement (r) and the bead width (w), in percentage. where p is the weld penetration [mm], t is the joint thickness [mm], r is the bead reinforcement [mm], w is the bead width [mm], S is the spattering rate [kg/h], D is the deposition rate [kg/h], Felect is the covered electrode fusion rate [kg/h], Miel is the initial mass of the covered electrode, before welding [g], Mfel is the final mass of the covered electrode, after welding [g], tarc is the arc duration time [s], Fwire is the wire fusion rate [kg/h], ϕ is the wire diameter [mm], fwire is the wire feed rate [m/min], γ is the steel density (7.85 x 103 g/mm3), Mfcp is the final mass of the test plate, after welding [g], Micp is the initial mass of the test plate, before welding [g] and de is the deposition efficiency [%].
Calculate Spatter Index Stability Index	CO2	Short Circuit	Spatter Index $F_{sp} = \frac{W_{SP}}{W_{SP}+W_{Bead}} *100$ (7) Index of stability (SI) was defined as: $SI = \frac{K}{F_{SP}}$ (8)	W_{sp} is the weight of the spatter collected in the box, and W_{Bead} is the weight of the weld bead. Where K is an arbitrary constant, which was considered by the author be equal to 100

Table 2.
Summary of transfer stability indexes.

below the optimum. In another study, models were developed for evaluating the spatter rate based on the conventional feed-forward multilayer perceptrons with the error back-propagation as the learning algorithm to estimated spatter rate.

Lastly, Fernandes et al. [34] propose a spatter index (Eq. 7, **Table 2**) relating in a mathematical equation of the weight of the spatter collected in the box and weight of the weld bead. Using the calculated value of the spatter index, they propose a new index of stability (Eq. 8, **Table 1**) that enhances the electrical stability of the process and the weight of spatter generated during welding. The proposed method is efficient as soon as the collection of spatters is carried out correctly. It is suitable for a laboratory environment but can hardly be implemented in the industry since it depends on the collection device. However, the results obtained can be generalized in an automatic learning model and implemented for the control of spatters.

2.3.6 Acoustic monitoring

According to Grad et al. [35], the acoustic signal contains information about the transfer mode and the behavior of the arc. It is also possible to identify changes in arc dimensions and geometry; changes in arc intensity; and metal transfer and oscillations of the molten pool.

Even according to Mota et al. [36], it is possible to observe that the sound signal accompanies the electrical signal, specifically the voltage, in relation to the moments of extinction and ignition of the arc. It is easy to see in **Figure 10** the sound pulses from the moments of the abrupt change in the voltage of the electric arc, and the time intervals between them follow the same pattern observed in the electric signal.

Grum et al. [37] use the sound signal and the light signal to detect even the smallest deviations of arc behavior, as well as large deviations due to the material transfer mode and excessive/inadequate weld penetration. They propose a mathematical model using sound and light values. The authors demonstrated the existence of a correlation between light signals and the energy provided to the system. With the monitoring of sound, it was possible to identify oscillations in the arc that indicated instability. The model was developed for the short-circuit transfer mode but was generalized for the spray transfer mode.

Cayo and Alfaro [38] use the sound to define the difference between the transfer modes on the GMAW process. They use sound pressure and current signals to identify changes in the transfer mode and identify defects. In the case of the spray

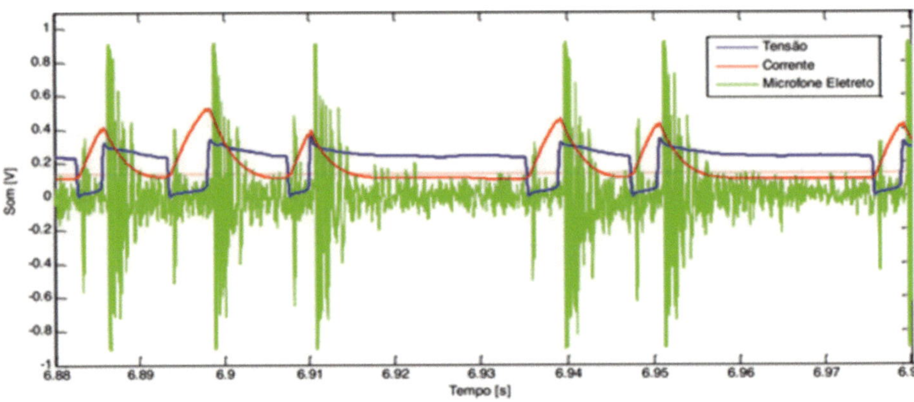

Figure 10.
Comparison between sound and current signals (modified from [36]).

transfer mode, the drops are small and practically imperceptible during the acoustic analysis. Already in the case of the short circuit, it is possible to monitor the occurrence of each short and the reignition of the arc.

Roca et al. [39] also applied acoustic monitoring, and the results obtained were used for the training of a neural network. To perform the analysis, they obtain the standard deviations of the peak amplitudes of the sound at the moment in which the short circuit is made, and they use as stability indicator. In Eq. 12, **Table 1** shows the stability index previously established. The combination of statistical technique, acoustic monitoring, and artificial intelligence allowed to use online monitoring, considering it an efficient and non-destructive technique.

It can be summarized that the electrical and acoustic signals are correlated mainly in the short-circuit transfer mode where it is possible to identify the detachment of the drop and the arc reignition. In addition, it is possible through sound monitoring to identify the transference modes. It is a method that is not expensive and that is feasible to implement in the industry. The combination of this method with machine learning techniques that allow prediction and classification is open for future works.

3. Synthesis of the study and future research directions

To synthesize the study, an analysis of the documentation was obtained, the metadata of the document collection was exported in Information Systems Research (RIS) format, and a bibliometric analysis was performed using the VOSViewer software. A graph with groups of the main authors and their relationship of co-authorship (taking five as a frequency of occurrence of the author's surname) is presented in **Figure 11**. It is possible to identify as the largest cluster the Chinese authors, followed by smaller groups of Brazilian and Indian authors, highlighting that there is little cooperation between those groups.

Figure 12 shows the most used terms in the area that can be defined as keywords.

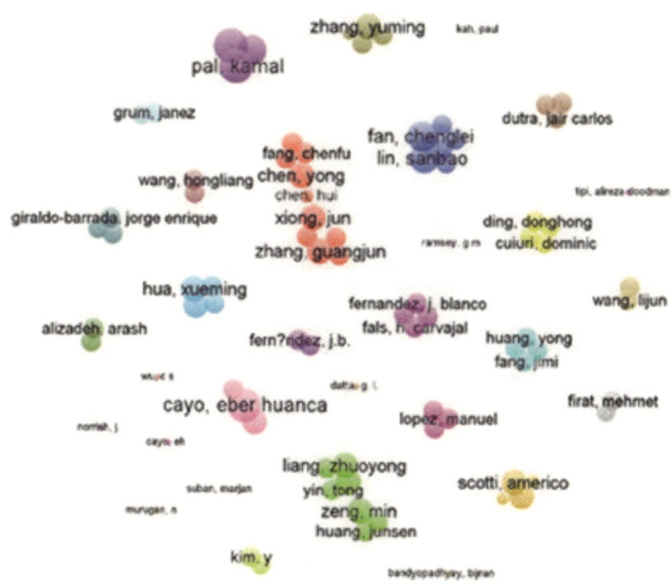

Figure 11.
Authors and their relationship of co-authorship (two as a frequency of occurrence of the author's surname).

Figures 13 and **14** show a summary of the signals and methods used to measure or estimate the indexes. Consequently, the current and voltage signals are widely used, as well as the camera in the image processing and the microphone for the analysis of acoustic signals.

Figure 15 summarizes the parameters and variables used in the studies showing that among the most influential in the stability of the process, current, voltage, wire feed speed, short-circuit time, arcing time, and short-circuit frequency can be mentioned.

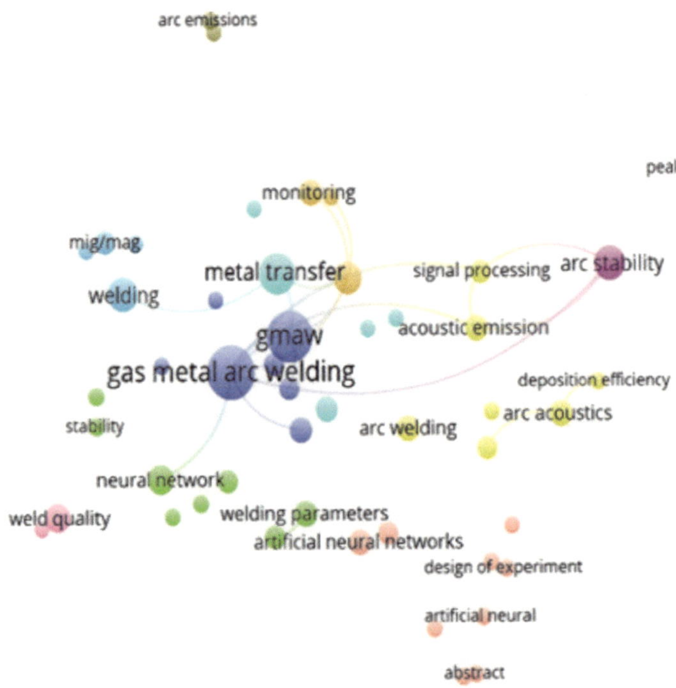

Figure 12.
Cluster of terms (two as a frequency of occurrence of the term).

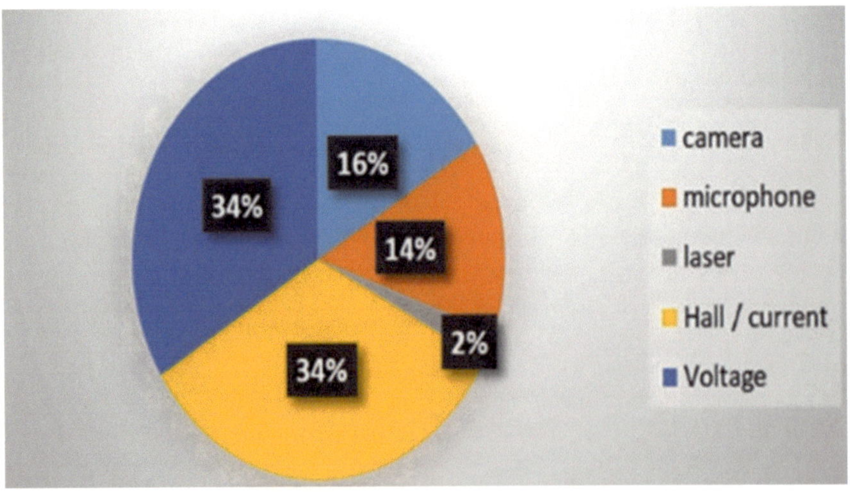

Figure 13.
Sensors used to measure.

Also, it is possible to classify the indexes into groups according to their purpose, those that are oriented to the monitoring of the metallic transfer, and the analysis of the stability of the arc and the process in general. **Figure 16** shows the percentage by group; **Figure 17** shows the technique used to develop the indexes for those groups. It is important to emphasize that these concepts are widely correlated.

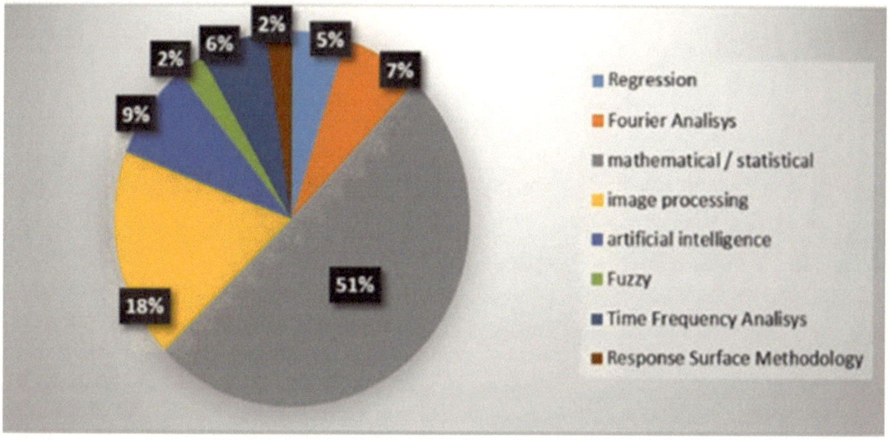

Figure 14.
Methods used to estimate the indexes.

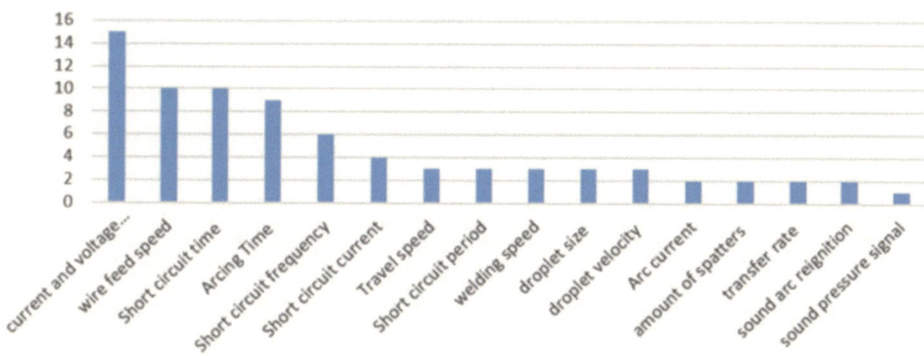

Figure 15.
Parameters and variables used in the studies.

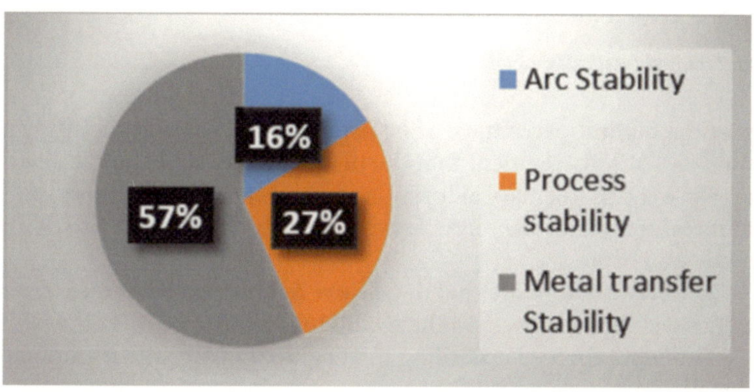

Figure 16.
Percentage by group.

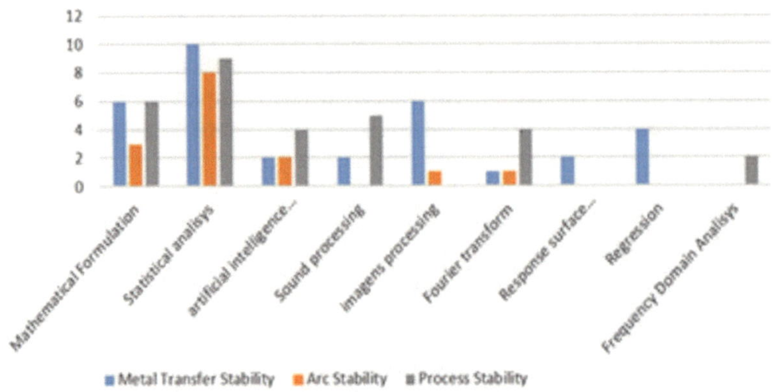

Figure 17.
Techniques used to develop the indexes classified by groups.

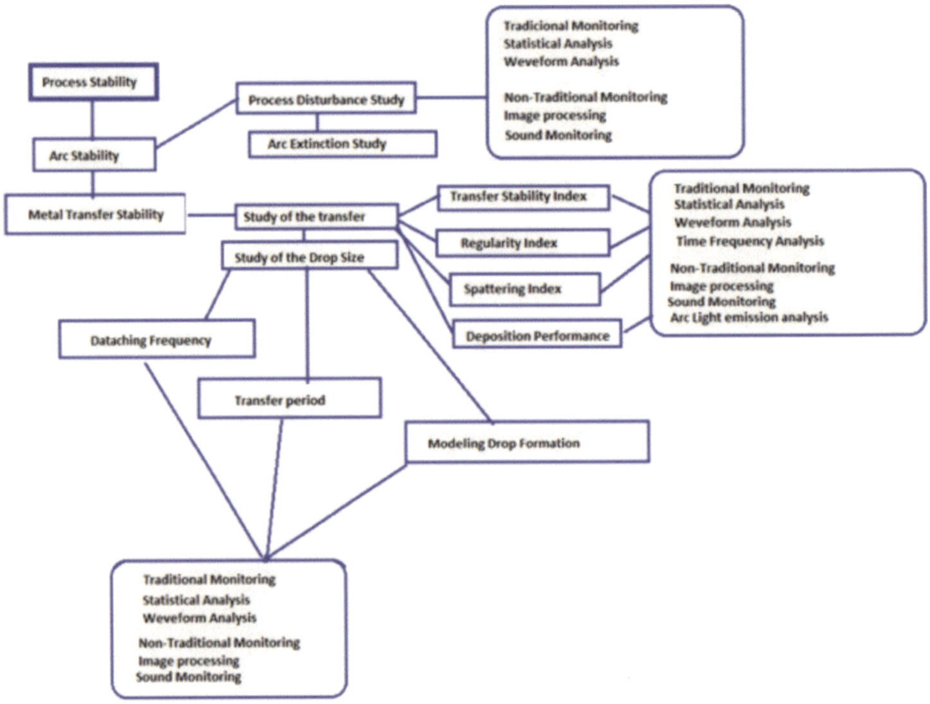

Figure 18.
Classification of the main methods used for monitoring.

Note that the highest percentage of investigation is aimed at the study of metal transfer stability. It is also evident that the main processing techniques to develop the indexes were the mathematical formulation and statistical methods. Although in the case of metal transfer, image processing is widely used, mainly to define the transfer mode and drop size.

Figure 18 shows a taxonomy that details the methods used to measure the stability of the welding process and the techniques associated with them. The techniques used were divided according to Weglowski [40] into traditional and nontraditional.

Finally, to find a trend and a possible vision of the direction of future studies, the following was analyzed:

3.1 Highlights of the works of the last 5 years

An analysis of the works in the field of stability in the last 5 years was made and allowed to find the following trends. There is a considerable increase in the study and application of works in pulsed GMAW (**Figure 19**). This increase is caused by the known improvements in quality and productivity with respect to regular metal gas arc welding (GMAW).

Another trend that could be identified is the increase in research that integrates classical statistics techniques and novel machine learning algorithms. It is well known that with the increase of the computing processing capacities, the data analysis, big data, and machine learning have had a significant boom since 2009. The welding area has not been oblivious to the use of such techniques, although it should be noted that in the area of stability, classical statistics is more commonly used as demonstrated in the present study.

Already in recent years, some interesting solutions have been presented. Alizadeh and Omrani [41] integrate successfully the Taguchi method with back-propagation neural network (BPNN) technique for controlling quality in offline mode. Gyasi et al. [42] are employing an artificial neural network (ANN) to predict geometric characteristics of the welded cord. Wan et al. [43] integrate multiple linear regression analysis and back-propagation neural network to estimate the weld quality. Yue-zhou et al. [44] use sound monitoring and develop a classification algorithm with SVM (support vector machine). Sumesh et al. [45] use machine learning algorithms for weld quality monitoring, acoustic signature, and the perform classification use J48 and random forest algorithms.

In addition, there has been an increase in the use of artificial intelligence algorithms and sensorial fusion. Two powerful techniques have enabled the monitoring and control of welding processes in real time. Also, and as expected, we already find in the literature novel proposals for applications of artificial intelligence and robotics.

Another area that has been highly developed in recent years and future perspectives is image processing. A great number of algorithms have been created for high performance in this subject. Thanks to these advances, the monitoring of the weld bead in real time is now a reality.

3.2 Innovative techniques

It is known that metal transfer has a direct influence on the stability of the process and on the final quality of the welding. Consequently, it has been widely studied as demonstrated in the present review of the literature.

But innovative techniques continue to appear in this field with future prospects of great interest. In this case, they were identified as laser-enhanced gas metal arc welding (GMAW), a modification of GMAW, used to control the metallic

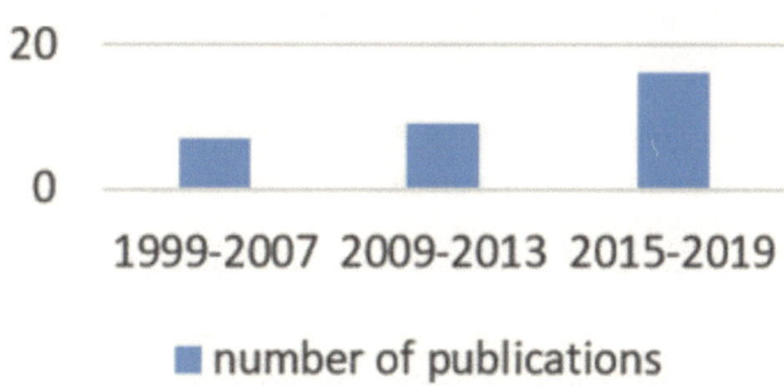

Figure 19.
Number of publications for years in the area of pulsed GMAW.

transfer. A low power laser helps to obtain greater strength in the detachment of the drop. It was also determined as a newly developed arc welding method and ultrasonic-wave-assisted arc welding. This new technique uses power ultrasound energy to radiate the arc and weld pool, modifying the speed with which the plasma heats and cools. According to Fan et al. [46], it allows increasing the stability of the process.

3.3 Areas where further study is required

Very correlated with the study of acoustic monitoring is the analysis of the arc light emission. It is used to control the metallic transfer. A low power laser helps to obtain greater strength in the detachment of the drop. Weglowski [47] demonstrate that light emission has a linear correlation with the current. By another hand, Shao et al. [48] affirm that the light signal indicates the arc radiation intensity and the arc radiation is proportional to the power released which has been a relevant indicator of the energy supplied into the weld.

To a lesser extent and with little representation in the scientific literature, there are works related to the quantification of the emission of fumes. Yamamoto et al. [49] conclude that a relation between heat content, fume emission rate, and molten metal transfer mode exists; consequently the fume emission rate per unit weight of consumed wire increases with the increase in heat content.

But Meneses et al. [50] proved that there is no correlation between the amount of spattering generated and fume generation rate, because the regularity of the transfer did not show influence on the morphology, size or composition of the fumes. Then the generation of fumes is not correlated with the stability of the process.

Finally, other techniques that can be developed are 3D computational modeling, simulation, spectroscopy, spectral analysis, and X-ray observation system.

4. Conclusions

- The chapter shows reliable and precise methods to measure the stability in the GMAW welding process.

- The greatest amount of effort in the area has been directed towards the study of metallic transfer.

- The transfer mode for which most indexes have been created is the short circuit.

- The indexes have been developed through the analysis of process signals as current, voltage, sound, and light.

- Techniques such as mathematical formulation, statistical analysis, image processing, and monitoring of acoustic signals and light spectrum are mentioned among the most useful.

- The methods of artificial intelligence and machine learning have been little used leaving an interesting path to be traveled in future research.

Acknowledgements

This work has been supported by the Brasilia University (UnB), the government research CAPES foundation and CNPQ.

Conflict of interest

Elina Mylen Montero Puñales and Sadek Crisótomo Absi Alfaro declare that they have no conflict of interest.

Author details

Elina Mylen Montero Puñales[1,2*] and Sadek Crisóstomo Absi Alfaro[3]

1 University of Brasilia, Campus Darcy Ribeiro, Brasilia, DF, Brazil

2 Department of Automatic Control, Institute of Cybernetics, Mathematics and Physics, La Habana, Cuba

3 Department of Mechanical, Mechatronic Engineering, Faculty of Technology, University of Brasilia, Campus Universitario Darcy Ribeiro, Brasilia, DF, Brazil

*Address all correspondence to: mylen88a@gmail.com

References

[1] Wu C, Polte T, Rehfeldt D. A fuzzy logic system for process monitoring and quality evaluation in GMAW. Welding Journal. 2001;**80**:33-38

[2] Ponomarev A. Arc Welding Process Statistical Analysis. Methodical Approaches, Analysis Conceptions, Experiences. Denmark: DTU-Helsingor, JOM-Instit.; 1997. pp. 59-100

[3] Meneses LYH. Modelagem E Simulação Da Transferência Metálica No Processo Gmaw-S Em Soldagem Orbital. Brasilia: UNB; 2013 p. 196

[4] Pinto Lopera JE. Uso de agentes inteligentes no controle simultâneo da largura e do reforço dos cordões de solda no processo GMAW-S. 2016

[5] Margarita E et al. Parameter Optimization in Gmaw Process With Solid and Metal-Cored Wires. Uberlandia, Brasil: Cobem, ABCM; 2013. pp. 5256-5266

[6] Bestard GA. Sensor Fusion and Embedded Devices to Estimate and Control the Depth and Width of the Weld Bead in Real Time. Brazil: Universidade de Brasília-UnB; 2017

[7] De Figueiredo KM. Mapeamento dos Modos de Transferência Metálica na Soldagem Mig de Alumínio. Brazil: Universidade Federal de Uberlândia; 2000

[8] Chávez JJM. Uma Metodologia Para Obtenção E Validação De Parâmetros Ótimos Em Soldagem Gmaw Pulsado. Brazil: UNB; 2014. p. 190

[9] Mita, Sakabe, Yokoo. Quantitative estimates of arc stability for CO_2 gas shielded arc welding. Welding International. 1988;**2**(2):152-159

[10] Adolfsson S, Bahrami A. On-line quality monitoring in short-circuit gas metal arc welding. Welding Journal. 1999;**78**(2):59S-73S

[11] Luksa K. Influence of weld imperfection on short circuit GMA welding arc stability. Journal of Materials Processing Technology. 2006; **175**(1–3):285-290

[12] Wu CS, Gao JQ, Hu JK. Real-time sensing and monitoring in robotic gas metal arc welding. Measurement Science and Technology. 2007;**18**(1): 303-310

[13] Hermans M, Den Ouden G. Process behavior and stability in short circuit gas metal arc welding. Welding Journal. 1999;**78**:137-141

[14] Ogunbiyi B, Norrish J. Monitoring indices for metal transfer in the GMAW process. Science and Technology of Welding and Joining. 1997;**2**(1):33-35

[15] Simpson SW. Signature image stability and metal transfer in gas metal arc welding. Science and Technology of Welding and Joining. 2008;**13**(2): 176-183

[16] Modenesi RCAPJ. The influence of small variations of wire characteristics on gas metal arc welding process stability. Journal of Materials Processing Technology. 1997;**86**:226-232

[17] de Rezende GMC, Liskévych O, Vilarinho LO, Scotti A. Um critério para determinar a regulagem da tensão em soldagem MIG/MAG por curto-circuito. Soldagem & Inspeção. 2011;**16**(2):98-103

[18] Souza D. Levantamento de Mapas Operacionais de Transferência Metálica para Soldagem MIG/MAG de Aço ao Carbono na Posição Plana. Brazil: Universidade Federal De Uberlândia; 2010

[19] Costa TF. Avaliação De Critérios De Estabilidade Do Processo Mig/Mag

Curto-Circuito Na Soldagem De Aços Inoxidáveis Austenítico E Duplex. Brazil: Universidade Federal De Uberlândia; 2014

[20] Huang Y, Wang K, Zhou Z, Zhou X, Fang J. Stability evaluation of short-circuiting gas metal arc welding based on ensemble empirical mode decomposition. Measurement Science and Technology. 2017;**28**(3):12

[21] Chu SPMYX, Hu SJ, Hou WK, Wang PC. Signature analysis for quality monitoring in short-circuit GMAW. Welding Journal. 2004;**83**(12):336S-343S (December)

[22] Cayo EH, Alfaro SCA. A non-intrusive GMA welding process quality monitoring system using acoustic sensing. Sensors (Basel). 2009;**9**(9): 7150-7166

[23] Macías EJ, Roca AS, Fals HC, Fernández JB, de la Parte MP. Time–frequency diagram applied to stability analysis in gas metal arc welding based on acoustic emission. Science and Technology of Welding and Joining. 2010;**15**(3):226-232

[24] Moinuddin SQ, Sharma A. Arc stability and its impact on weld properties and microstructure in anti-phase synchronised synergic-pulsed twin-wire gas metal arc welding. Materials and Design. 2015;**67**: 293-302

[25] Cayo E. Weld transference modes identification through sound pressure level in GMAW process. Journal of Achievements in Materials and Manufacturing Engineering. 2008; **29**(1):57-62

[26] Suban M. Determination of stability of MIG/MAG welding processes. Quality and Reliability Engineering International. 2001;**17**(5):345-353

[27] Mousavi M, Haeri M. Welding current and arc voltage control in a GMAW process using ARMarkov based MPC. Control Engineering Practice. 2011;**19**(12):1408-1422

[28] Kulkarni SD, Mallya UD, Chaturvedi RC. Droplet detachment and plate fusion characteristics in pulsed current gas metal arc welding. Welding Journal. 1998;**6**(June):254-269

[29] Soderstrom EJ, Mendez PF. Metal transfer during GMAW with thin electrodes and Ar-CO_2. Welding Research. 2008;**87**(May):124-133

[30] Pal K, Pal SK. Effect of pulse parameters on weld quality in pulsed gas metal arc welding: A review. Journal of Materials Engineering and Performance. 2011;**20**(6):918-931

[31] Silva CR, Ferraresi VA, Scotti A. A quality and cost approach for welding process selection. Journal of the Brazilian Society of Mechanical Sciences and Engineering. 2000;**22**(3): 389-398

[32] Kang, Rhee. The statistical models for estimating the amount of spatter in the short circuit transfer mode of GMAW. Welding Journal. 2001;**80**:1-8

[33] Kang MJ, Kim Y, Ahn S, Rhee S. Spatter rate estimation in the short-circuit transfer region of GMAW. Welding Journal. 2003;**82**(9):238s-247s

[34] Fernandes DB, de Menezes Júnior LC, Vilarinho LO, Scotti A. Ensaio para medição de nível e alcance de respingos em soldagem MIG/MAG. Soldag. e Insp. 2010;**15**(2):150-151

[35] Grad L, Grum J, Polajnar I, Slabe JM. Feasibility study of acoustic signals for on-line monitoring in short circuit gas metal arc welding. International Journal of Machine Tools and Manufacture. 2004;**44**(5):555-561

[36] Mota CP, Mendes R, Neto F, Vilarinho LO. Do Por Curto-Circuito. 2013

[37] Grum J, Bergant Z, Polajnar I. Monitoring the GMAW process by detection of welding current, light intensity and sound. In: Defectoscopie 2012, Czech Republic: Czech Society for Nondestructive Testing; 2012

[38] Cayo, Alfaro. Welding Quality Measurement Based on Acoustic Sensing. Brazil: COBEM, ABCM; 2008

[39] Roca AS, Fals HC, Fernández JB, Macías EJ, De La Parte MP. Artificial neural networks and acoustic emission applied to stability analysis in gas metal arc welding. Science and Technology of Welding and Joining. 2009;**14**(2):117-124

[40] Weglowski M. Modeling and analysis of the arc light Spectrum in GMAW. Welding Journal. 2008;**87** (August):212-218

[41] Alizadeh A, Omrani H. An integrated multi response Taguchi-neural network- robust data envelopment analysis model for CO_2 laser cutting. Measurement. 2019;**131**: 69-78

[42] Gyasi EA, Kah P, Wu H, Kesse MA. Modeling of an artificial intelligence system to predict structural integrity in robotic GMAW of UHSS fillet welded joints. 2017;1139-1155

[43] Wan X, Wang Y, Zhao D, Huang Y, Yin Z. Weld quality monitoring research in small scale resistance spot welding by dynamic resistance and neural network. Measurement. 2017;**99**:120-127

[44] Ma YZ, Qu M, Chen JH. SVM classifier for wire extension monitoring using arc sound signal in GMAW. Transactions of the China Welding Institution. 2006;**27**(5):21-26

[45] Sumesh A, Rameshkumar K, Mohandas K, Babu RS. Use of machine learning algorithms for weld quality monitoring using acoustic signature. Procedia Computer Science. 2015;**50**: 316-322

[46] Fan C, Xie W, Yang C, Lin S, Fan Y. Process stability of ultrasonic-wave-assisted gas metal arc welding. Metallurgical and Materials Transactions A: Physical Metallurgy and Materials Science. 2017;**48**(10): 4615-4621

[47] Węglowski MS, Huang Y, Zhang YM. Relationship between wire feed speed and metal transfer in GMAW. Journal of Achievements in Materials and Manufacturing Engineering. 2008;**29**(2):191-194

[48] Shao Y, Wang Z, Zhang Y. Monitoring of liquid droplets in laser-enhanced GMAW. International Journal of Advanced Manufacturing Technology. 2011;**57**(1–4):203-214

[49] Yamamoto E, Yamazaki K, Suzuki K, Koshiishi F. Effect of flux ratio in flux-cored wire on wire melting behaviour and fume emission rate. Welding World. 2013;**54**(5–6):R154-R159

[50] De Meneses VA, Gomes JFP, Scotti A. The effect of metal transfer stability (spattering) on fume generation, morphology and composition in short-circuit MAG welding. Journal of Materials Processing Technology. 2014;**214**(7):1388-1397

Chapter 2

Data Analysis and Modeling Techniques of Welding Processes: The State-of-the-Art

Rogfel Thompson Martinez and Sadek Crisóstomo Absi Alfaro

Abstract

Information contributes to the improvement of decision-making, process improvement, error detection, and prevention. The new requirements of the coming Industry 4.0 will make these new information technologies help in the improvement and decision-making of industrial processes. In case of the welding processes, several techniques have been used. Welding processes can be analyzed as a stochastic system with several inputs and outputs. This allows a study with a data analysis perspective. Data mining processes, machine learning, deep learning, and reinforcement learning techniques have had good results in the analysis and control of systems as complex as the welding process. The increase of information acquisition and information quality by sensors developed at present, allows a large volume of data that benefits the analysis of these techniques. This research aims to make a bibliographic analysis of the techniques used in the welding area, the advantages that these new techniques can provide, and how some researchers are already using them. The chapter is organized according to some stages of the data mining process. This was defined with the objective of highlighting evolution and potential for each stage for welding processes.

Keywords: data mining, deep learning, welding process, machine learning

1. Introduction

One of the most important processes of joining metals is welding process, like the one that appears in [1]. It is used in simple structures fabrication, nuclear and petroleum industries, as well as chemical components.

In a typical fusion welding process of metals, such as resistance welding, arc welding, electron beam welding, laser welding, a heat source is applied locally to the interfaces of the two metals to be joined. The interface can be metals' surfaces, where faces of each other are joined by a nugget (e.g., spot or resistance welding). In arc welding, the interface will be the weld seam. However, complex physical phenomena and processes occur due to the heating/melting and cooling/solidifying. This may produce adverse effects on weld properties and base metal properties [2]. In order to reduce adverse effects and obtain desired results, many studies have been developed to monitor, predict, or control welding processes. All these studies are based on the optimal welding parameters' adjustment, but all of these are adjustable.

All adjustable welding parameters, such as current or current waveform, heat input, wire feed speed, travel speed, and arc voltage, may be used as system inputs and be designed to assure the required outputs. For that reason and the interrelations-parameters complexity, welding process can be analyzed like a stochastic system, which has input and output parameters and several disturbances [2]. Chen's article [3] was related to the need to improve the information acquired from these welding parameters and identify characteristics in order to improve and control the welding process results. Chen defined new objectives of modern welding manufacturing technology to show the way for better welding processes. It exposes some problems of the intelligentized welding manufacturing technology (IWMT), which are shown in **Figure 1**.

Other science areas present the potential to solve these problems. Computer science areas have had great results with new technique applications of data analysis, learning models, and intelligent control. Data analysis objective indicates nontrivial features on a large amount of data. Due to the increase and complexity of data, more efficient data analysis techniques have been developed. Welding process can be analyzed with this point of view. So, the welding process analysis with new techniques is nothing more than a continuity in the development of welding analysis processes. This interdisciplinarity is one of the necessary contributions proclaimed by the so-called Industry 4.0, like the one shown in [4–6].

The fourth industrial revolution refers to the next manufacturing generation, where automation technology will be improved by self-optimization and intelligent feedback [7]. For this reason, the application of the most recent data analysis techniques and processes can contribute to a better control and monitoring of welding processes. These techniques can be joined in machine learning techniques [8–12], data mining process [13–17], and control process [18–20]. The interrelation of these areas and their origins are presented in **Figure 2**. Machine learning is a growing area in computer science, with far-reaching applications, for data analysis [21]. Machine learning uses computer theory and statistics for building mathematical models with the goal of making inference from a sample [22]. One branch in machine learning with fast growing is deep learning. These methods are an essential part of the research on speech recognition in the state of the art [23], image recognition [24–26], object detection [27, 28], videos [29, 30], and sound [31, 32] analysis. Interesting patterns come out from such a machine learning techniques. One important process is data mining. Data mining puts strong emphasis on different aspects, like efficiency, effectiveness, and validity of process [33]. Data mining

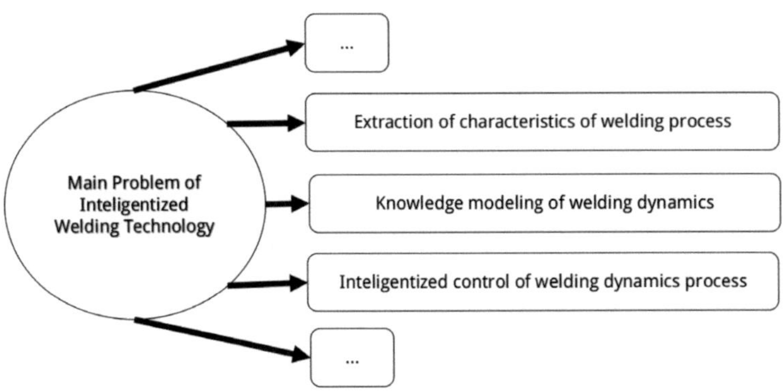

Figure 1.
Some technical problems in IWMT [3].

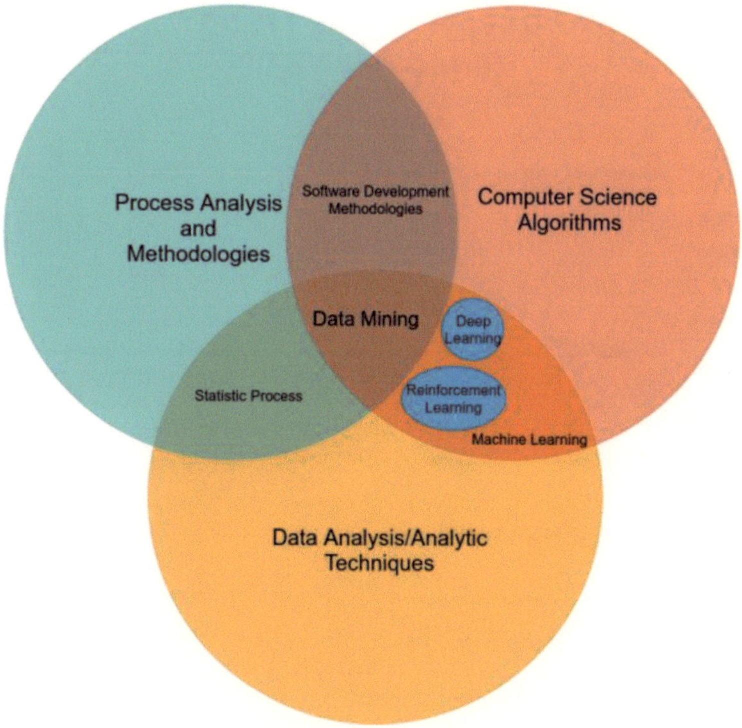

Figure 2.
Origin diagram of the new data analysis techniques.

processes define several stages and methodologies to achieve these objectives, as exposed by Marbán in [34]. An important objective of data analysis is to reveal and indicate diverse, nontrivial features in data. For this reason, welding process can be analyzed with this point of view.

A search conducted in the Web of Science from 2011 to October 3, 2018, shows the growing trend of these new data analysis techniques and processes in welding process researches **Figure 3**, but when comparing with the investigations on models welding process, growth is almost imperceptible, as appearing in **Figure 4**.

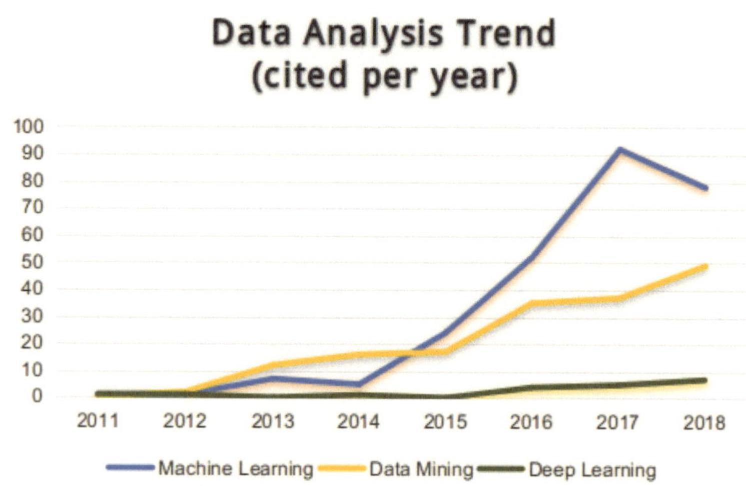

Figure 3.
Cited per year on welding (Web of Science [35]).

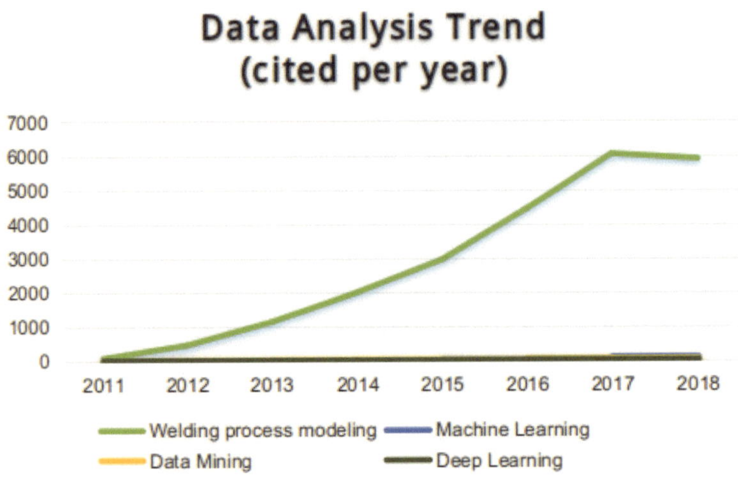

Figure 4.
Cited per year on welding (Web of Science [35]).

These demonstrate the need for this review to show these techniques, the advantages in their applications, and the increasing trend of their utilization. This review can be resumed in following stages:

1. Welding process—understanding of welding processes being analyzed.

2. Sensors—analysis of some principal sensors in welding process.

3. Data processing—analysis of technique to transform sensors information to welding process dataset.

4. Modeling welding process—analysis of some modeling techniques in welding process.

5. Intelligent control of welding process—analysis of some intelligent control techniques in welding process.

These stages has a close relationship with data mining processes as a sample [34].

2. Welding process

American Welding Society (AWS) definition for a welding process is:

> *"a materials joining process which produces coalescence of materials by heating them to suitable temperatures with or without the application of pressure or by the application of pressure alone and with or without the use of filler material"* [36].

AWS defines groups of welding techniques depending on the energy transfer mode. The processes analyzed in this chapter are grouped as shown in **Table 1**.

These groups present different parameters and characteristics that were analyzed in the articles presented in this chapter.

Group	Welding process
Arc welding	Gas metal arc welding (GMAW)
	Gas tungsten arc welding (GTAW)
	Plasma arc welding (PAW)
	Shielded metal arc welding (SMAW)
	Submerged arc welding (SAW)
	Variable polarity plasma arc welding (VPPAW)
	Rotating arc narrow gap MAG welding (RANGMW)
	Girth welds
Resistance welding	Resistance spot welding (RSW)
	Large scale RSW (LSRSW)
Other welding processes	Laser beam welding (LBW)

Table 1.
Welding processes group.

2.1 Arc welding

The group arc welding is characterized with electric arc. The electric arc is the heat source most commonly used in fusion welding of metallic materials. The welding arc comprises a relatively small region of space characterized by high temperatures (similar to or even higher than the sun's surface), strong generation of light and ultraviolet radiation, intense flow of matter, and large gradients of physical properties. It has an adequate concentration of energy for localized base metal fusion, ease of control, low relative cost of equipment, and an acceptable level of health risks to its operators. The study of the arc is of special interest in areas such as astrophysics and the electrical and nuclear industries [37]. The electric arc generates a complex interrelation of thermal, electrical, and magnetic parameters. These are hampering much of their studies based on definite theoretical formulations. Despite many studies, the electric arc is quite complex and the knowledge so far allows a partial understanding of the phenomenon [1].

2.2 Resistance welding

Resistance welding is the joining of metals by applying pressure and passing current for a length of time through the metal area that is to be joined. Its principal advantage is no other materials are needed to create the bond; this reason makes this process extremely cost effective. Resistance welding is applied in a wide range of automotive, aerospace, and industrial applications. Among the main parameters are welding time, welding force, contact resistance, materials properties [1]. Resistance spot welding, like all resistance welding processes, creates welds using heat generated by resistance to the flow of welding current between the faying surfaces, as well as force to push the workpieces together, applied over a defined period of time. Resistance spot welding uses the electrode face geometries to focus the welding current at the desired location, and apply force to the workpieces. Once sufficient resistance is generated, the materials set down and combine, and a weld nugget is formed [36]. The process is fast and effective, and it is also complicated due to complex interactions between electrical, mechanical, thermal, and metallurgical processes. The heat generation in RSW is due to the resistance of the parts being welded to the flow of a localized electric current, based on Joule's law. The quality of the joint in RSW is influenced by the welding parameters. These parameters mainly include welding current, welding time, electrode force, and electrode geometry [38]. Large scale resistance spot welding (LSRSW), as mentioned in **Table 1**, is generally adopted in the automotive industry. It is an automotive structure that includes thousands of spot welds. It presents the same parameters and complexity as RSW; only the parameters related and influenced by its scalability are increased [39].

2.3 Other welding processes

In this group, AWS presents various welding processes. Laser welding is the only one belonging to this group, which is found in the analyzed articles.

Laser beam welding is one of the most technically advanced welding processes. Laser welding is in general a keyhole fusion welding technique which is achieved with the very high power density obtained by focusing a beam of laser light to a very fine spot [40]. This light ray heats metals up quickly so that the two pieces fuse together into one unit. The light beam is very small and focused, so the metal weld also cools very quickly. Laser welding operates in two fundamentally different modes: conduction limited welding and keyhole welding. The mode in which the laser beam will interact with the material is welding; it will depend on the power density of the focused laser spot on the work piece [41].

Other parameters that are present in these processes are those of final welding geometry, which behave differently in different processes and under different conditions. The parameters of the respective sources generate their influence on the final result of each welding process.

Welding is a complex process, so it requires more intelligent techniques in its analysis, monitoring, and production quality improvement. The use of sensors allows the acquisition of process parameters. The new artificial intelligence techniques will allow a better study, modeling, and control of these processes.

3. Sensors

Several sensors have been applied in the welding process for monitoring. The weld bead and the weld-pool indirect sensing technologies can be classified like exposed in [42] and in **Figure 5**.

Infrared vision techniques have been widely applied in the welding process [43–50]. One of the problems of this technique is that the environment where it is applied can interfere in the precision of the data obtained from a process. This may be due to the own heat emission of the technologies utilized.

3.1 Sound sensor

Sound may indicate conditions that generate weld defects. Acoustic information plays a relevant role for expert welders, as described in [51]. Sound signature produced by GMAW contains information about arc column behavior, the molten metal, and the metal transfer mode. High-speed data acquisition and computer-aided analysis of sound signature may indicate conditions that generate weld defects [52, 53]. Di Wu, in 2016 [54], tried to monitor penetration and keyhole with acoustic signals and image analysis. Lv et al. [55], proposed a recognition model to analyze the relationship between penetration state and arc sound. In 2017, Lv et al. [56] again presented a welding quality control in pulse gas tungsten arc welding

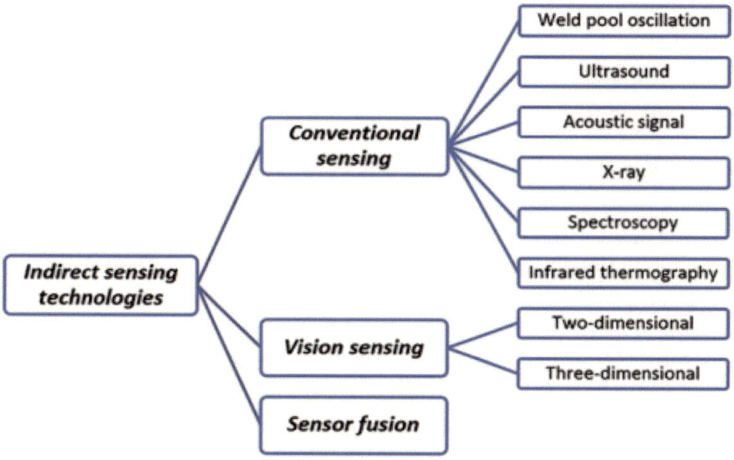

Figure 5.
Some indirect monitoring technologies in welding process [42].

(P-GTAW). The welding acoustic signal was used to analyze the design of an automated welding penetration control system.

In welding, it is easy to capture sound, but it is very difficult to analyze the noises and differences of intensities that are sometimes generated. This is not a problem to sound deep learning technique like present [32, 57]. To understand the welding sound analysis with deep learning techniques, it is necessary make an image arc correlation to know what happens in welding arc.

3.2 Vision sensor

Vision sensor is largely utilized in welding process to analyze weld-pool process [58, 59], arc-welding process [60, 61], and weld bead geometry [62, 63]. The more light generated by arc can be difficult for the image obtention. Some techniques are utilized. One of them was utilized by Chen in 2010 [64].

He made monitoring and control of the hybrid laser-gas metal arc welding process with an economical sensor system, and a coaxial vision system, which was integrated from a relatively inexpensive industrial vision system and a personal computer (PC). Another visualization technique is Shadowgraphy, applied in Esdras Ramos investigation, in 2013 [65, 66]. This is based on process shadow arc with laser source.

In [60], a laser illumination was utilized. To reduce the arc light, a narrow band interference filter was applied. For precise measurements, an image-analysis technique was used. This technique can be used to obtain high quality images but only it can be used in processes without material transfer.

Chen et al. [67] utilized a visual double-sided sensing system. In one frame, the weld-pool geometry parameters in GTAW process were determined.

With high speed illumination laser in [68], great quality images are obtained. This technique is more recent one but it needs a laser with more potentiality than Shadowgraphy technique. This technique is more expensive too.

4. Data processing

Some papers define their own image processing technologies, like Hong Yue in 2009 [69], where the weld image processing adopts the classic techniques such as

Laplacian, Gaussian, neighborhood mean filters, and threshold segmentation. Yanling Xu, in 2014 [70], proposed the Canny edge detection algorithm for detecting edges and extracting pool and seam characteristic parameters. Qian-Qian Wu in [71] researched to find out the optimal algorithm to filter. He made a comparison of Wiener filter, Gaussian filter, and Median filter on welding seam image. In the classic image processing, it is very difficult to generalize a filter or algorithm, because it depends on the conditions and characteristics of camera parameters and light.

Another problem with these algorithms mentioned above is that the real-time analysis has an insufficient response time to be utilized in a process control despite recent developments in computational resources.

Deep learning techniques have efficient result in real-time executions [28] and classifications [24, 25] despite classifications on new images. One example applied in welding process is [62, 63]. It utilizes autoencoder deep learning technique to extract features of images process in laser welding. Another example of recent application of deep learning technique is [72]. It presents a method based on deep learning aims to extract information from photographs on spot welding. This monitoring system on the spot welding productive line shown better performance than the previous images analysis.

Not focused on welding arc analysis, but with good results, the work [73] proposed an automatic detection for weld defects in X-ray images. A classification model on deep neural network was developed. The accuracy rate of the proposed model was 91.84%. This was one more example of the potential of these techniques in welding area on images processing.

5. Modeling welding process

Today's manufacturing environments has a rapid advancement on demand for quality products. Many techniques and methods are applied to correlate between process parameters and bead geometry. One of them is response surface methodology (RSM). It was applied by Sen in 2015 [74]. He made to evaluate the correlations between process parameters and weld bead geometry in double-pulsed gas metal arc welding (DP-GMAW). Santhana Babu [75] with the same technique got good results for predicting and controlling the weld bead quality in GTAW process. The problem of this method is that the researcher can find the equation, called response surface, by test and error. This can be very difficult. Many theoretical models have been defined to determine the process that occurs in the welding arc, including [76]. The main problems of these models were that they lose precision because it was very difficult to obtain a formula that contains all the complexity of these processes, as well as affirmed by Hang Dong in [77]. Mathematical models, based on machine learning techniques, have better results in problems as complex as this one. In the same paper, Hang Dong expressed the potential of these models.

One of the well-known and utilized regression algorithms is the least squares method. It was utilized in [78] to predict the seam position under strong arc light influence. Other work is [79] a LR model that is utilized to analyze the pool image centroid deviation and weld based on visual weld deviation measurement in GTAW process. The other technique is Gaussian process (GP) regression (GP), which was utilized in [77] to predict better performance in arc welding process of GTAW process.

An interesting method, utilized in [80], was Mahalanobis Distance Measurement (MDM). It was employed to determine welding faults occurrences. The same method was utilized in 2017 by Khairul Muzaka [81] on GMAW process to optimize welding current on a vertical-position welding. One problem of this method is that only correlate in function one input.

Bai and Lubecki [82] proposed a Localized Minimum and Maximum (LMM) analysis method in real time for welding monitoring system. The problem of LMM is that it exposes a simple function to measure the quality than not defining the complexity of the system. That is why, this work is limited only to the short-circuit transfer mode.

In 2017 by Junheung Park [83], a SVM was proposed with bootstrap aggregating that reduced the noisy on RSW data with computational efficiency. In this framework, other techniques as Generalized Regressive Neural Networks (GRNN) and Genetic algorithms for optimization were joined. This article demonstrates an increase in more complex computer science techniques for better analysis of welding processes. But the only way to know if all this was necessary is comparing with other techniques.

5.1 Artificial neural network models

Some researchers already had this reference of advantages of these algorithms. Bo Chen in 2009 [84] utilized ANN to training the experimental obtaining data. The good result of ANN prediction was validated by D-S evidence theory information fusion. They have also been utilized for different purposes and in different welding processes such as in SAW process [85] and GMAW cold metal transfer (CMT) process [86], for predicting weld bead geometry; in GTAW process, for predicting the angular distortion considering the bead geometry [87]; in girth welded pipes process, for predicting residual stresses [88]; and in underwater wet welding process, for predicting the weld seams, geometric parameters [89].

For better results, ANNs have been mixed with other techniques. One example is [90], where ANN and Support Vector Machine (SVM) are utilized for welded defect detecting and monitoring on a laser welding process. The other technique is by Bo Chen and Shanben Chen [91] for predicting the penetration in GTAW process. But they used different ANNs to process information from different sensors, and finally, they used the predictive fuzzy integral method.

Another example is [92], for predicting bead height and width in GMAW process using ANN Fuzzy ARTMAP, like monitoring task.

The increase in computational resources has allowed an increase in the complexity of ANN architectures. These are called Deep Neural Networks (DNN). They, bit by bit, begin to be applied in the welding process. One of them utilized was in [93]. The model is based on a DNN architecture to make a study of the estimation of weld bead parameters. This article mixed data from different welding processes. This is a risk for results analysis since different processes can have different outcomes with the same input parameters.

Rao et al. [94] utilized Generalized Regressive Neural Networks (GRNN) technique for estimating and optimizing the vibratory assisted welding parameters to produce quality welded joints. But in this case, it does not have comparison with other algorithms.

Di Wu, in 2017 [95], wrote a paper that addresses to perform Variable Polarity Plasma Arc Welding (VPPAW) process. Deep Belief Network (DBN), DNN variant, and t-Stochastic Neighbor Embedding (t-SNE) were studied for monitoring and identifying the penetration values. Experimental comparisons and verifications expose better performance for DBN, 97.62% exactly. This reaffirms the good results offered by the learning models developed with these algorithms. This work does not take the advantage of DNN algorithms to analyze both images and sound in real time.

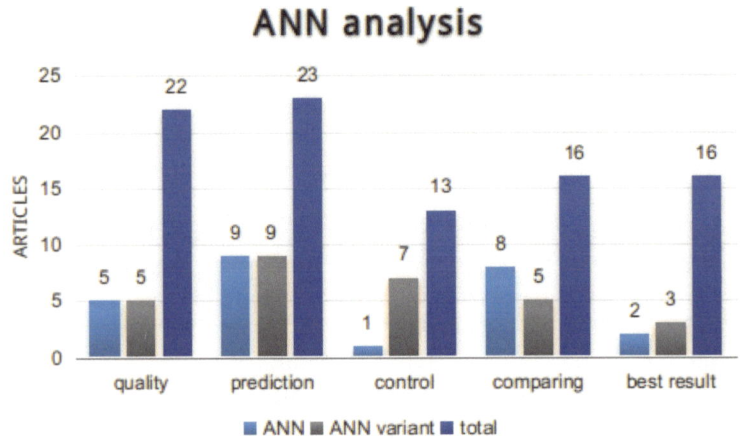

Figure 6.
Comparison between ANNs and ANN variations.

Figure 6 shows a summary of articles analyzed. It shows that ANNs are one of the most used techniques, but they do not always offer the best result. This demonstrates the need to make comparisons between various modeling techniques in order to define the best result, in terms of efficiency and computational cost.

5.2 Comparison of different models

As it has been expressed in the previous sections, there are new techniques to analyze very complex systems. But they require expensive computational resources for their construction and sometimes for their execution. A comparison between models will allow to know which model has better results and which model can be the most effective to be utilized. This effectivity is measured in function of problem necessity, like the one shown in data mining (DM) methodologies and processes [16, 17].

An interesting comparison is Support Vector Machine (SVM) and ANN model, to identify weld groove state and weld deviation extraction in rotating arc narrow gap MAG welding (RANGMW) [96]. It presented SVM models with better results than ANN model.

One comparison with focus on time optimized was [97]. It utilized an ANN and ANN with differential evolutionary algorithm (DEA) separately. The results obtained by ANN using DEA were closer to ANN, but the computational time of ANN using DEA was shorter.

In the article [98], Response Surface Methodology (RSM) was compared with linear isotonic regression, regression (LR), regression trees, ANN, GP, and SVM, to evaluate mechanical properties in GMAW process. The results present that the DM models have poorer generalization on this research, because DM techniques require, to obtain acceptable results, a large amount dataset.

Sumesh in 2015 [99] compared Decision Trees (DT), ANN, Fuzzy Logic, SVM, and Random forest technique Weld Quality Monitoring in SMAW. The most efficient technique was Random forest. This shows that not always the most complex techniques offer the best results.

One of the few comparative analysis algorithms is Kumar's paper in 2016 [100]. This paper explores Self-Organizing Maps (SOM) using as a mechanism for performing unsupervised learning, for comparing performance characteristics of various welding parameters which include welding power supplies and welders.

Results obtained using SOM has been compared with the Probability Density Distributions (PDDs) obtained during statistical analysis. Voltage and current data analyzed using the SOM technique can also be utilized to evaluate the arc welding process. These studies demonstrate that there are other potential algorithms for welding process analysis. For that reason, it is necessary to evaluate and compare several of them to be agreed upon in a real-time process.

Other comparison in 2016 by Di Wu is [54]. The article compared a prediction model for Plasma Arc Welding based on Extreme Learning Machine (ELM) with ANN and SVM techniques. The ELM model had better generalization performance and was faster than others. This potentiality was established too by Nandhitha in

Author	Year	Welding process	Sensors	Data preparations	Modeling	Online	Compare
Saini [52]	1998	GMAW	Sound	Classic	No	Yes	No
Yue [69]	2009	Pipeline welding	Visual	Classic	Theoretical model	No	No
Chen [64]	2010	LBW/ GMAW	Visual	Classic		Yes	No
Horvat [53]	2011	GMAW	Sound	Classic	No	Yes	No
Gao [78]	2011	GTAW	Visual	Classic	LR-ANN	No	No
Feng [80]	2012	GMAW	Standard	Classic	MDM	Yes	No
Fidali [45]	2013	GMAW	Infrared	Classic	Statistical analysis	Yes	No
Sreedhar [48]	2013	GTAW	Infrared	Classic	Statistical analysis	Yes	No
Kalaichelvi [101]	2013	GMAW	Standard	Classic	GA-Fuzzy	Yes	No
Kumar [97]	2014	GMAW	Visual	Classic	ANN, ANN-DEA	Yes	Yes
Deyong You [90]	2015	Laser welding	Photodiode, spectrometer	WPD-PCA	FFANN-SVM	Yes	No
Sumesh [99]	2015	SMAW	Sound	Classic	Some DM (RF)	Yes	Yes
Kumar [100]	2016	SMAW	Standard	Classic	PDDs, SOM	No	Yes
Muzaka [81]	2016	GMAW	Standard	Classic	MDM	Yes	No
Bai [82]	2016	GMAW	Standard	Classic	LMM	Yes	No
Park [83]	2017	RSW	Standard	Classic	GRNN-SVM	Yes	No
Wan [102]	2017	LSRSW	Standard	Classic	ANN (BP), ANN (Prob)	Yes	Yes
Huang [103]	2017	P-GTAW	Visual	Classic	DM, EMD	No	Yes
Petković [104]	2017	Laser welding	Multiples	Classic	SVM, ANN, GP	Yes	Yes
Muniategui [72]	2017	RSW	visual	DL, classic	Fuzzy	Yes	Yes
Wan [105]	2017	GTAW	visual	Classic	ANN and fuzzy	Yes	No

Table 2.
Table articles with quality objective

2016 [106]. He utilized GRNN and Radial Basis Networks (RBN) for torch current prediction in GTAW process. The torch current deviation was 98.95 % accuracy for the best result of GRNN.

In 2016 too, Kyoung-Yun Kim [107] discusses that in Resistance Spot Welding (RSW) process. He examined the prediction performance with GRNN and k-Nearest Neighbor (kNN) algorithms. The results indicate that with smaller k of kNN, the prediction performance measured by mean acceptable error has increased.

Author	Year	Welding process	Sensors	Data preparations	Modeling	Online	Compare
Bo Chen [84]	2009	GTAW	Multiples	Classic	ANN-DS	No	No
Bo Chen [91]	2010	GTAW	Multiples	Classic	ANN-fuzzy	No	No
Seyyedian [108]	2012	GTAW	Standard	Classic	ANN	Yes	No
Li [79]	2014	GTAW	Visual	Classic	LR	No	No
Bo Chen [89]	2014	UWW	Visual	Classic	ANN	Yes	No
Li [96]	2014	RANGMW	Visual	Classic	SVM, ANN	Yes	Yes
Escribano-García [98]	2014	GMAW	Standard	Classic	RSM, some DM	Yes	Yes
Sen [74]	2015	DP-GMAW	Standard	Classic	Taguchi-RSM	No	No
Keshmiri [93]	2015	SAW, GMAW, GTAW	Standard	Classic	DNN	Yes	No
Wu [54]	2016	VPPAW	Sound	Classic	ELM, ANN, SVM	Yes	Yes
Lv [55]	2016	GTAW	Sound	Classic	BP-Adaboost	Yes	Yes
Dong [77]	2016	GTAW	Standard	Classic	GPR	Yes	No
Sarkar [85]	2016	SAW	Standard	Classic	MRA and ANN	Yes	Yes
Rong [87]	2016	GTAW	Standard	Classic	ANN	Yes	No
Rios-Cabrera [92]	2016	GMAW	Visual	Classic	ANN fuzzy ARTMAP	Yes	No
Nandhitha [106]	2016	GTAW	Thermography	Classic	ELM, RBN, GRNN	Yes	Yes
Kim [107]	2016	RSW	Standard	Classic	kNN, GRNN	Yes	Yes
Aviles-Viñas [109, 110]	2016	GMAW	Visual	Classic	ANN-fuzzy	Yes	No
Pavan Kumar [86]	2017	GMAW CMT	Standard	Classic	ANN	Yes	No
Mathew [88]	2017	Girth welds	Standard	Classic	ANN	Yes	No
Di Wu [95]	2017	VP-PAW	Visual, sound	Classic	t-SNE and DBN	No	No

Table 3.
Table articles with prediction objective.

Other quality welding article was Xiaodong Wan in 2017 [102]. It proposed a Probabilistic Neural Network (PNN) model for quality prediction in large scale RSW process. In this case, the PNN model was more appropriate in quality level classification than the Back Propagation Neural Network.

The one of the last articles with direct DM techniques and welding relation is of Yiming Huang in 2017 [103]. This is an investigation of porosity on pulsed gas tungsten arc welding (P-GTAW) with an X-ray image analysis. To detect, an Empirical Mode Decomposition (EMD) and Spectral Analyses were made based on DM.

In 2017, Petković [104] predicted the laser welding quality by training data for the computational intelligence methodologies and support vector regression (SVR). SVR is a novel variant of SVM for regression task. This article made a comparison between SVR, ANN, and GP. It is another example that in certain problems, less complex algorithms can offer better results.

Table 2 presents a series of articles that were based on the monitoring and quality of the welding processes. The column *Preparation* defines the technique of processing the data obtained by the sensors; *Classic* for processes that do not use the latest techniques of image processing and DL for the use of deep learning; *Online* defines if the model was executed in real time; *Compare*, if in the research carried out in the article, a comparison is made between several algorithms; and *Modeling* defines the algorithms used in specific article. When a comparison exits, the first model before coma was the best quality result. As **Tables 2–4**, the best algorithm does not always match.

Defining which of the techniques is more effective for our problem also helps in the effectiveness of a future process of intelligent control.

Author	Year	Welding process	Sensors	Data preparations	Modeling	Online	Compare
Chen [66]	2000	P-GTAW	Double-visual	Classic	ANN-learning control	Yes	Yes
Chen [111]	2009	GTAW	Visual	Classic	ANN-fuzzy	Yes	No
Malviya [112]	2011	GMAW	Standard	Classic	ANN-PSO	Yes	No
Hailin [105]	2012	GMAW	Visual	Classic	ANN and fuzzy	Yes	No
Cruz [113]	2015	GMAW	Visual	Classic	ANN and fuzzy	Yes	No
Günther [63]	2016	Laser welding	Visual	DL	DL-RL	Yes	No
Santhana [75]	2016	GTAW	Standard	Classic	RSM	Yes	No
Sharma [114]	2016	SAW	Standard	Classic	RSM and fuzzy	Yes	No
Moghaddam [115]	2016	GMAW	Visual	Classic	ANN-PSO	Yes	No
Lv [56]	2017	GTAW	Sound	Classic	ANN	Yes	No
Rao [94]	2017	Vibratory Welding	Standard	Classic	GRNN	Yes	No
Pengfei Hu [116]	2017	GMAW	Standard	Classic	Math-model—fuzzy	Yes	No

Table 4.
Table articles with control objective.

6. Intelligent control of welding process

The intelligent control approach offers interesting perspectives since it is able to provide methodologies that allow to perform automatically some of the tasks typically performed by humans [117]. This combines with data mining models.

One intelligent control tendency utilized is a fuzzy method with ANN model. Example of this was [111] on GTAW process for predicting the dynamic of the weld pool; and in [105] for GMAW pipe-line welding, to improve the welding quality.

Another example was [113], on GMAW process, for modeling and control of weld bead width. Other example of fuzzy methods but different model techniques was [114]. It was applied for better control purpose of bead geometry parameters in submerged arc welding (SAW) process. This article proposed the response of a fuzzy logic approach with surface methodology (RSM). Demonstrating that any model obtained from a welding process can be integrated into a control system. As long as it meets time demands.

Conventional and intelligent control methods were investigated by [67] in P-GTAW process. This work made a comparison with PID control, fuzzy control, and neuron self-learning PSD control. It had better performance. This article highlights the advantage of learning-based control.

Other optimization based in learning was [115]. It proposed ANN model with a Particle Swarm Optimization (PSO) algorithm to optimize weld bead geometry characteristics on the GMAW process. The ANN-PSO model obtained an efficient optimization and multi-criteria modeling.

An emerging learning-based control system was used by Günther in [62, 63] for laser welding control. This technique is called reinforcement learning (RL). It is a machine learning branch. It is focused on decision-making by learning process [118]. Control learning can be an optimization-based method like Q-learning algorithm. It can be used to solve optimal control problems like expressed in [119].

Günther's study [63] is one of the few RL studies for laser welding system. This makes this work an important contribution to welding process engineering. RL is a new technique open now in welding process with noble success in other areas like appearing in [120–123].

7. Future perspective

These techniques of data analysis based on learning, as appearing in this article, is not yet widespread in welding process area. A bibliometric analysis among the authors studied in this research, presents a very little relationship between them. **Figure** 7 exposes this. The small dimensions of the authors' clouds (articles with welding process and new data analysis techniques) and their relationships (joint publications) show little maturity in the interrelation of these areas.

Some of the works demonstrate a small approximation between the areas, fulfilling the interdisciplinarity that Industry 4.0 advocates. Achieving this interdisciplinarity implies new study processes, defining new methodologies that unify the potential of these two areas. The needs of the modern world are going to make this happen in a short time. The new data analysis conception in welding processes area will be an acceleration in obtaining new and better models, more efficient predictions, and controls.

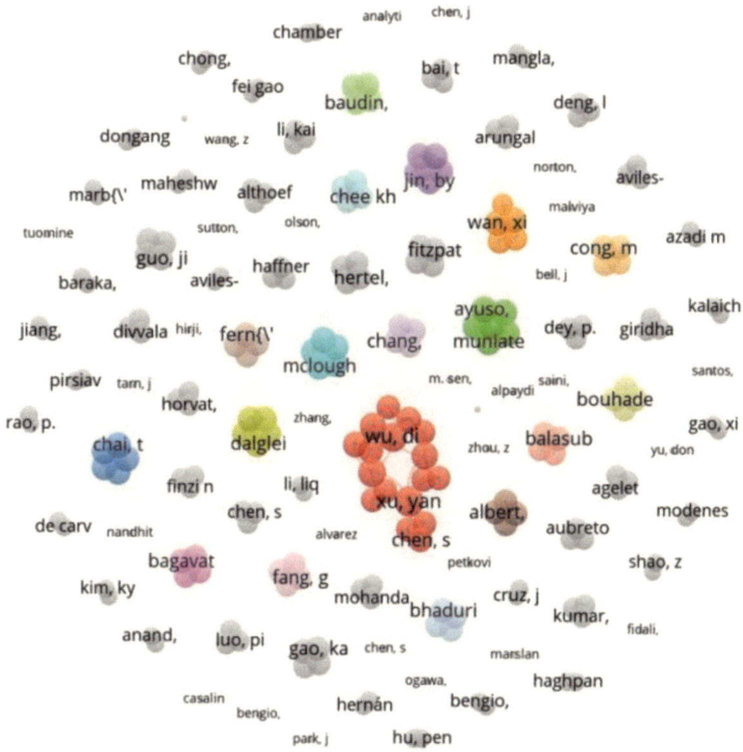

Figure 7.
Bibliometric analysis: authors' interrelationship.

8. Conclusions

Several articles about the welding process were analyzed. These allowed to determine for each data mining stage how it is possible to optimize the results to obtain a good result of process analysis. Several analysis algorithms of the welding process were shown, and it was demonstrated that the comparison between them can make the process analysis more efficient and less expensive. The potential of learning-based techniques was described, because computational resources are becoming cheaper, and more quality information of welding process can be obtained. All these premises aligned with the so-called Industry 4.0, where a set of technologies that allow a fusion of physical and digital world, create a more intelligent and dynamic system.

Acknowledgements

The authors would like to acknowledge IntechOpen, Brasilia University, CNPq, CAPES, and PPMEC-UnB, and also to professors Alysson Martin Silva, and Guillermo Albarez Bestard.

Author details

Rogfel Thompson Martinez and Sadek Crisóstomo Absi Alfaro*
Department of Mechanical Engineering, Brasilia University, Brazil

*Address all correspondence to: sadek@unb.br

References

[1] Villani P, Modenesi PJ, Bracarense AQ. Soldagem: Fundamentos e Tecnologia. Brasil: Elsevier; 2016

[2] Zhang YM. Institute of Materials, Real-time Weld Process Monitoring. Woodhead Pub. and Maney Pub. on behalf of the Institute of Materials, Minerals and Mining; 2008. Available from: http://www.sciencedirect.com/science/book/9781845692681

[3] Chen SB, Lv N. Research evolution on intelligentized technologies for arc welding process. Journal of Manufacturing Processes. 2014;**16**(1):109-122

[4] Haffner O, Kucera E, Kozak S, Stark E. Proposal of system for automatic weld evaluation. In: 2017 21st International Conference on Process Control (PC). IEEE; 2017. pp. 440-445. Available from: http://ieeexplore.ieee.org/document/7976254/

[5] Jiang C, Zhang F, Wang Z. Image processing of aluminum alloy weld pool for robotic VPPAW based on visual sensing. IEEE Access. 2017;**5**:21567-21573. Available from: http://ieeexplore.ieee.org/document/8064625/

[6] Chong L, Ramakrishna S, Singh S. A review of digital manufacturing-based hybrid additive manufacturing processes. The International Journal of Advanced Manufacturing Technology. 2018;**95**(5-8):2281-2300. Available from: http://link.springer.com/10.1007/s00170-017-1345-3

[7] Tuominen V. The measurement-aided welding cellgiving sight to the blind. The International Journal of Advanced Manufacturing Technology. 2016;**86**(1-4):371-386. Available from: http://link.springer.com/10.1007/s00170-015-8193-9

[8] Hernandez Orallo J, Ramirez Quintana MJ, Ferri Ramirez C. Introduccion a la Mineria de Datos. NJ, USA: Pearson Prentice Hall; 2004

[9] Marsland S. Machine Learning, An Algorithmic Perspective. USA: CRC Press; 2015

[10] Bell J. Machine Learning: Hands-On for Developers and Technical Professionals. Indianapolis, IN, USA: John Wiley & Sons, Inc.; 2015

[11] Casalino G. [INVITED] Computational intelligence for smart laser materials processing. Optics & Laser Technology. 2018;**100**:165-175. Available from: https://linkinghub.elsevier.com/retrieve/pii/S0030399217303286

[12] Yu D, Deng L. Deep learning and its applications to signal and information processing [exploratory DSP]. IEEE Signal Processing Magazine. 2011;**28**(1):145-154. Available from: http://ieeexplore.ieee.org/document/5670617/

[13] Hirji KK. Discovering data mining: From concept to implementation. SIGKDD Explorations Newsletter. 1999;**1**(1):44-45. Available from: http://doi.acm.org/10.1145/846170.846181

[14] Norton MJ. Knowledge discovery in databases. Library Trends. 1999;**48**(1):9-21. Available from: https://search.proquest.com/docview/220463919?accountid=26646

[15] Olson DL, Delen D. Advanced Data Mining Techniques. 1st ed. NY, USA: Springer Publishing Company, Incorporated; 2008

[16] Piatetsky G. CRISP-DM, still the top methodology for analytics, data mining, or data science projects. 2014. [Online].

Available from: http://www.kdnuggets.com/2014/10/crisp-dm-top-methodology-analytics-data-mining-data-science-projects.html. [Accessed: 27 July 2017]

[17] Chambers M, Doig C, Stokes-Rees I. Breaking Data Science Open. 1st ed. CA, USA: O'Reilly Media, Inc; 2017

[18] Huang Z, Xu X, He H, Tan J, Sun Z. Parameterized batch reinforcement learning for longitudinal control of autonomous land vehicles. IEEE Transactions on Systems, Man, and Cybernetics: Systems. 2019;**49**(4): 730-741

[19] Chi R, Hou Z, Jin S, Huang B. An improved data-driven point-to-point ilc using additional on-line control inputs with experimental verification. IEEE Transactions on Systems, Man, and Cybernetics: Systems. 2019;**49**(4): 687-696

[20] Woods AC, La HM. A novel potential field controller for use on aerial robots. IEEE Transactions on Systems, Man, and Cybernetics: Systems. 2019;**49**(4):665-676

[21] Shalev-Shwartz S, Ben-David S. Understanding Machine Learning: From Theory to Algorithms. New York, USA: Cambridge University Press; 2014

[22] Alpaydin E. Introduction to Machine Learning. USA: Massachusetts Institute of Technology; 2010

[23] Mesnil G, He X, Deng L, Bengio Y. Investigation of recurrent-neural-network architectures and learning methods for spoken language understanding Iterspeech. In: Bimbot F, Cerisara C, Fougeron C, Gravier G, Lamel L, Pellegrino F, et al. ISCA. 2013. pp. 3771-3775

[24] Zhu Z, Luo P, Wang X, Tang X. Multi-View Perceptron: A Deep Model for Learning Face Identity and View Representations. 2014. pp. 217-225

[25] Pachitariu M, Packer AM, Pettit N, Dalgleish H, Hausser M, Sahani M. Extracting regions of interest from biological images with convolutional sparse block coding. 2013. pp. 1745-1753

[26] Yang J, Price B, Cohen S, Lee H, Yang M-H. Object contour detection with a fully convolutional encoder-decoder network. Cvpr 2016. 2016. Available from: http://arxiv.org/abs/1603.04530

[27] Pachauri D, Kondor R, Sargur G, Singh V. Permutation Diffusion Maps (PDM) with Application to Image Association Problem in Computer Vision. 2014. pp. 541-549

[28] Redmon J, Divvala S, Girshick R, Farhadi A. You only look once: Unified, real-time object detection. Cvpr 2016. 2016. pp. 779-788

[29] Vondrick C, Pirsiavash H, Torralba A. Anticipating visual representations from unlabeled video. In: IEEE Conference on Computer Vision and Pattern Recognition. 2015. Available from: http://arxiv.org/abs/1504.08023

[30] Zheng S, Dongang W, Shih-Fu C. Temporal action localization in untrimmed videos via multi-stage CNNs. In: IEEE Conference on Computer Vision and Pattern Recognition (CVPR), 2016. 2016. pp. 1049-1058

[31] Luo S, Zhu L, Althoefer K, Liu H. Knock-Knock: Acoustic object recognition by using stacked denoising autoencoders. Neurocomputing. 2017; **267**:18-24. Available from: http://linkinghub.elsevier.com/retrieve/pii/S092523121730509X

[32] McLoughlin I, Zhang H, Xie Z, Song Y, Xiao W, Phan H. Continuous robust sound event classification using time-frequency features and deep learning. PLoS ONE. 2017;**12**(9):

e0182309. Available from: http://dx.plos.org/10.1371/journal.pone.0182309

[33] Zhou Z-H. Three perspectives of data mining. Artificial Intelligence. 2003;**143**(1):139-146. Available from: http://citeseerx.ist.psu.edu/viewdoc/download?doi=10.1.1.2.3790&rep=rep1&type=pdf

[34] Marbán Ó, Mariscal G, Segovia J. A data mining & knowledge discovery process model. Data Mining and Knowledge. 2009;(February):1-17. Available from: http://cdn.intechopen.com/pdfs/5937/InTech-Adataminingampknowledgediscoveryprocessmodel.pdf

[35] C. Analytics. Web of Science. 2018. Available from: http://webofknowledge.com

[36] AWS. Welding Inspection Handbook 3rd Edition. 2000

[37] Modenesi PJ. Introdução à Física do Arco Elétrico e sua Aplicação na Soldagem dos Metais. Dep. Eng. Met. e Mater. Univ. Fed. Minas Gerais—UFMG. 2004. p. 159

[38] Abdullahi I, Hamza MF. A review on the application of resistance spot welding of automotive sheets. 2015;(December)

[39] Ouisse M, Cogan S. Robust design of spot welds in automotive structures: A decision-making methodology. Mechanical Systems and Signal Processing. 2010;**24**(4):1172-1190

[40] Dawes CT. Laser Welding: A Practical Guide. 1992

[41] Mazmudar CP, Patel K. Effect of laser welding process parameters on mechanical properties of stainless steel-316. 2014;**1**(5):1-11

[42] Alvarez Bestard G. Sensor fusion and embedded devices to estimate and control the depth and width of the weld bead in real time [Ph.D. thesis, Ph.D. dissertation]. 2017. Available from: http://repositorio.unb.br/handle/10482/31429

[43] Nagarajan S, Nagarajan S, Banerjee P, Banerjee P, Chen W, Chen W, et al. Control of the welding process using infrared sensors. Society. 1992;**8**(1):86-93

[44] Mota CP, Machado MVR, Finzi Neto RM, Vilarinho LO. Sistema de visão por infravermelho próximo para monitoramento de processos de soldagem a arco. Soldagem & Inspeção. 2013;**18**(1):19-30

[45] Fidali M, Jamrozik W. Diagnostic method of welding process based on fused infrared and vision images. Infrared Physics & Technology. 2013;**61**:241-253

[46] Bagavathiappan S, Lahiri BB, Saravanan T, Philip J, Jayakumar T. Infrared thermography for condition monitoring—A review. Infrared Physics & Technology. 2013;**60**:35-55

[47] Vilarinho LO, Mota CP, Machado MVR, Finzi Neto RM. Near-infrared vision system for arc-welding monitoring. In: DebRoy T, David SA, JN DP, Koseki T, Bhadeshia HK, editors. Trends in Welding Research: Proceedings of the 9th International Conference. Proceedings Paper. ASM Int. 9503 Kinsman Rd, Materials Park, OH 44073 USA: ASM International; 2013. pp. 1029-1037

[48] Sreedhar U, Krishnamurthy CV, Balasubramaniam K, Raghupathy VD, Ravisankar S. Automatic defect identification using thermal image analysis for online weld quality monitoring. Journal of Materials Processing Technology. 2012;**212**(7):1557-1566

[49] Vasudevan M, Chandrasekhar N, Maduraimuthu V, Bhaduri AK, Raj B.

Real-time monitoring of wield pool during gtaw using infra-red thermography and analysis of infra-red thermal images. Welding in the World. 2011;**55**(7-8):83-89

[50] Benoit A, Paillard P, Baudin T, Klosek V, Mottin JB. Comparison of four arc welding processes used for aluminium alloy cladding. Science and Technology of Welding and Joining. 2015;**20**(1):75-81

[51] Tarn J, Huissoon J. Developing psycho-acoustic experiments in gas metal arc welding. IEEE International Conference Mechatronics and Automation. 2005, 2014;**2**(January): 1112-1117. Available from: http://ieeexplore.ieee.org/document/1626707/

[52] Saini BYD. An Investigation of Gas Metal Arc Welding Sound Signature for On-Line Quality Control. 1998. pp. 172-179. Available from: http://files.aws.org/wj/supplement/WJ199804s172.pdf

[53] Horvat J, Prezelj J, Polajnar I, Čudina M. Monitoring gas metal arc welding process by using audible sound signal. Strojniški Vestnik Journal of Mechanical Engineering. 2011;**2011**(03): 267-278

[54] Wu D, Chen H, He Y, Song S, Lin T, Chen S. A prediction model for keyhole geometry and acoustic signatures during variable polarity plasma arc welding based on extreme learning machine. Sensor Review. 2016;**36**(3):257-266

[55] Lv N, Xu YL, Fang G, Yu XW, Chen SB. Research on welding penetration state recognition based on BP-Adaboost model for pulse GTAW welding dynamic process. In: Proceedings of IEEE Workshop on Advanced Robotics and its Social Impacts, ARSO. Vol. 2016. IEEE; 2016. pp. 100-105. Available from: http://ieeexplore.ieee.org/document/7736264/

[56] Lv N, Xu Y, Li S, Yu X, Chen S. Automated control of welding penetration based on audio sensing technology. Journal of Materials Processing Technology. 2017;**250**:81-98. Available from: http://linkinghub.elsevier.com/retrieve/pii/S0924013617302777

[57] LeCun Y, Bengio Y, Hinton G. Deep learning. Nature. 2015;**521**(7553): 436-444. Available from: http://www.nature.com/doifinder/10.1038/nature14539

[58] Xu Y, Yu H, Zhong J, Lin T, Chen S. Real-time image capturing and processing of seam and pool during robotic welding process. Industrial Robot—An International Journal. 2012; **39**(5):513-523

[59] Liu Y-K, Huang N, Zhang Y-M. Modeling of human welder response against 3D weld pool surface using machine-human cooperative virtualized welding platform. In: Tarn TJ, Chen SB, Chen XQ, editors. Robotic Welding, Intelligence and Automation, RWIA'2014, Ser. Advances in Intelligent Systems and Computing. Proceedings Paper. Vol. 363. Heidelberger Platz 3, D-14197 Berlin, Germany: Springerverlag Berlin; 2015. pp. 451-457

[60] Ogawa Y. High speed imaging technique. Part 1—High speed imaging of arc welding phenomena. Science and Technology of Welding and Joining. 2011;**16**(1):33-43

[61] Gao F, Chen Q, Guo L. Study on arc welding robot weld seam touch sensing location method for structural parts of hull. In: 2015 International Conference on Control, Automation and Information Sciences (ICCAIS). IEEE; 2015. pp. 42-46

[62] Günther J, Pilarski PM, Helfrich G, Shen H, Diepold K. First steps towards an intelligent laser welding architecture

using deep neural networks and reinforcement learning. Procedia Technology. 2014;**15**:474-483

[63] Günther J. Intelligent laser welding through representation, prediction, and control learning: An architecture with deep neural networks and reinforcement learning. Mechatronics. 2016;**34**:1-11. Available from: http://linkinghub.elsevier.com/retrieve/pii/S0957415815001555

[64] Chen JZ, Farson DF. Hybrid welds coaxial vision monitoring of LBW/GMAW hybrid welding process. Materials Evaluation. 2010;**68**(12):1318-1328

[65] Ramos EG, de Carvalho GC, Absi Alfaro SC. Analysis of weld pool oscillation in P-GMAW by means of shadowgraphy image processing. Soldagem & Inspeção. 2013;**18**(1):39-49

[66] Siewert E, Wilhelm G, Haessler M, Schein J, Hanson T, Schnick M, et al. Visualization of gas flows in welding arcs by the Schlieren measuring technique. Welding Journal. 2014;**93** (January):1-5

[67] Chen SB, Lou YJ, Wu L, Zhao DB. Intelligent methodology for sensing, modeling and control of pulsed GTAW: Part I—Bead-on-plate welding. Welding Journal. 2000;**79**(6):151s-163s

[68] Ma G, Li L, Chen Y. Effects of beam configurations on wire melting and transfer behaviors in dual beam laser welding with filler wire. Optics and Laser Technology. 2017;**91**(April): 138-148. DOI: 10.1016/j.optlastec.2016.12.019

[69] Yue H, Li K, Zhao HW, Zhang Y. Vision-based pipeline girth-welding robot and image processing of weld seam. Industrial Robot—An International Journal. 2009;**36**(3): 284-289. Available from: http://www.emeraldinsight.com/doi/10.1108/01439910910950568

[70] Xu Y, Fang G, Chen S, Zou JJ, Ye Z. Real-time image processing for vision-based weld seam tracking in robotic GMAW. International Journal of Advanced Manufacturing Technology. 2014;**73**(9-12):1413-1425

[71] Wu Q-Q, Lee J-P, Park M-H, Park C-K, Kim I-S. A study on development of optimal noise filter algorithm for laser vision system in GMA welding. In: Xavior MA, PKDV Y, editors. 12th Global Congress on Manufacturing and Management (GCMM—2014), ser. Procedia Engineering. Proceedings Paper. Vol. 97. VIT Univ, Sch Mech & Bldg Sci; Queensland Univ Technol. Sara Burgerhartstraat 25, PO BOX 211, 1000 AE Amsterdam, Netherlands: Elsevier Science BV; 2014. pp. 819-827

[72] Muniategui A, Hériz B, Eciolaza L, Ayuso M, Iturrioz A, Quintana I, et al. Spot welding monitoring system based on fuzzy classification and deep learning. In: 2017 IEEE International Conference on Fuzzy Systems (FUZZ-IEEE). IEEE; 2017. pp. 1-6. Available from: http://ieeexplore.ieee.org/document/8015618/

[73] Hou W, Wei Y, Guo J, Jin Y, Zhu C. Automatic detection of welding defects using deep neural network. Journal of Physics: Conference Series. 2018;**933**: 012006

[74] Sen M, Mukherjee M, Pal TK. Evaluation of correlations between DP-GMAW process parameters and bead geometry. Welding Journal. 2015; (July):265-279

[75] Santhana Babu AV, Giridharan PK, Ramesh Narayanan P, Narayana Murty SVS. Prediction of bead geometry for flux bounded TIG welding of AA 2219-T87 aluminum alloy. Journal of Advanced Manufacturing Systems. 2016;**15**(02):69-84. Available from: http://www.worldscientific.com/doi/abs/10.1142/S0219686716500074

[76] Boutaghane A, Bouhadef K, Valensi F, Pellerin S, Benkedda Y. Theoretical model and experimental investigation of current density boundary condition for welding arc study. European Physical Journal-Applied Physics. 2011;**54**(1):13

[77] Dong H, Cong M, Liu Y, Zhang Y, Chen H. Predicting characteristic performance for arc welding process. In: 2016 IEEE International Conference on Cyber Technology in Automation, Control and Intelligent Systems (CYBER). IEEE; 2016. pp. 7-12

[78] Gao X, Ding D, Bai T, Katayama S. Weld-pool image centroid algorithm for seam-tracking vision model in arc-welding process. IET Image Processing. 2011;**5**(5):410-419

[79] Li Z, Gao X. Study on regression model of measuring weld position. In: Choi SB, Yarlagadda P, AbdullahAlWadud M, editors. Sensors, Mechatronics and Automation, Ser. Applied Mechanics and Materials. Proceedings Paper. Vol. 511-512. Laublsrutistr 24, CH-8717 Stafa-Zurich, Switzerland: Trans Tech Publications Ltd; 2014. pp. 514-517

[80] Feng S, Lin G, Ma B, Hu S. A novel measurement and qualification method of GMAW welding fault based on digital signals. In: Chen WZ, Xu XP, Dai PQ, Chen YL, editors. Advanced Manufacturing Technology, Pts 1-4, Ser. Advanced Materials Research. Proceedings Paper. Vol. 472-475. Fujian Univ Technol; Xiamen Univ; Fuzhou Univ; Huaqiao Univ; Univ Wollongong; Fujian Mech Engn Soc; Hong Kong Ind Technol Res Ctr. Laublsrutistr 24, CH-8717 Stafa-Zurich, Switzerland: Trans Tech Publications Ltd; 2012. pp. 1201-1205

[81] Muzaka K, Park MH, Lee JP, Jin BJ, Lee BR, Kim WYIS. A study on prediction of welding quality using mahalanobis distance method by optimizing welding current for a vertical-position welding. Procedia Engineering. 2017;**174**:60-67. Available from: http://linkinghub.elsevier.com/retrieve/pii/S1877705817301431

[82] Bai F, Lubecki TM. Robotic arc welding with on-line process monitoring based on the LMM analysis of the welding process stability. In: 2016 IEEE International Conference on Advanced Intelligent Mechatronics (AIM). 2016. pp. 566-571. Available from: http://ieeexplore.ieee.org/document/7576828/

[83] Park J, Kim K-Y. Prediction modeling framework with bootstrap aggregating for noisy resistance spot welding data. Journal of Manufacturing Science and Engineering. 2017;**139**(10): 101003

[84] Chen B, Wang J, Chen S. Prediction of pulsed GTAW penetration status based on BP neural network and D-S evidence theory information fusion. International Journal of Advanced Manufacturing Technology. 2010;**48** (1-4):83-94

[85] Sarkar A, Dey P, Rai R, Saha S. A comparative study of multiple regression analysis and back propagation neural network approaches on plain carbon steel in submerged-arc welding. Sadhana—Academy Proceedings in Engineering Sciences. 2016;**41**(5):549-559

[86] Pavan Kumar N, Devarajan PK, Arungalai Vendan S, Shanmugam N. Prediction of bead geometry in cold metal transfer welding using back propagation neural network. The International Journal of Advanced Manufacturing Technology. 2017;**93** (1-4):385-392. Available from: http://link.springer.com/10.1007/s00170-016-9562-8

[87] Rong Y, Huang Y, Zhang G, Chang Y, Shao X. Prediction of angular

distortion in no gap butt joint using BPNN and inherent strain considering the actual bead geometry. International Journal of Advanced Manufacturing Technology. 2016;**86**(1-4):59-69. Available from: http://link.springer.com/10.1007/s00170-015-8102-2

[88] Mathew J, Moat R, Paddea S, Fitzpatrick M, Bouchard P. Prediction of residual stresses in girth welded pipes using an artificial neural network approach. International Journal of Pressure Vessels and Piping. 2017;**150**: 89-95

[89] Chen B, Feng J. Modeling of underwater wet welding process based on visual and arc sensor. Industrial Robot—An International Journal. 2014; **41**(3):311-317

[90] You D, Gao X, Katayama S. WPD-PCA-based laser welding process monitoring and defects diagnosis by using FNN and SVM. IEEE Transactions on Industrial Electronics. 2015;**62**(1):628-636

[91] Chen B, Chen S. Multi-sensor information fusion in pulsed GTAW based on fuzzy measure and fuzzy integral. Assembly Automation. 2010; **30**(3):276-285

[92] Rios-Cabrera R, Morales-Diaz AB, Aviles-Viñas JF, Lopez-Juarez I. Robotic GMAW online learning: Issues and experiments. International Journal of Advanced Manufacturing Technology. 2016;**87**(5-8):2113-2134

[93] Keshmiri S, Zheng X, Feng LW, Pang CK, Chew CM. Application of deep neural network in estimation of the weld bead parameters. In: IEEE International Conference on Intelligent Robots and Systems. Vol. 2015. 2015. pp. 3518-3523. Available from: http://arxiv.org/abs/1502.04187

[94] Rao PG, Srinivasa Rao P, Deepak BB. GRNN-immune based strategy for estimating and optimizing the vibratory assisted welding parameters to produce quality welded joints. Engineering Journal. 2017;**21**(3):251-267

[95] Wu D, Huang Y, Chen H, He Y, Chen S. VP-PAW penetration monitoring based on fusion of visual and acoustic signals using t-SNE and DBN model. Materials and Design. 2017; **123**:1-14. Available from: http://linkinghub.elsevier.com/retrieve/pii/S0264127517302721

[96] Li W, Gao K, Wu J, Hu T, Wang J. SVM-based information fusion for weld deviation extraction and weld groove state identification in rotating arc narrow gap MAG welding. International Journal of Advanced Manufacturing Technology. 2014;**74**(9-12):1355-1364

[97] Kumar GS, Natarajan U, Veerarajan T, Ananthan SS. Quality level assessment for imperfections in GMAW. Welding Journal. 2014;**93**(3): 85S-97S

[98] Escribano-García R, Lostado-Lorza R, Fernández-Martínez R, Villanueva-Roldán P, Mac Donald BJ. Improvement in manufacturing welded products through multiple response surface methodology and data mining techniques. Advances in Intelligent Systems and Computing. 2014;**299**: 301-310

[99] Sumesh A, Rameshkumar K, Mohandas K, Babu RS. Use of machine learning algorithms for weld quality monitoring using acoustic signature. Procedia Computer Science. 2015;**50**: 316-322. Available from: http://linkinghub.elsevier.com/retrieve/pii/S1877050915005438

[100] Kumar V, Albert SK, Chandrasekhar N, Jayapandian J, Venkatesan MV. Performance analysis of arc welding parameters using self organizing maps and probability density

distributions. In: 2016 IEEE First International Conference on Control, Measurement and Instrumentation (CMI). IEEE; 2016. pp. 196-200

[101] Kalaichelvi V, Karthikeyan R, Sivakumar D. Analysis of gas metal arc welding process using GA tuned fuzzy rule based system. Journal of Intelligent & Fuzzy Systems. 2013; **25**(2):429-440

[102] Wan X, Wang Y, Zhao D, Huang Y. A comparison of two types of neural network for weld quality prediction in small scale resistance spot welding. Mechanical Systems and Signal Processing. 2017;**93**:634-644

[103] Huang Y, Wu D, Lv N, Chen H, Chen S. Investigation of porosity in pulsed GTAW of aluminum alloys based on spectral and X-ray image analyses. Journal of Materials Processing Technology. 2017;**243**:365-373

[104] Petković D. Prediction of laser welding quality by computational intelligence approaches. Optik. 2017; **140**:597-600. Available from: http://linkinghub.elsevier.com/retrieve/pii/S0030402617304928

[105] Hailin H, Jing L, Fang L, Wei Z, Heqiang P. Neural-fuzzy variable gap control method for GMAW pipe-line welding with CCD camera. In: Zhao H, editor. Mechanical and Electronics Engineering III, Pts 1-5, Ser. Applied Mechanics and Materials. Proceedings Paper. Vol. 130-134. Hefei UnivTechnol. Laublsrutistr 24, CH-8717 Stafa-Zurich, Switzerland: Trans Tech Publications Ltd; 2012. pp. 2358-2363

[106] Nandhitha NM. Artificial Neural Network Based Prediction Techniques for Torch Current Deviation to Produce Defect-Free Welds in GTAW Using IR Thermography. 2016. pp. 137-142. Available from: http://link.springer.com/10.1007/978-81-322-2538-614

[107] Kim KY, Park J, Sohmshetty R. Prediction measurement with mean acceptable error for proper inconsistency in noisy weldability prediction data. Robotics and Computer-Integrated Manufacturing. 2017;**43**:18-29

[108] Seyyedian Choobi M, Haghpanahi M, Sedighi M. Prediction of welding-induced angular distortions in thin butt-welded plates using artificial neural networks. Computational Materials Science. 2012; **62**:152-159

[109] Aviles-Viñas JF, Rios-Cabrera R, Lopez-Juarez I. On-line learning of welding bead geometry in industrial robots. International Journal of Advanced Manufacturing Technology. 2016;**83**(1-4):217-231

[110] Wan X, Wang Y, Zhao D, Huang YA, Yin Z. Weld quality monitoring research in small scale resistance spot welding by dynamic resistance and neural network. Measurement: Journal of the International Measurement Confederation. 2017;**99**:120-127

[111] Chen SB, Wang WY, Ma HB. Intelligent control of arc welding dynamics during robotic welding process. In: Chandra T, Wanderka N, Reimers W, Ionescu M, editors. Thermec 2009, PTS 1-4, Ser. Materials Science Forum. Proceedings Paper. Vol. 638-642. Minerals, Met & Mat Soc. Laublsrutistr 24, CH-8717 Stafa-Zurich, Switzerland: Trans Tech Publications Ltd; 2010. pp. 3751-3756

[112] Malviya R, Pratihar DK. Tuning of neural networks using particle swarm optimization to model MIG welding process. Swarm and Evolutionary Computation. 2011;**1**(4):223-235. Available from: https://www.sciencedirect.com/science/article/abs/pii/S221065021100040X

[113] Cruz JG, Torres EM, Alfaro SCA. A methodology for modeling and control of weld bead width in the GMAW process. Journal of the Brazilian Society of Mechanical Sciences and Engineering. 2015;**37**(5):1529-1541

[114] Sharma SK, Maheshwari S, Rathee S. Multi-objective optimization of bead geometry for submerged arc welding of pipeline steel using RSM-fuzzy approach. Journal for Manufacturing Science and Production. 2016;**16**(3):141-151

[115] Azadi Moghaddam M, Golmezergi R, Kolahan F. Multivariable measurements and optimization of GMAW parameters for API-X42 steel alloy using a hybrid BPNNPSO approach. Measurement. 2016;**92**: 279-287

[116] Wang Z. Monitoring of GMAW weld pool from the reflected laser lines for real-time control. IEEE Transactions on industrial informatics. 2014;**10**(4): 2073-2083

[117] Santos M. Un enfoque aplicado del control inteligente. RIAI—Revista Iberoamericana de Automatica e Informatica Industrial. 2011;**8**(4): 283-296. Available from: http://linkinghub.elsevier.com/retrieve/pii/S1697791211000501

[118] Sutton R, Barto A. Reinforcement learning: An introduction. Trends in Cognitive Sciences. 1999;**3**(9):360

[119] Li J, Chai T, Lewis FL, Fan J, Ding Z, Ding J. Off-policy Q-learning: Set-point design for optimizing dual-rate rougher flotation operational processes. IEEE Transactions on Industrial Electronics. 2018;**65**(5): 4092-4102

[120] Chincoli M, Liotta A. Self-learning power control in wireless sensor networks. Sensors. 2018;**18**(2):375. Available from: http://www.mdpi.com/1424-8220/18/2/375

[121] Ramanathan P, Mangla KK, Satpathy S. Smart controller for conical tank system using reinforcement learning algorithm. Measurement: Journal of the International Measurement Confederation. 2018;**116**: 422-428

[122] Yin L, Yu T, Zhou L. Design of a novel smart generation controller based on deep Q learning for large-scale interconnected power system. Journal of Energy Engineering. 2018;**144**(3): 04018033

[123] Hu P, Huang J, Zeng M. Application of fuzzy control method in gas metal arc welding. The International Journal of Advanced Manufacturing Technology. 2017;**92**(5-8):1769-1775. Available from: http://link.springer.com/10.1007/s00170-017-0245-x

Chapter 3

Automatic Control of the Weld Bead Geometry

Guillermo Alvarez Bestard and Sadek Crisostomo Absi Alfaro

Abstract

Automatic control of the welding process is complex due to its nonlinear and stochastic behavior and the difficulty for measuring the principal magnitudes and closing the control loop. Fusion welds involve melting and subsequent solidification of one or more materials. The geometry of the weld bead is a good indicator of the melting and solidification process, so its control is essential to obtain quality junctions. Different sensing, modeling, estimation, and control techniques are used to overcome this challenge, but most of the studies are using static single-input/single-output models of the process and focusing on the flat welding position. However, theory and practice demonstrate that dynamic models are the best representation to obtain satisfactory control performance, and multivariable techniques reduce the effect of interactions between control loops in the process. Also, many industrial applications need to control orbital welding. In this chapter, the above topics are discussed.

Keywords: arc welding, feedforward decoupling, multivariable control, weld bead geometry, welding control

1. Introduction

All fusion welds involve the melting and subsequent solidification of the base metal. The geometry of the weld bead is a good indicator of the melting and solidifying process. Generally, weld inspection starts by evaluating this weld bead geometry and is followed by further inspection of the mechanical properties and metallurgical structures [1]. Many resources and time are employed in the final inspection of the weld bead, which is conducted when the part is finished. At that time, the problem usually has no solution or the solution is very expensive. This is one of the reasons why the control of the weld pool geometry is imperative to improve the quality in the weld and to reduce the cost of welded components.

Research and development in this area have increased in the last five decades, starting from simple control methods and analogical devices, as shown in [2], to complex algorithms and digital devices and computers, as shown in [3, 4]. But, in the literature analyses made in [5], it is possible to observe that most of the cases (90%) of the developments work in horizontal position and only 10% work in orbital welding, despite the importance of orbital welding for the industry. It is important to note that this type of welding imposes strong challenges in the use of sensors due to orbital movement (which can be quite irregular) around the piece. The main challenges are in the size and portability of the sensor, flexibility in the

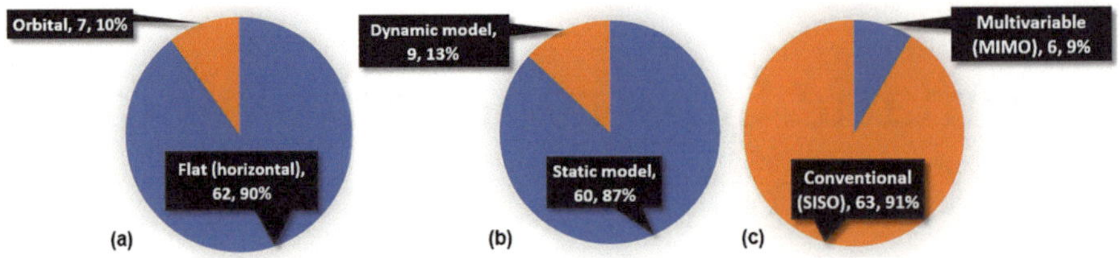

Figure 1.
Statistical information of publications about control systems of the weld bead geometry: (a) welding positions; (b) model types; (c) control type.

communication lines, continuous changes in position and lighting conditions (for optical sensors), and the effects of the force of gravity. These statistical data are shown in **Figure 1a**.

In addition, most cases apply static models, do not control all parameters of the bead geometry, and do not apply multivariable techniques, as shown in **Figure 1b** and **c**. The dynamic models can be a better representation of the process, producing better prediction results. The research where these models were used represents only 13% of the total, as analyzed in [5]. The black box model approach is widely used. Because of the complex characteristics of the process, a physic model approach is very difficult and needs extensive research and resources.

In this chapter the principal control loops and techniques used for online control of the weld bead geometry are discussed. The more usual variables, control, and techniques to modeling, used in welding power sources and welding robotic systems, are critically discussed. Some examples of singled and multivariable control loops are shown. A decoupling technique for multivariable loops is also explained. The dead time and disturbances that can affect the processes and some techniques to determine them are also explained. A special topic about the embedded systems in the welding process was included.

This chapter aims to create a knowledge base necessary to understand the main control systems in welding processes before addressing more complex control techniques. Its main contributions are the exhaustive literature review that is critically discussed and the solutions provided for the control of each part of the process, especially the control of the weld bead geometry for electric arc welding processes.

2. Control loops in welding processes

A typical automatic closed-loop control is composed of a controller or control system, an actuator system, and a measurement system or sensor, as shown in **Figure 2a**. The controller calculates the control law based on the control error, which is the difference between the set point value and the measurement of the controlled variable. The actuator modifies the process state, based on controller output (manipulate variable), to bring the controlled variable to the desired value. In sequence, to close the loop, the measurement system obtains the value of the controlled variable and sends it to the controller. An open-loop control does not have a measurement system, or the controller does not use its feedback as shown in **Figure 2b**.

The selection of variables to the control loop is a very important task. For this, it is necessary to analyze and quantify the influence of all process variables on the variable to be controlled. A statistical tool to quantify these relationships is the cross-correlation, using experimental data series of these variables and the variable

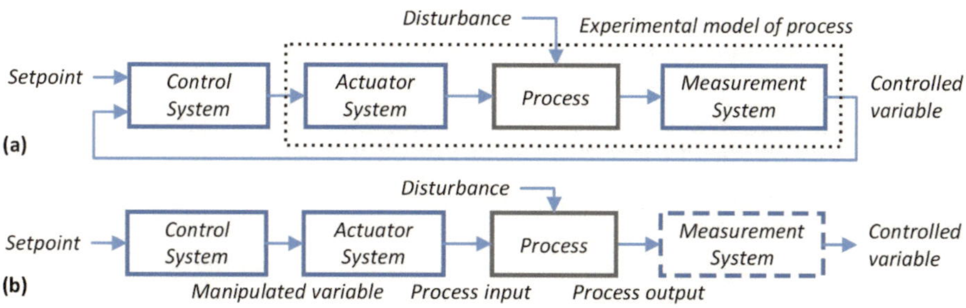

Figure 2.
Representation of a typical automatic control system: (a) closed-loop control; (b) open-loop control.

to be controlled. Other factors must also be considered, such as the actual possibility of modifying each of these variables and how these modifications affect other control loops. These correlations are mentioned in the literature, but often they are not totality quantified, indicating that there is still a wide-open field of research in modeling these processes.

The dead time (transportation lag or time delay) is another important parameter that can be obtained from cross-correlation analysis too [6]. The sampling time for digital control systems must be selected based on the process dynamic, and auto-correlation techniques can be useful.

An important input to be considered on the control loop is the disturbance. This signal (or signals) can affect the process response and must be compensated by the controller. The open-loop controller cannot compensate for the disturbance action because it does not have a feedback signal. An example of disturbance is small variations in the height of the base metal surface product of heat. These variations change the contact tip to work distance and consequently the arc conditions. If the disturbance is measurable, some techniques can be used to improve the response of the controller, as shown in [6].

The model or process can be obtained using white or black box techniques. The first modeling technique required great knowledge of the process and its manufacture parameters to be able to create the equations that satisfactorily describe the process, actuator, and sensor. It is important to keep in mind that these parameters may change during the life of the equipment. Then if you use the parameters defined by the manufacturer, you can make the model inaccurate or useless. Because of this aspect and because of the development of powerful methods and tools for modeling, the black box techniques are more used.

To create a black box model of process for simulation, estimation, prediction, or control purposes, it is necessary that an actuator system and a measurement system modify the state of the process and see its response. With the process input and output values, it is possible to obtain statics or dynamic models of the process, but the actuator and sensor are included in the process model, as shown in **Figure 2a**. This becomes more evident in dynamic models when the dynamic response of the actuator and sensor affects the dynamic of the set. In static models, the use of a different sensor (or actuator), but with the same static gain, does not affect the model.

Other types of control algorithms are the logic control; those are classified as combinational and sequential. The response of the combinational logic algorithm depends on the inputs on current sampling time only, for example, the torch travel limit was reached, the robot must stop, and the arc must close. On sequential control the response depends on the previous sampling time inputs also, for example, in seam tracking control, it is important to know the last positions reached by the torch to calculate the next position.

In welding processes it is possible to find several control loops with different complexities and purposes. Each control loop has a set point or desired value of the controlled variable (supply by the operator or by a higher-level controller), a controlled variable (obtained from measurement system), a manipulated variable (supplied to the actuator by the control system), and disturbances. For example, on GMAW conventional welding power sources with the constant voltage, you can find an arc controller loop that tries to keep the voltage, the wire feed speed controller, and the gas flow controller (commonly included in the sequential logic controller) constant. The more complex processes have other control loops and sequential controllers to generate the arc signal form.

In welding robotic systems, you can find several control loops too but related with the torch or piece position and torch or piece travel. The combination of several controllers makes possible the control of the geometry of the weld bead, for example, a loop that changes the set point of voltage input in the welding power source to control the weld bead width based on profile sensor or video camera. Another example is a control loop that changes the wire feed speed in the welding power source and the welding speed in the welding robot to control the weld bead penetration.

These several controllers and actuators, which modify the same process at the same time to reach the manufacture objectives, will also cause interactions between the loops and its strongly coupled variables. A change in a control loop may affect other loops as disturbance and turns the process unstable. In these cases, a multivariable control loop must be considered. In the next sections, examples of control loops used in arc welding processes are shown.

2.1 Control of welding speed and seam tracking

If the torch trajectory is known, two open-loop controls can be used to govern the torch movement in two axes of the flat welding robotic system that uses stepper motors as actuator elements. In this system, the X axis is then governed by the welding speed controller and the Y axis by the seam tracking controller. The first loop keeps the welding speed set point, and the second loop applied the torch trajectory correction as shown in **Figure 3**.

The welding speed controller reads a lookup table with the speed set points ($WSsp$) for each torch X position and applies the control signals $SM1$ to stepper motor driver to reach the desired welding speed. These control signals are pulses with a time interval that corresponds to the motor speed (step time), and the pulse count is equivalent to distance traveled. The stepper motor driver sends the signal necessary for each coil to the stepper motor, and the stepper motor generates the rotation movement so that the gear transforms it into linear torch movement. The nT parameter is the time of the sample, made up of the sample time (T) and the sample number (n). The $nT - T$ is the previous sample time.

On the other hand, the seam tracking controller receives the X torch position too and finds the correction to the Y axis on another lookup table. Then, the controller

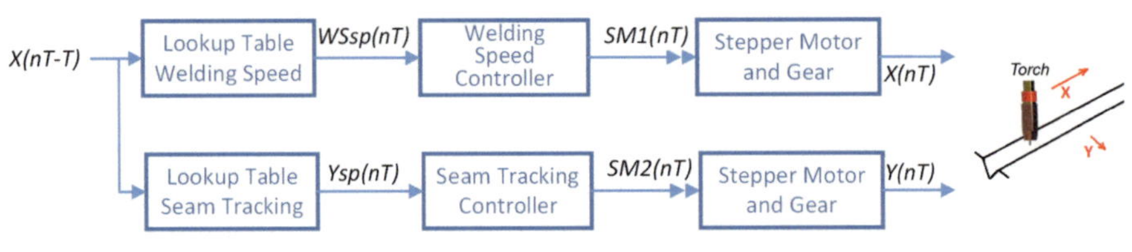

Figure 3.
Blocks diagram of the welding speed and seam tracking control loops.

moves the torch to this *Y* position. These systems do not need feedback because of stepper motor accuracy in normal working conditions.

To reduce the amount of data in the lookup tables, it is possible to save only the significant changes of welding speed and trajectory and hold the last value in the output until the new value is found.

To obtain the correct torch trajectory, the weld joint can be scanned before the welding process starts and the center joint can be calculated in all the points of the torch trajectory. With this data, it is possible to define the correct trajectory for the torch and move the seam tracking stepper motor accordingly.

The seam tracking loop can be transformed on closed-loop, but it is necessary for a profile sensor (e.g., a laser profilometer) to obtain online the joint profile and to make the trajectory analysis. The same control strategy can be used to substitute the lookup table by the algorithm that analyzes the profilometer data. This control strategy is better if the pieces are expected to move or deform during welding, but more calculation resources are needed.

2.2 Control of contact tip to work distance

The contact tip to work distance (*CTWD*) can be controlled in a closed-loop by a proportional-integral-derivative (PID) controller [6]. The CTWD is measured with a laser profilometer and calculated with the algorithm described in the chapter "On-line measurements in welding processes" and compared with the set point (*CTWDsp*). The controller manipulates the step time and the step counter (embedded in the SM3 signal) of the stepper motor associated with the CTWD movement in a robotic flat welding system. The $CTWD(0)$ is the initial value for the controller algorithm. **Figure 4** shows this closed control loop.

Due to the advanced position of the sensor, the CTWD measurement is in advanced time (θ). These future values are saved in a memory element that implements a first input-first output list to supply the correct value to the controller. This loop can use the same profile sensor with that of the seam tracking loop when the weld joint is scanned before the welding process starts.

2.3 Control of width, reinforcement, and penetration of the weld bead

The principal variables that affect the bead geometry in the conventional arc welding process are the welding voltage, the welding current, the wire feed speed, the contact tip to work distance, and the welding speed. The most common relation found in the scientific literature shows that increasing the electric current by increasing the wire feed speed, for the same welding speed, results in a greater weld bead depth and welding pools with greater volume and production. An increase in electric current, accompanied by a proportional increase of the welding speed (*wire feed speed/welding speed = constant*), also results in greater penetration, but the welding pool keeps the same volume. Then, the same welding joint volume can be filled (maintaining production) and ensure its integrity by full penetration (good quality) at the same time [7].

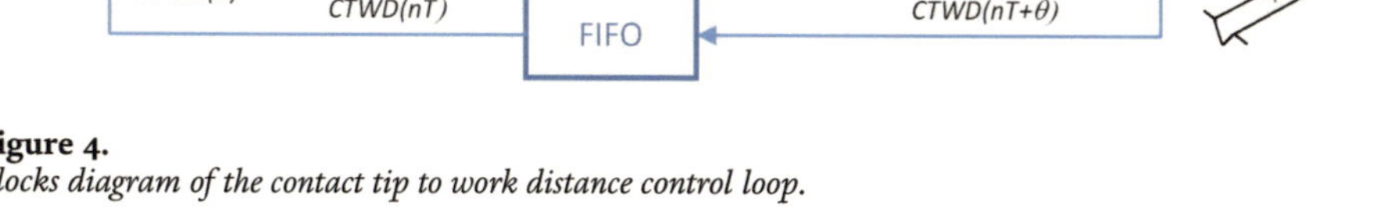

Figure 4.
Blocks diagram of the contact tip to work distance control loop.

On the other hand, laser welding has an additional set of parameters, such as the laser power and optical adjustments of the laser beam, but it is restricted by the availability of the equipment and difficult to make online adjustment in welding parameters. Shortly, the controllable parameters will become diversified, but right now the focus adjustment can be made by changing the position of focus lens inside the laser head. A review of these topics is shown in [8].

It is important to note that the relations between variables are more complex and multivariable techniques are necessary to describe them. With the multivariable techniques, it is possible to consider the interactions between variables in the process and reduce their effect in the control loop, but the implementation is difficult because of the complexity of the modeling and the control algorithm adjustment. In the literature analysis made in [5], only 9% of the papers use multi-variable control loops.

A generic diagram of a multivariable control loop of the weld bead geometry for the GMAW process is shown in **Figure 5**. It used the main variables that affect the process, to control the bead geometry, but it should be noted that due to the interdependence between them, setting the controller parameters and the use of uncoupling become difficult.

The GMAW conventional process can be represented as a multiple-input multiple-output (MIMO) process with three inputs and two outputs, as shown in **Figure 6**. The pairing of controlled and manipulated variables is shown too, and the other variables are considered as disturbances or controlled by other control loops. This selection is based on the bibliographic review and requires the analysis of the relative gains defined in [9].

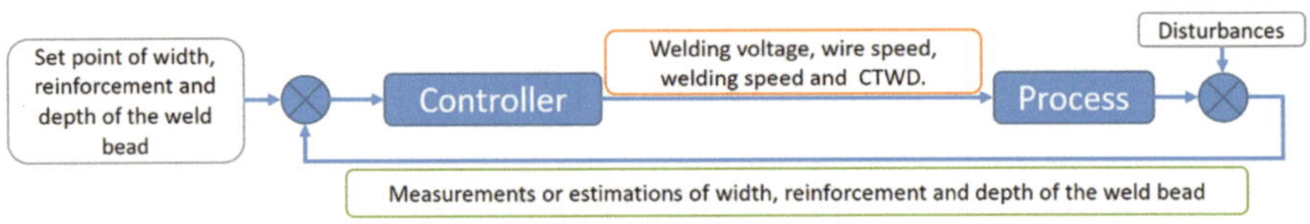

Figure 5.
Main variables used in the weld bead geometry control in GMAW process.

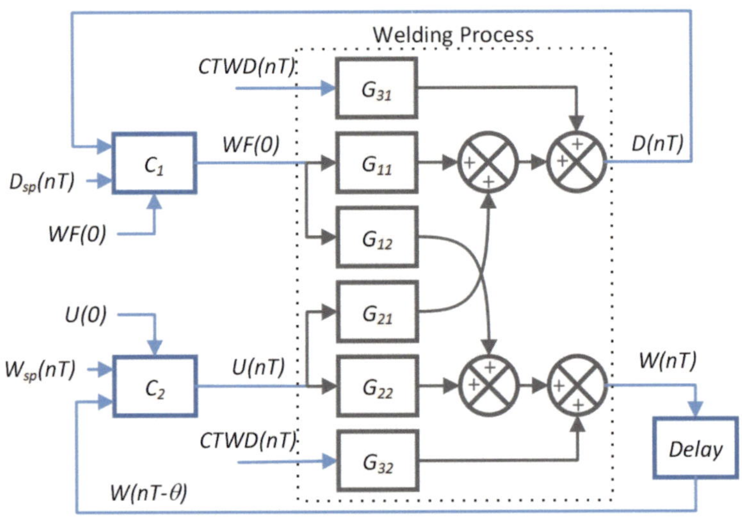

Figure 6.
Arc welding process represented as a multiple-input multiple-output system, with the depth and width of the weld bead controlled by PID algorithms.

It is important to note that the orbital welding adds more complexity to control, due to the effect of gravity on the transfer of material, the weld pool, and the weld bead formation, in addition to other requirements. So, it is important to consider the orbital angle as a measurable disturbance.

Control strategies proposed in [5] are based on a PID controller improved with a decoupling method, a Smith predictor, and a fuzzy self-adaptive algorithm. In the PID control strategy, shown in **Figure 6**, the welding voltage (U) and wire feed speed (WF) are manipulated by two independent control loops that control the weld bead width (W) and the weld bead depth (D). The $CTWD$ can be considered a disturbance or a manipulated variable, depending on the control strategy. The quantity of material deposited (deposition rate) is indirectly controlled by the relation between WS and WF, and the weld bead reinforcement depends on the relation between bead width and quantity of material deposited. The values $U(0)$ and $WF(0)$ are initial values of manipulated variables.

The weld bead depth value (D) is estimated using the algorithms described in the chapter "On-line measurements in welding processes," and the weld bead width (W) can be estimated too or measured using a profile sensor. These values are feedback to the controller and the control errors, which are calculated using the set point of the weld bead depth (Dsp) and width (Wsp) values. If a laser profilometer is used, the real width and reinforcement of the weld bead are calculated. Note that the measurements have a dead time (θ) and these values are zero during the first sampling times.

While many control algorithms have been proposed, in which approaches are theoretically elegant, most of the industrial processes nowadays are still controlled by proportional-integral-derivative controllers [10–12]. Conventional PID controllers have been widely applied in industrial process control for about half a century because of their simple structure and convenience of implementation [13]. However, a conventional PID controller can have poor control performance for nonlinear and complex systems for which there are no precise mathematical models. Numerous variants of conventional PID controller, such as self-tuning, auto-tuning, and adaptive PID controllers, have been developed to overcome these difficulties. Several online self-tuning or adaptive algorithms are based on fuzzy inference systems that were developed in [11, 12, 14–19].

The weld bead width measure dead time problem can be solved using a modified Smith predictor as shown in [5]. In the same work, the nonlinear is solved using a fuzzy self-adaptive algorithm for PID tuning. But the control strategy still has a problem, the interdependence between the process variables.

To improve the controller behavior, the decoupling structures can be incorporated for reducing the interaction between the loops. In this case, the decoupling used is based on feedforward control. These decoupling techniques are useful when the process is affected by strong measurable disturbances. This strategy can help improve the behavior of the controller in the face of this disturbance, but it cannot replace the feedback control [20]. The typical feedforward control is shown in **Figure 7**.

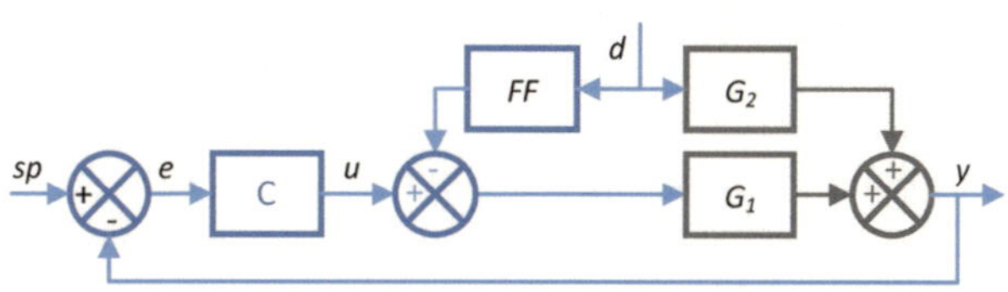

Figure 7.
Typical feedforward control in a single control loop. Adapted from [5].

The feedforward control tries to anticipate the effect of the disturbance (d). The control action is applied directly to the control loop drive element, before the disturbance can affect the controlled variable [6, 20]. Eqs. (1) and (2) usually define the necessary conditions to it, while the feedforward model is obtained in Eq. (3).

$$\frac{\Delta y}{\Delta v} = \frac{G_2(z^{-1})}{1 - CG_1(z^{-1})} + \frac{FFG_1(z^{-1})}{1 - CG_1(z^{-1})} = 0 \tag{1}$$

$$\frac{\Delta y}{\Delta v} = \frac{G_2(z^{-1}) + FFG_1(z^{-1})}{1 - CG_1(z^{-1})} = 0 \tag{2}$$

$$FF = \frac{G_2(z^{-1})}{G_1(z^{-1})} \tag{3}$$

If $G_1(z^{-1})$ and $G_2(z^{-1})$ are quite close to first-order transfer function with delay, as it is shown in

$$G_1\left(z^{-1}\right) = \frac{K_1 z^{p_1}}{1 - \alpha_1 z^{-1}} \tag{4}$$

$$G_2\left(z^{-1}\right) = \frac{K_2 z^{p_2}}{1 - \alpha_2 z^{-1}} \tag{5}$$

where the terms α_1 and α_2 are related with the sample time (T_s) and process time constants T_1 and T_2, as shown in the next equations,

$$\alpha_1 = e^{-\frac{T_s}{T_1}} \tag{6}$$

,

$$\alpha_2 = e^{-\frac{T_s}{T_2}} \tag{7}$$

then the feedforward transfer function is

$$\frac{\Delta u}{\Delta v} = FF\left(z^{-1}\right) = -\frac{K_2 z^{-p_2}\left(1 - \alpha_1 z^{-1}\right)}{K_1 z^{-p_1}\left(1 - \alpha_2 z^{-1}\right)} \tag{8}$$

$$\frac{\Delta u}{\Delta v} = FF\left(z^{-1}\right) = -K_{FF} z^{(p_1 - p_2)} \frac{1 - \alpha_1 z^{-1}}{1 - \alpha_2 z^{-1}} \tag{9}$$

where

$$K_{FF} = \frac{K_2}{K_1} \tag{10}$$

Sometimes only the steady value is compensated, and the dynamic is depreciated in which case the transfer function of the feedforward block is K_{FF}. This simplifies the modeling work and the model structure. It is important to note that the dead time in the principal channel must be less or equal than the disturbance channel ($p_1 \leq p_2$) for the dynamic compensation to be effective.

If the interactions between loops are considered as disturbances, the same feedforward scheme can be used to decoupling the weld bead width and penetration control loop in our welding process. **Figure 8** shows the block diagram with decouples.

Decouples FF_1 and FF_2 are designed to minimize the interaction between loops and improve the controller behavior. The calculation of the transfer function of decouples is similar to what is explained in the feedforward transfer function. This strategy is similar to inverse decoupling shown in [21, 22].

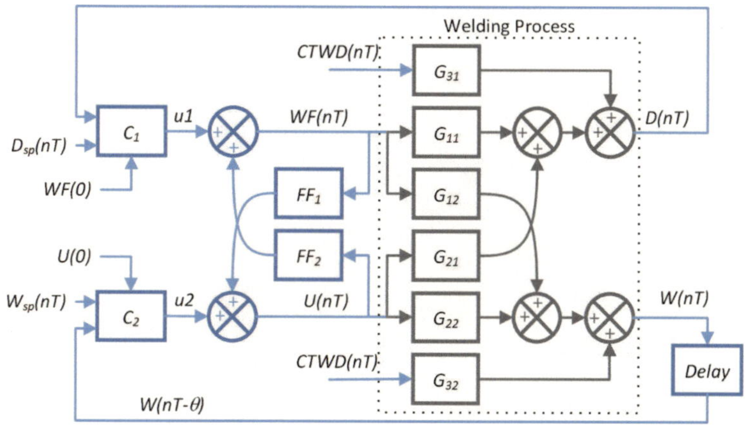

Figure 8.
Weld bead width and depth penetration controller with decouples.

The design and synthesis of PID controllers and fuzzy algorithms in FPGA or microcontroller devices are possible, and the resource consumption is very low, as is shown in [23]. But many other scientific researchers are being developed and tested to solve and improve the control of the welding process. In the next section, some of them will be described.

3. Methods or techniques used for modeling and control of the geometry of weld bead

The welding process is characterized as multivariable, nonlinear, and time-varying, with stochastic behavior and having a strong coupling among welding parameters. For this reason, it is very difficult to find a reliable mathematical model to design an effective control scheme by conventional modeling and control methods.

Due to these characteristics, the use of adaptive techniques has spread in the last decades with favorable results, only overcome by a proportional-integral-derivative controller. The adaptive control has been implemented in some researchers to cope with the problem of the high dependence of process parameters to its operating condition. The main drawback of this method is that it requires online estimation or tuning of the parameters, which is usually a time-consuming operation. The single implementation of PID and low computational resources make it still the most used, as in the rest of industrial applications. A graphic summary is shown in **Figure 9**, and **Table 1** shows the document references analyzed.

Neural networks, fuzzy methods, and their combinations also stand out. Note that the magnificent behavior of the neural network can be clouded by a slow convergence because of the excessive quantity of neurons or hidden layers. Many research efforts use this approach but neglected the need for the quick response of the control system.

Statistical methods, such as the classic autoregressive moving-average and expert systems, are represented too. Other less used techniques include state space, model-free adaptive, first- and second-order model, support vector machine, and finite elements.

3.1 Some scientific research about the geometry control in arc welding processes

Developments in the field of automatic control of the geometry in the arc welding process have been intense in the last four decades. The most representative

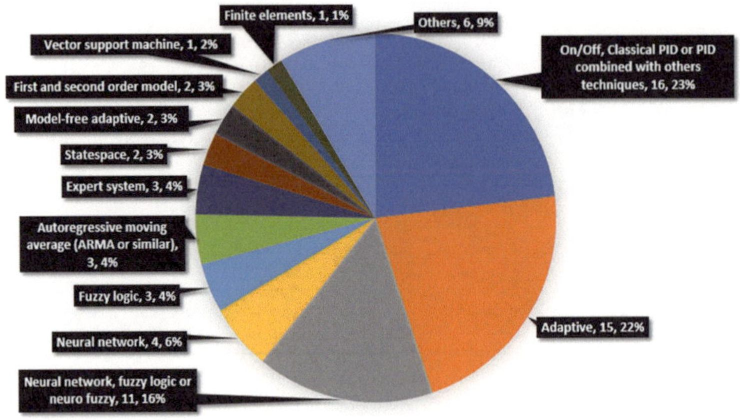

Figure 9.
Graphic summary about the main methods or techniques used for control of the weld bead geometry, including the quantity of references and the percentage they represent in the total papers consulted.

Control technique or control model	References
On/off, classic PID, or PID combined with other method	[2, 24–37]
Adaptive	[26, 29, 38–50]
Neural network and fuzzy logic or neuro-fuzzy	[3, 4, 51–60]
Neural network	[27, 61–63]
Fuzzy logic	[39, 64, 65]
Autoregressive moving average (ARMA or similar)	[29, 66, 67]
Expert systems	[38, 53, 40]
State space	[1, 68]
Model-free adaptive	[68, 69]
First- or second-order model	[70, 71]
Support vector machine	[64]
Finite elements	[52]

Table 1.
References classified according to the methods or techniques used for control of the weld bead geometry.

research efforts found in the literature are in this area in the last two decades. In this section, some relevant works are discussed in chronological order.

In 1986, Nied and Baheti [46] registered a patent in which a robotic welding system has an integrated vision sensor to obtain images and analyze the welding scene in real-time. It used an adaptive feedback control to assure the full penetration, the weld puddle area, and the maximum width in the TIG process. The adaptive control system determines a puddle geometry error and uses the nominal welding current to change the heat input to the weld pool, regulating the combination of puddle area and width. Arc voltage is modulated to reflect changes in welding current and maintain a constant arc length.

In 1989 Edmund R. Bangs and others [38] registered a patent that described a system for real-time welding adaptive control using infrared images, artificial intelligence (expert system), and distributed processors.

Already in 1990, Andersen designed a control system for the GMAW process [24]. As shown in **Figure 10**, a neural network set point selector defines the start

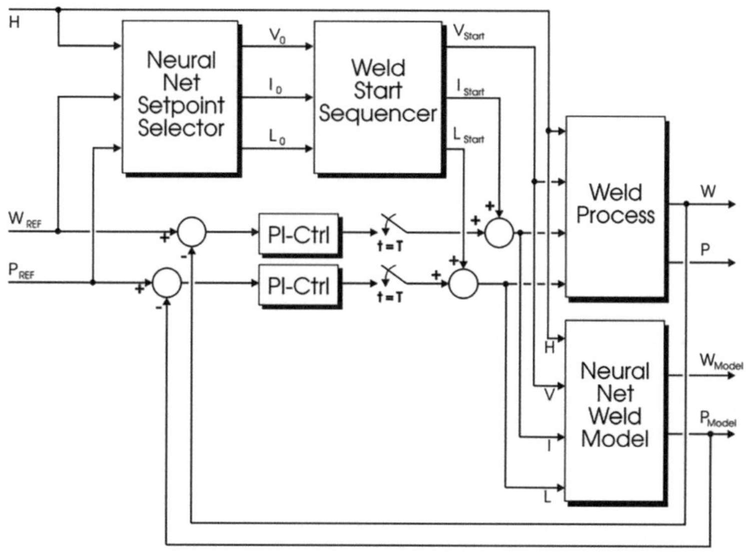

Figure 10.
Control loop proposed for Andersen [24] when the initial set point conditions are defined by neural network and the penetration are estimated. Adapted from [73].

parameters of welding speed (V), current (I), and arc length (L). Another neural network estimates the width (W) and weld depth penetration (P). Two independently closed-loop PIs adjust continuously the start parameters in function of weld bead width and penetration control error. The experimental results are satisfactory. In [72], a similar structure was proposed for the control of weld bead width, without the estimator and using a fuzzy controller. The same philosophy is used in [54] for width and reinforcement control in independent control loops.

Zhang et al. [74] developed an adaptive control of full penetration for GTAW processes, based on the generalized predictive control (GPC) algorithm presented by Clarke [75–77]. The controlled variables are the width and reinforcement of the weld bead, measured by a vision system and a laser stripe. The control or manipulated variables are the welding current and arc length. The sampling interval of the control system was 0.5 s. The process has been described by a moving-average model with a predictive decoupling algorithm. The system is not controlling the penetration directly but has a satisfactory performance in the weld quality control.

Brown et al. [26] developed a control loop with a PID controller and an adaptive dead-time compensator for GMAW processes in a horizontal welding position. The controlled variable is the weld pool width and the welding speed is manipulated. The simulation results show an acceptable response.

A dynamic model, based on neural network, was designed in [53] to predict the maximum backside width in pulsed GTAW processes. Also, a self-learning fuzzy neural network controller was developed for controlling the maximum backside width, and the fuzzy rules were modified online. Another intelligent multivariable controller, type double-input and double-output (DIDO), based on a neuro-fuzzy algorithm and combined with an expert system, was developed to control the maximum backside width and the shape of the weld pool. Experimental results showed the best behavior using the DIDO intelligent controller.

In [55], a neuro-fuzzy controller was designed for the GTAW process. This method overcomes the dependency of human experts for the generation of fuzzy rules and the non-adaptive fuzzy set. The adaptation of membership function and the fuzzy rule self-organization are carried out by the self-learning and competitiveness of the neural network with three hidden layers. The simulation test got promising behavior.

Chin [28] developed a system for infrared image sense and PID control of the SAW process. Several tests were executed with diverse conditions in the control variables. Similarly, an infrared point sensor (pyrometer) is used in [36] to estimate the weld bead depth in GTAW and SAW processes. The penetration is controlled indirectly (using the infrared emission) by a PI regulator, which changes the welding current in the GTAW process and the welding voltage and welding speed in the SAW process. A satisfactory result is obtained under laboratory conditions.

Moon and Beattie [45] developed an adaptive fill control for multi-torch multi-pass SAW processes. The system measures the joint geometry with laser seam tracking and calculates the total area of the joint, computing by numerical integration based on the actual joint profile. With the area ratio of the joint, the welding current/voltage combination is obtained, and the welding speed is adjusted inversely in proportion to the area to be filled. The control significantly improves weld bead quality. This technology has been used in the manufacture of longitudinal and spiral pipes and pressure vessels.

In [66], two simultaneous but independent control loops are used. The first loop monitors the temperature with an infrared camera and controls the trajectory of the torch. The second loop monitors the geometric profile with the laser stripe and manipulates the welding speed and the wire feed speed. A variable delay Smith predictor is used for reducing the dead-time effect of laser strip sensor, as shown in **Figure 11**. The author tested a SISO closed loop with a PI controller and a MIMO GOSA. The last one obtains the best result.

An H-infinity robust control system in [1] is proposed to control the length and width of the weld pool, manipulating welding current and weld speed. The simulation shows an effective robust method to control processes with large uncertainties in the dynamics. However, the complexity of the H-infinity control algorithm can make the implementation of embedded devices difficult. Also, it needs an effective description of the uncertainties of the welding process dynamics, and it is difficult under conditions of the real processes.

A weld pool imaging system, with a LaserStrobe high shutter speed camera, is used in [58] to obtain contrasting images and eliminate the arc interference. Image processing algorithms, based on edge detection and connectivity analysis, extract information about the weld pool length and width online. A neuro-fuzzy controller, based on human experience and experimental results, manipulate the welding voltage and speed in real-time based on weld pool dimensions. A welding speed closed-loop control is used to reach the set point of the weld pool geometry. The simulation results are satisfactory, but the response may be slow due to the time required for image processing and fuzzy calculation.

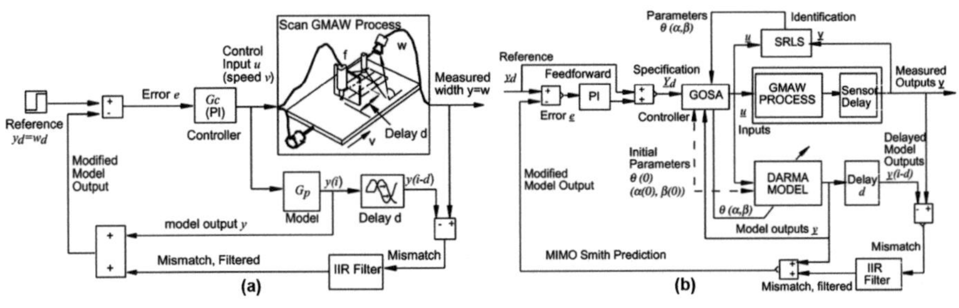

Figure 11.
SISO (a) and DIDO (b) control loops that employ infrared information. Adapted from [66].

In [31], the difficulty to tune the PI controller parameters to achieve the desired performance, across the entire range of the process operation, is shown. Therefore, the design and implementation of more complex controllers are required to obtain better control of the welding process.

A method developed by Casalino in [52] defines an automated methodology for selecting the weld process parameters based on artificial intelligence. While there are many methods available to improve the reliability of traditional open-loop control schemes, these can only be used with a particular welding power source and a specific welding arrangement.

Lu et al. [43] developed a non-transferred plasma charge sensor-based vision system to measure the depth of the weld pool surface in orbital GTAW processes. The sensor measures the welding voltage when the welding current is off and calculates the arc distance by an inversely linear relation. An insulated gate bipolar transistor (IGBT) power module is utilized to temporarily switch off the main power supply. During this period, the large arc pressure associated with the main arc is not present, the depth of the weld pool surface decreases, and the sensor output increases, obtaining the arc measurement. An adaptive interval algorithm controls the depth of the weld pool surface, regulating the main-arc-on period. Four experiments were executed under different conditions and show a satisfactory response.

Smith et al. [34] use two independent PIs to control a pulsed TIG process in a horizontal position. Both controllers use the control error of the top face weld pool width as input. The independent outputs are the welding current and the wire feed speed. The active adjustment of the welding current and the wire feed speed allows compensating variations on the weld pool size. A camera and image processing algorithm measures the top face weld pool width. The controllers and image processing algorithms run concurrently on a PC. The controllers use CAN serial communication protocol to adjust the outputs in two distributed actuator nodes, based on a microcontroller system. A step-change in plate thickness was used to test the controller system. The experimental results produced welds with a more consistent profile when faced with variations in process conditions.

In [63], a three-dimensional vision system is used for obtaining the geometrical parameters of the weld joint. A closed-loop neuronal-network controller is developed to control the width and depth of the weld joint, by regulating the welding voltage, welding speed, and wire feed speed. The neuronal model has two hidden layers with six and four neurons, respectively, as shown in **Figure 12**. The experimental results, using the neural network learning data and the error range of width and depth, are within 3%.

Fuzzy and PID controllers are employed in [65] to control the geometry stability in a GTAW process, by regulating the welding current (ΔI) and wire feed rate (v). A real-time image analysis algorithm was developed for seam tracking and seam gap measuring using passive vision. The control was applied in a teach and playback robotic welding system, which helps the robot track the seam with the gap variation. The quality of the products obtained reached the standard of first-order welding seam (QJ2698-95) in terms of dimensions and soundness, which was verified with an X-ray inspection. **Figure 13** shows the block diagram.

In [78] two SISO subsystems are developed to control a double-electrode GMAW process. The control structure is selected for convenient implementation and design. One system controls the welding current of the main torch by manipulating the wire feed speed. The other system controls the welding voltage of the bypass torch by manipulating the welding current. Two interval models have been obtained, based on experimental data from step-response experiments under different manufacturing conditions. These simple controllers show an acceptable behavior to control this relatively complex process.

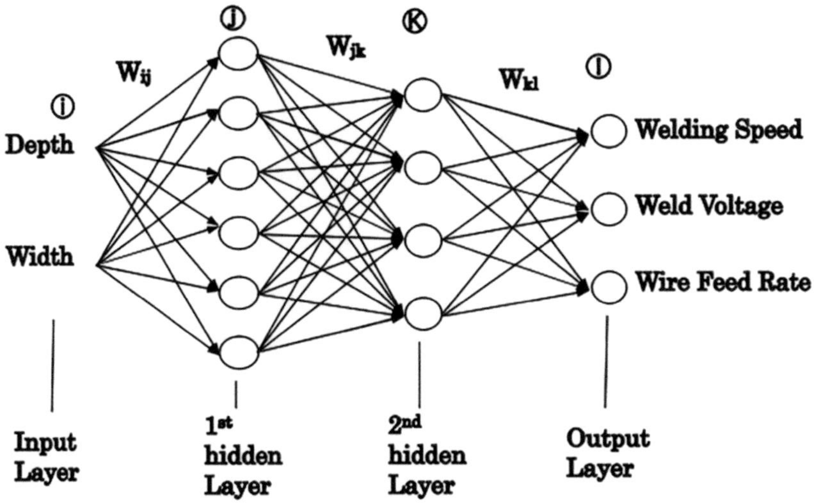

Figure 12.
Neural network to determine welding parameters. Adapted from [63].

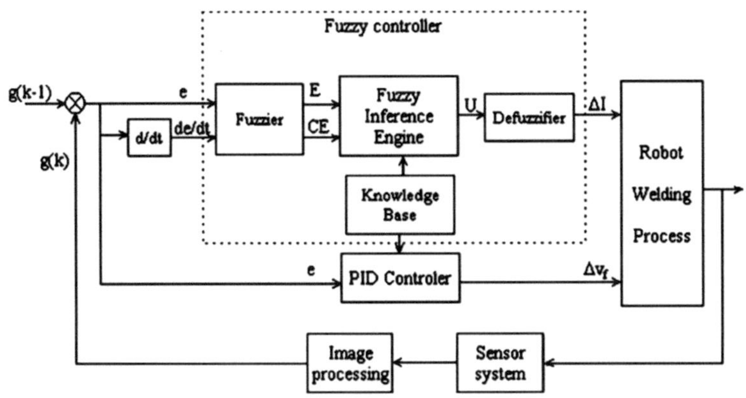

Figure 13.
Block diagram of the geometry stability controller for GTAW process, developed in [65].

Chen solved the full penetration detection, in orbital the GTAW process, with a vision system over the topside pool and a neural network model for estimating the backside weld bead width in [39]. The neural network has 17 neurons in the input layer, 40 in the hidden layer, and 1 in the output. An adaptive controller receives the backside weld bead width, estimated by the neural network, and regulates the peak current. A fuzzy controller received the gap state and controlled the wire feed speed. The experimental results, examined in X-ray, have shown a uniform backside of the weld bead.

An adaptive inverse control based on support vector machine-based fuzzy rules acquisition system is proposed by [64] for pulsed GTAW process. This method extracts the control rules automatically from the process data, using an adaptive learning algorithm and a support vector machine to adjust the fuzzy rules. The controlled variable is the backside weld width, and the manipulated variable is the peak current of the pulse. A double-side visual sensor system captures the topside and backside images of the weld pool simultaneously. The data for model identifying is obtained experimentally. The control is simulated and shows satisfactory results.

Lü et al. [69] developed a MISO algorithm for the control of the width of the weld pool backside in the GTAW process. The vision sensing technology and model-free adaptive control (MFC) are used. The welding current and wire feed speed are selected as the manipulated variables, and the backside width of the weld pool is the controlled variable. The main difficulty was the availability of computational resources to maintain the control period and the image processing speed within acceptable values. It has the disadvantage of using complex optical systems for obtaining the image of the back- and the front side of the weld pool.

In [68] a space state model of the GMAW process is obtained to compare the behavior of three controllers: ARMarkov-PFC (based MPC controller), PI, and feedback linearization based on the PID (FL-PID). The controller outputs are the wire feed motor voltage and the welding voltage. The controlled variables are the welding current and voltage. The simulation results show that the ARMarkov-PFC outperformed other controllers from the viewpoints of the transient response, desired output tracking, and robustness against the process parameter uncertainties but require more computational resources. The FL-PID controller was sensitive to the process parameter variations, presence of noise and disturbance, and results in an improper performance. The PI controller produced an inappropriate transient response and inadequate interaction reduction despite good tracking performance, low sensitivity to parameter variations, and low computational resource requirements.

The main advantages of MPC over structured PID controllers are its ability to handle constraints, non-minimum phase processes, changes in system parameters (robust control), and its straightforward applicability to large, multivariable, or multiple-input multiple-output processes. Despite having many advantages, a noticeable drawback of an MPC is that it requires higher computation capability, as is shown in [79].

The change in arc voltage during the peak current is used in [49] to estimate the weld penetration depth in the pulsed GMAW process. An adaptive interval model control system is implemented, but, contrary to the author's comment, the control accuracy is not good.

Lui et al. [60] developed a model-based predictive control for the orbital GTAW process. The control input is the backside weld bead width, and the outputs are the welding current and welding speed. A nonlinear neuro-fuzzy (ANFIS) model is utilized to estimate the backside weld bead width (related with weld depth penetration) using the front-side weld pool characteristic. **Figure 14** shows a block diagram of this approach.

In [67] the GMAW LAM process is modeled using the recursive least squares algorithm for system identification. An image processing algorithm is employed for obtaining the nozzle to top surface distance. An adaptive controller adjusts the deposition rate and keeps the nozzle to top surface distance constant. The precision range of the control system is limited within ± 0.5 mm.

A segmented self-adaptive PID controller was developed in [30] for controlling the arc length and monitoring the arc sound signal in the pulsed GTAW process for the flat and arched plate. The experiments show that the controller has acceptable accuracy.

Lui and Zhang [4] developed a machine-human cooperative control scheme to perform welder teleoperation experiments in the orbital GTAW process. They developed an ANFIS model of the human welder's adjustment on the welding speed. The welder sees the weld pool image overlaid with an assistant visual signal and moves the virtual welding torch accordingly. The robot follows the welder's movement and completes the welding task. The experimental data is used to obtain the model. Later, they transfer this model to the welding robot controller to perform

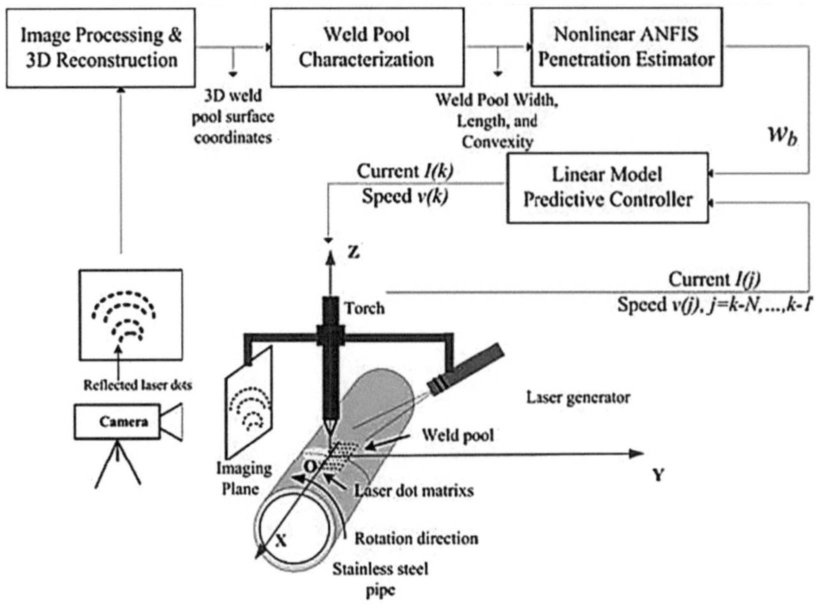

Figure 14.
Model-based predictive control to orbital GTAW process. Adapted from [60].

automated welding. The controller receives the three-dimensional weld pool characteristic parameters (weld pool length, width, and convexity) and changes the welding speed.

In other similar work [3], a human intelligence model based on a neuro-fuzzy algorithm is proposed to implement an intelligent controller to maintain a full penetration manipulating the welding current. These works establish a method to rapidly transform human welder intelligence into welding robots by using three-dimensional weld pool surface sense, fitting the human welder response to the information through a neuro-fuzzy model, and using the neuro-fuzzy model as a replacement for human intelligence in automatic systems. In previous works [56, 57], the skilled human welder response to the fluctuating three-dimensional weld pool surface is correlated and compared with a novice welder.

3.2 Embedded devices in welding process control

Embedded systems, especially the FPGA and system on chip (SoC), are used in a multitude of technological processes in various industries, covering hazardous areas such as medical, aerospace, and military or even the most common household appliances. With the increase in processing capabilities of these systems, based on microcontrollers and new processor generation, it is possible to obtain remarkably improved measurement and control systems with the use of advanced algorithms for processing information provided by the sensors. The parallel processing capabilities of the FPGA (into the SoC) allow lower execution times than in processors or microcontrollers. These capabilities are important to estimators based on neural networks (parallel execution) and to control systems in real-time that need to attend several sensors and actuators.

The FPGA has numerous digital inputs and outputs, with the possibility of adding several analogs and other peripherals. Many of them have a hard processor, with one or more cores and various peripherals for communication, video, sound,

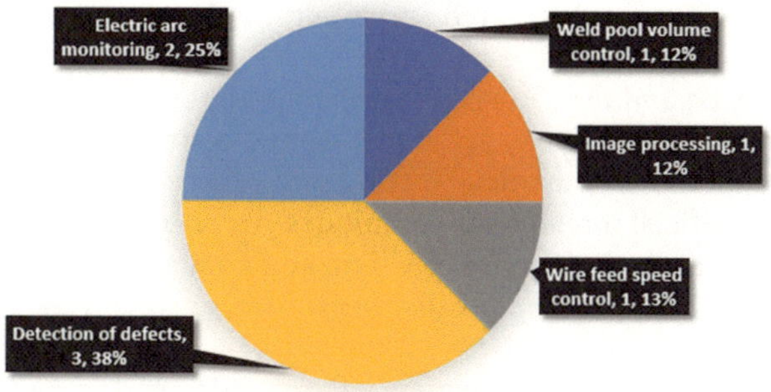

Figure 15.
Graphical summary of the applications of embedded devices in welding processes found in the literature review.

and large random access memory (RAM) capacity, where you can run a standard operating system interconnected with the FPGA. These features and the small size, low power consumption, low heat dissipation, and reconfigurable architecture make it an ideal tool for monitoring and control systems with real-time requirements. For all these good reasons, we must pay special attention to these devices.

The modern welding power sources are controlled by embedded systems. These systems provide communication, data acquisition, and control functions for different welding processes, but their most important specialization is the control of the electric arc, laser signal, and other methods to transfer the energy to the base material. This specialization permitted the substitution of big transformer and switches to select the welding parameters such as the voltage, current, and inductance, by a smaller transformer and high-frequency switching semiconductors governed by a microcontroller. With these changes it was possible to generate various types of waveforms on the output of the welding power source, improving the conventional processes and creating new processes and new control algorithms.

In these systems, the embedded controller emulates the impedance of the old transformer and keeps the set points of welding voltage and current with more accuracy. The configuration of welding parameters, data acquisition, and set point changes can be made using a communication protocol, defined by the manufacturer. The integration with a supervisory or higher control system is possible using this communication protocol.

The literature review does not show an extensive application of FPGA to bead geometry control in the arc welding process. But other works have shown applications related with wire feed control [80], defect detection [81–83], and arc signal monitoring [84, 85]. A graphical summary is shown in **Figure 15**.

4. Conclusions

Based on the literature review and the experimental experience about the control of the welding bead geometry, it is possible to observe the great complexity of the welding process and the many efforts to control it. The proposed solutions range from simple open-loop controllers to complex intelligent control algorithms, highlighting the legendary PID combined with other techniques, adaptive methods, and the neural network and fuzzy algorithms.

Despite the arduous research efforts, few of these algorithms are being applied in the industry, in some cases due to its complexity but others due to commercial interests and its cost of implementation. For these reasons, the control of welding processes is an open topic for research and especially for the development of feasible solutions to be used in the industry.

Scientific research and the slow but continuous application of its results in the welding industry show a tendency for modeling and control of these processes to be carried out using methods of artificial intelligence. These methods, in addition to including classic artificial intelligence techniques, are incorporating bioinspired algorithms, deep learning techniques, big data and data mining for the analysis of the measurements, the adjustment of the controllers, and even the implementation of the controller itself.

Undoubtedly, the current development of embedded systems and the small and smart sensors is allowing the implementation of many algorithms proposed decades ago and new algorithms that make extensive use of the calculation capabilities of these systems. The use of multivariable control and dynamic models of the process will be possible and will allow a notable improvement in the quality of the welds and the number of parts rejected in the production process.

But the advantages of these technologies will not be accepted and exploited efficiently without adequate training of the technical staff that directs and operates the industries. Many of these modeling and control techniques are still unknown or their advantages are poorly disclosed. This is a problem when it is compared in terms of ease of use and productivity against classical techniques with decades of use in the industry. In this sense, we try to contribute to the dissemination of this knowledge throughout this chapter.

Author details

Guillermo Alvarez Bestard*† and Sadek Crisostomo Absi Alfaro†
University of Brasilia, Brasilia, Brazil

*Address all correspondence to: guillermo@unb.br

† These authors contributed equally.

References

[1] Jou M. A study on development of an H-infinity robust control system for arc welding. Journal of Manufacturing Systems. 2002;**21**(2):140-150

[2] Iceland WF, O'Dor ME. Weld Penetration Control. United States: North American Rockwell Corporation; 1971. p. 7

[3] Liu Y, Zhang W, Zhang Y. Dynamic neuro-fuzzy-based human intelligence modeling and control in GTAW. IEEE Transactions on Automation Science and Engineering. 2015;**12**(1):324-335

[4] Liu Y, Zhang Y. Iterative local ANFIS-based human welder intelligence modeling and control in pipe GTAW process: A data-driven approach. IEEE/ASME Transactions on Mechatronics. 2015;**20**(3):1079-1088

[5] Bestard GA. Sensor fusion and embedded devices to estimate and control the depth and width of the weld bead in real time [PhD thesis]. Universidade de Brasília; 2017

[6] Smith CA, Corripio AB. Principles and Practice of Automatic Process Control. 2nd ed. New York: John Wiley; 1997

[7] Scotty A, Ponomarev V. Soldagem MIG-MAG: melhor entendimento, melhor desempenho. São Paulo: Artibler Editora Ltda.; 2008

[8] You DY, Gao XD, Katayama S. Review of laser welding monitoring. Science and Technology of Welding and Joining. 2014;**19**(3):181-201

[9] Bristol EH. On a new measure of interaction for multivariable process control. IEEE Transactions on Automatic Control. 1966;**11**(1):133-134

[10] Astrom KJ, Eykhoff P. System identification—A survey. Automatica. 1971;**7**(2):123-162

[11] Ko CN, Wu CJ. A PSO-tuning method for design of fuzzy PID controllers. Journal of Vibration and Control. 2008;**14**(3):375-395

[12] Kumar V, Nakra BC, Mittal AP. A review on classical and fuzzy pid controllers. International Journal of Intelligent Control Systems. 2011;**16**(3): 170-181

[13] Bennett S. Development of the PID controller. IEEE Control Systems. 1993; **13**(6):58-62

[14] Jantzen J. Tuning of fuzzy PID controllers. Fuzzy Information and Engineering. 1998;**871**(98-H):1-22

[15] Karasakal O, Guzelkaya M, Eksin I, Yesil E, Kumbasar T. Online tuning of fuzzy PID controllers via rule weighing based on normalized acceleration. Engineering Applications of Artificial Intelligence. 2013;**26**(1):184-197

[16] Kazemian HB. Comparative study of a learning fuzzy PID controller and a self-tuning controller. ISA Transactions. 2001;**40**(3):245-253

[17] Liu WH, Xie Z. Design and simulation test of advanced secondary cooling control system of continuous casting based on fuzzy self-adaptive PID. Journal of Iron and Steel Research International. 2011;**18**(1):26-30

[18] Soyguder S, Karakose M, Alli H. Design and simulation of self-tuning PID-type fuzzy adaptive control for an expert HVAC system. Expert Systems with Applications. 2009;**36**(3 PART 1): 4566-4573

[19] Wu Y, Jiang H, Zou M. The research on fuzzy PID control of the permanent magnet linear synchronous motor. Physics Procedia. 2012;**24**: 1311-1318

[20] Behar AA. Identificacion y Control Adaptativo. Mexico: Pearson Prentice Hall; 2003

[21] Bharathi M, Selvakumar C. Interaction reducer for closed-loop control of multivariable systems. International Journal of Engineering Trends and Technology. 2012;**3**(4):1-15

[22] Garrido J, Vázquez F, Morilla F. Diseño de Sistemas de Control Multivariable por Desacoplo con Controladores PID. In: X Simposio CEA de Ingeniería de Control, September. 2012. pp. 64-71

[23] Duboué RC, Bestard GA. Diseño e implementación de un PID Profesional y un PID Difuso utilizando un microcontrolador PIC18F4550. In: Jornada Científica ICIMAF 2014. 2014

[24] Andersen K, Cook GE, Karsai G, Ramaswamy K. Artificial neural networks applied to arc welding process modeling and control. IEEE Transactions on Industry Applications. 1990;**26**(5):824-830

[25] Boughton P, Rider G, Smith EJ. Feedback of weld penetration. In: Adisances Weld. Proc. 1978. pp. 203-209

[26] Brown LJ, Meyn SP, Weber RA. Adaptive dead-time compensation with application to a robotic welding system. IEEE Transactions on Control Systems Technology. 1998;**6**(3):335-349

[27] Cheng SB, Zhang Y, Lin Y, Qiu T, Wu T. Welding robotic systems with visual sensing and real time control of dynamic weld pool during pulsed GTAW. International Journal of Robotics and Automation. 2004;**19**(1): 28-35

[28] Chin BA. Welding GAP control using infrared sensing. In: Technical Report. Office of Naval Research, Auburn University; 2001

[29] Doumanidis CC. Multiplexed and distributed control of automated welding. IEEE Control Systems. 1994; **14**(4):13-24

[30] Lv N, Zhong J, Wang J, Chen S. Automatic measuring and processing system of audio sensing for real-time arc height control of pulsed GTAW. Sensor Review. 2014;**34**(1):51-66

[31] Naidu DS, Ozcelik S, Moore KL. Modeling, Sensing and Control of Gas Metal Arc Welding. Oxford: Elsevier; 2003

[32] Nomura H, Satoh Y, Tohno K, Satoh Y, Kuratori M. Arc light intensity controls current in SA welding system. I. Welding Metal Fab. United Kingdom: Institute of Welding; 1980. pp. 457-463

[33] Smith CJ. Self-adaptive control of penetration in tungsten inert gas weld. Advances in Welding Processes. 1974: 272-282

[34] Smith JS, Balfour C. Real-time top-face vision based control of weld pool size. Industrial Robot: An International Journal. 2005;**32**(4):334-340

[35] Vorman AR, Brandt H. Feedback control of GTA welding using puddle width measurement. Welding Journal. 1978:742-746

[36] Wikle HC III, Kottilingam S, Zee RH, Chin BA. Infrared sensing techniques for penetration depth control of the sub-merged arc welding process. Journal of Materials Processing Technology. 2001;**113**:228-233

[37] Zi B, Sun H, Zhu Z, Qian S. The dynamics and sliding mode control of multiple cooperative welding robot manipulators. International Journal of Advanced Robotic Systems. 2012;**9**:1-10

[38] Bangs ER, Longinow NE, Blaha JR. Using Infrared Image to Monitor and Control Welding. United States: IIT

Research Institute; 1989. p. 167. Available from: http://illinois.patentlibrary.us/us-4877940.html

[39] Chen H, Lv F, Lin T, Chen S. Closed-loop control of robotic arc welding system with full-penetration monitoring. Journal of Intelligent and Robotic Systems: Theory and Applications. 2009;**56**(5):565-578

[40] Doumanidis CC, Hardt DE. Simultaneous in-process control of heat affected zone and cooling rate during arc welding. Welding Journal. 1990;**69** (5):186s-196s

[41] Henderson DE, Kokotovic PV, Schiano JL, Rhode DS. Adaptive control of an arc welding process. IEEE Control Systems. 1993:49-53

[42] Kovacevic R, Zhang YM, Ruan S. Sensing and control of weld pool geometry for automated GTA welding. Journal of Engineering for Industry. 1995;**117**(2):210

[43] Lu W, Zhang YM, Zhang C, Walcott BL. Robust sensing and control of weld pool surface. IFAC Proceedings Volumes (IFAC-Papers Online). 2005; **16**:301-306

[44] Lv N, Zhong J, Chen H, Lin T, Chen S. Real-time control of welding penetration during robotic GTAW dynamical process by audio sensing of arc length. International Journal of Advanced Manufacturing Technology. 2014;**74**(1-4):235-249

[45] Moon H, Beattie RJ. Development of adaptive fill control for multitorch multipass submerged arc welding. International Journal of Advanced Manufacturing Technology. 2002;**19**: 867-872

[46] Nied HA, Baheti RS. Arc welding adaptive process control system. 1986

[47] Ozcelik S, Moore KL, Naidu SD. Application of MIMO direct adaptive control to gas metal arc welding. In: American Control Conference. 1998. pp. 1762-1766

[48] Suzuki A, Hardt DE, Valavani L. Application of adaptive control theory to on-line GTA weld geometry regulation. Transactions of ASME Journal of Dynamic Systems, Measurement, and Control. 1991:93-103

[49] Wang Z, Zhang Y, Wu L. Adaptive interval model control of weld pool surface in pulsed gas metal arc welding. Automatica. 2012;**48**(1):233-238

[50] Zhang YM, Liu YC. Modeling and control of quasi-keyhole arc welding process. Control Engineering Practice. 2003;**11**(12):1401-1411

[51] Banerjee P, Chin BA. Front side sensor based dynamic weld penetration control in robotic GTAW. In: Proceedings of the ASME Japan/USA Symposium on Flexible Automation. 1992

[52] Casalino G, Hu SJ, Hou W. Deformation prediction and quality evaluation of the gas metal arc welding butt weld. Institution of Mechanical Engineers, Part B: Journal of Engineering Manufacture. 2003; **217**(11):1615-1622

[53] Chen SB, Zhao DB, Wu L, Lou J. Intelligent methodology for sensing, modeling and control of pulsed GTAW. Part 2. Butt joint welding. Welding Research Supplement. 2000;**June**:164s-174s

[54] Cruz JAG. Uma metodologia para modelagem e controle da altura do reforço e da largura do cordão de solda no processo gmaw [Master thesis in Mechatronic Systems]. Universidade de Brasília; 2014

[55] Di L, Srikanthan T, Chandel RS, Katsunori I. Neural-network-based self-organized fuzzy logic control for arc welding. Engineering Applications of

Artificial Intelligence. 2001;**14**(2): 115-124

[56] Liu YK, Zhang YM, Kvidahl L. Skilled human welder intelligence modeling and control: Part 1. Modeling. The Welding Journal. 2014;**93** (February):46s-52s

[57] Liu YK, Zhang YM, Kvidahl L. Skilled human welder intelligence modeling and control: Part II. Analysis and control applications. The Welding Journal. 2014;**93**(May):162-170

[58] Luo H, Devanathan R, Wang J, Chen X, Sun Z. Vision based neurofuzzy logic control of weld pool geometry. Science and Technology of Welding and Joining. 2002;7(5):321-325

[59] Wu CS, Gao JQ. Vision-based neuro-fuzzy control of weld penetration in gas tungsten arc welding of thin sheets. International Journal of Modelling, Identification and Control. 2006;**1**(2):126-132

[60] Liu YK, Zhang YM. Model-based predictive control of weld penetration in gas tungsten arc welding. IEEE Transactions on Control Systems Technology. 2014;**22**(3):955-966

[61] Torres EMM. Uma metodologia para modelagem e controle da largura do cordão de solda no processo GMAW [master thesis in Mechatronic Systems]. Universidade de Brasília; 2013

[62] Yan ZH, Zhang GJ, Wu L. Simulation and controlling for weld shape process in P-GMAW based on fuzzy logic. In: Proceedings of the 2011 International Conference on Mechatronics and Automation. 2011. pp. 2078-2082

[63] Yang S-M, Cho M-H, Lee H-Y, Cho T-D. Weld line detection and process control for welding automation. Measurement Science and Technology. 2007;**18**(3):819-826

[64] Huang X, Gu W, Shi F, Chen S. An adaptive inverse control method based on SVM-fuzzy rules acquisition system for pulsed GTAW process. International Journal of Advanced Manufacturing Technology. 2009;**44**(7-8):686-694

[65] Shen H-y, Wu J, Lin T, Chen S-b. Arc welding robot system with seam tracking and weld pool control based on passive vision. International Journal of Advanced Manufacturing Technology. 2008;**39**:669-678

[66] Doumanidis C, Kwak YM. Multivariable adaptive control of the bead profile geometry in gas metal arc welding with thermal scanning. International Journal of Pressure Vessels and Piping. 2002;**79**(4):251-262

[67] Xiong J, Zhang G. Adaptive control of deposited height in GMAW-based layer additive manufacturing. Journal of Materials Processing Technology. 2014; **214**(4):962-968

[68] Mousavi MA, Haeri M. Welding current and arc voltage control in a GMAW process using ARMarkov based MPC. Control Engineering Practice. 2011;**19**(12):1408-1422

[69] Lü F, Chen H, Fan C, Chen S. A novel control algorithm for weld pool control. Industrial Robot: An International Journal. 2010;**37**(1):89-96

[70] Domfeld DA, Tomizuka M, Langari G. Modelling and Adaptive Control of Arc Welding Processes. New York: Meus Control Batch Manufact.; 1982. pp. 53-64

[71] Hale M. Multivariable geometry control of welding. In: ASME's Winter Annual Meeting, Symposium for Manufacturing Proceedings on Modeling Control. 1990

[72] Anderson Pereira C. Um Projeto De Controle De Movimentação Veicular Projetado Em Um Processador

Embarcado Em Fpga Com Ambiente De Simulação Usando Instrumentação Virtual [PhD thesis]. Universidade de Brasília; 2007

[73] Cook G, Barnett R, Hartman D, Strauss A. Neural network systems techniques in weld modeling and control. In: Computer-Aided Design, Engineering, and Manufacturing. CRC Press; 2000

[74] Zhang YM, Kovacevic R, Li L. Adaptive control of full penetration gas tungsten arc welding. IEEE Transactions on Control Systems Technology. 1996; **4**(4):394-403

[75] Clarke DW, Mohtadi C, Tuffs PS. Generalized predictive control—Part I. The basic algorithm. Automatica. 1987; **23**(2):137-148

[76] Clarke DW, Mohtadi C, Tuffs PS. Generalized predictive control—Part II. Extensions and interpretations. Automatica. 1987;**23**(2):149-160

[77] Clarke DW, Mohtadi C. Properties of generalized predictive control. Automatica. 1989;**25**(6):859-875

[78] Li K, Zhang YM. Interval model control of consumable double-electrode gas metal arc welding process. IEEE Transactions on Automation Science and Engineering. 2009;**7**(4): 1-14

[79] Abu-Ayyad M, Dubay R. Real-time comparison of a number of predictive controllers. ISA Transactions. 2007; **46**(3):411-418

[80] Yang F, Zhang S, Zhang WM. Design of GTAW wire feeder control system based on Nios II. Applied Mechanics and Materials. 2013;**397-400**: 1909-1912

[81] Hurtado RH, Sadek C, Alfaro A, Llanos CH. Discontinuity welding detection using an embedded hardware system. In: ABCM Symposium Series in Mechatronics. Vol. 5. 2012. pp. 879-888

[82] Llanos CH, Hurtado RH, Alfaro SCA. FPGA-based approach for change detection in GTAW welding process. Journal of the Brazilian Society of Mechanical Sciences and Engineering. 2016;**38**(3):913-929

[83] Velasco RHH. Detecção on-line de descontinuidades no processo de soldagem GTAW usando sensoriamento infravermelho e FPGAs [master thesis in Mechatronic Systems]. Universidade de Brasília; 2010

[84] Machado MVR, Mota CP, Neto RMF, Vilarinho LO. Sistema Embarcado para Monitoramento Sem Fio de Sinais em Soldagem a Arco Elétrico com Abordagem Tecnológica. Soldag. Insp. 2012;**17**(2):147-157

[85] Millán R, Quero JM, Franquelo LG. Welding data acquisition based on FPGA. IC's for Instrumentation and Control. Sevilla, Spain; 1997. pp. 2-6. DOI: 10.13140/2.1.4318.3044. Available from: http://hdl.handle.net/11441/23596

Chapter 4

Online Measurements in Welding Processes

Guillermo Alvarez Bestard

Abstract

The online measurement of principal magnitudes in welding processes is important to close the control loop and meet the project requirements. But, it is difficult because of the adverse environmental conditions that exist near the weld pool. Some conventional measurement techniques are used, but under these conditions, indirect sensing techniques are a better option. Sensor fusion algorithms and indirect sensing techniques allow estimate magnitudes that are impossible to measure directly. Sensor fusion is used to describe the static and dynamic behavior of process variables and is based on several areas of knowledge, such as statistics and artificial intelligence. By combining different sensing technologies to take advantage of each one, it is possible to obtain better sensing results. In this chapter selected sensing techniques and estimation algorithms used online for collecting values on the welding process are shown. Special attention is given to sensor fusion techniques. Some real applications and innovative research results are discussed.

Keywords: estimation, online measurement, sensor fusion, welding

1. Introduction

The welding process is used by many manufacture companies in a wider range of applications. Many studies have been carried out to improve the quality and to reduce the cost of welded components. Part of the overheads is employed in the final inspection, which begins with a visual inspection, followed by destructive and nondestructive testing techniques. In addition to cost raises, a final inspection is conducted when the part is finished only. When a defect occurs during welding, it can be reflected in the physical phenomena involved: magnetic field, electric field, temperature, sound pressure, radiation emission, and others. Thus, if a sensor monitors one of these phenomena, it is possible to build a system to monitor weld parameters and quality.

For years, much has been done to predict problems in welding to make it a stable process capable of making union parts with minimal human interference. Despite various sensors used in welding processes, there is still no effective option able to identify, directly, the weld bead characteristics obtained during the process. This is a limiting factor in the process control because weld bead characteristics can only be determined after the completion of welding through testing (destructive or not) when no control action can be taken.

In the last decade, measurements of multiple sensors are used to estimate some geometric parameters of the weld bead, such as the weld penetration, reinforcement, and width, and to classify the process stability. These estimations can be made online[1] and can be used to control the weld bead geometry formation. More recently, research has shown the possibility of estimating some microstructural characteristics and thermally affected zone dimensions, as shown in [1]. These estimations can be used to limit the control actions so as not to affect the desired characteristics of the weld bead. These techniques do not eliminate destructive testing but greatly reduce its use and prevent the defect formation or loss of characteristics.

In the welding processes, many variables can be used to control the geometry and quality of the final product. For this, other variables need to be measured or estimated. Some variables can be measured or modified online, directly in the welding power source, such as welding voltage and welding electric current intensity. Others can be measured using noncontact sensing methods, such as width and reinforcement of weld bead, drop volume, and electrical stick-out. But several variables, necessary to close the control loops, are difficult to measure online. This difficulty is due to the extreme conditions in the arc zone, because the electric arc is a powerful radiation emitter, in a long range of the frequency spectrum, including high temperature and visible light, generating steam and droplets of molten metal, coming from the electrode and the base metal. For these latter cases are necessary estimation methods based on models of welding processes that may include the power source and the robotic system used to move the torch or piece.

For the automation and control of complex manufacturing systems, a great deal of progress came up in the last decade, for precision and online documentation (bases for the quality control). With the advent of electrically driven mechanical manipulators and later the whole, relatively new, multidisciplinary mechatronics engineering, the need for information acquisition has increased. The acquisition is, in many cases, distributed through the system, with strong interaction between the robot and its environment. The design objective is to attain flexible and lean production. The requirement of real-time processing of data from multi-sensor systems with robustness, in an industrial environment, shows the need for new concepts on system integration.

Technology advancements seek to meet the demands for quality and performance through product improvements and cost reductions. An important area of research is the optimization of applications related to welding and the resultant cost reduction. The use of nondestructive tests and defect repair are slow processes. To avoid this, online monitoring and control of the welding process can favor the correction and reduction of defects before the solidification of the melted/fused metal, reducing the production time and cost [2].

Developments in microelectronics have led to rapid advances in welding power sources. With the use of fast microelectronic circuits, the speed of welding process control and welding parameter adjustment has been increased tremendously, and dynamic control over the arc and molten metal transfer have become possible. Research and development carried out by manufacturers of welding power sources focus on rapid optimal control of the welding parameters during welding. Modern welding sources are equipped with special control functions of arc and molten metal transfer, focusing on two basic areas: The first area of focus is on welding thin metal sheets (0.53 mm), and the second is on high-productivity thick metal sheet welding (over 5 mm) [3].

---

[1] When the measurement or control action is made inside the process flow and during the process execution, this task is classified as online. When the measurement or control action is made outside the process flow and before or after the process execution, this task is classified as offline.

With continuing advancements in digital and sensor technology, new methods with relatively high accuracy and quick response time for the identification of perturbations during the welding process have become possible. Arc position, part placement variations, surface contaminations, and joint penetration are key variables that must be controlled to ensure satisfactory weld production [4]. The techniques related to welding process optimization are based on experimental methodologies. These techniques are strongly related to experimental tests and seek to establish relations between the welding parameters and welding bead geometry.

Researches related to adaptive systems for welding seek the improvement of welding bead geometry with direct (if based on monitoring sensors) or indirect monitoring techniques. The indirect monitoring systems are the more used, looking to link elements such as welding pool vibrations, superficial temperature distribution, and acoustic emissions to size, geometry, or welding pool depth [5]. The most used approaches in welding control are infrared monitoring, acoustic monitoring, welding pool vibrations, and welding pool depression monitoring as shown in [6]. The literature analysis made in [7] shows a similar conclusion, but adds vision techniques with the same level of use with infrared measurement.

The majority of modeling methods used in the past were based on the physics and chemical characteristics of the process components. This method needs great knowledge about welding processes, and, because of its complexity, these magnitudes are difficult to obtain and keep constant throughout the process, causing inaccuracies in the models. With the development of information processing capabilities, black box modeling methods were successfully used, as such the statistics and probabilities, but needs very complex models because of the nonlinearity of the welding process and the correlation between variables. Nowadays, the black box methods based on intelligent artificial algorithms are predominant as shown in [7], where 29% of models analyzed were obtained using intelligent artificial algorithms, 18% used image processing (which can include intelligent artificial algorithms such as deep learning methods), 12% used statistical methods, and only 2% used physics and chemical characteristics. Also, recently, big data and data mining algorithms are to be introduced in industrial applications.

This chapter focuses on the online measurement of main magnitudes in welding processes to close the control loop and allow the welding power source and robot parameters to be adjusted. These actions make it easier to meet the project requirements, reduce the cost of welding production, and increase productivity by reducing the number of parts rejected in final quality inspection. The next sections discuss the sensing and analysis techniques used to measure or estimate important variables in welding processes, with emphasis on the conventional gas metal arc welding (GMAW) process. A summary of its evolution is also shown, and some examples of algorithms to measure data processing are discussed.

Also, a novel modeling method, based on a sensor fusion algorithm, is shown. This method uses dynamic information of welding processes to improve the model response, rather than relying on static models used in research found in the literature. The method uses arc welding measurements and thermographic information to estimate the weld bead penetration. The algorithm obtains the thermographic features and supplies information about the amount and spatial distribution of the energy in the workpiece, minimizing the errors when multiple inflection points are found because it does not use the second derivative for the calculation of thermographic width. Besides, volume calculations are performed using the actual thermographic curve instead of the ideal Gaussian curve used in most research as an approximation value and using more complex equations. This approach uses only addition operations, simplifies calculations, and improves model accuracy, allowing its implementation in an embedded device. The actual application area is automatic control, arc welding, and sensor fusion research.

2. Measurements in welding processes

Automatic welding is affected by disturbances in the production parameters and welding conditions. Variations in components' position and dimensions, weld joint misalignment, oil in the surface and other improprieties in materials, instability in welding wire speed, and shielding gas flow are examples of disturbances. However, by adding measuring systems, the control system may have a better performance when disturbances are found. For this purpose, the disturbances and affected variables must be measured or estimated online, to counteract the effects of the disturbances through the control actions. These variables can be different in each process type or change your significance in the process. Some variables can be measured directly from the welding power source or using indirect measuring techniques, and others need to be estimated from measured variables. This last group includes the weld bead depth or penetration.

In welding processes, the variables can be classified depending on whether they can be measured or modified and if these actions can be done online or offline. Responding to this classification, the variables in welding processes can be divided into five basic groups [8]:

- Fixed, which cannot be modified by the operator but defined in the process design
- Adjustable online, which can be modified during the process
- Adjustable offline, which can be modified only before starting the process
- Quantifiable online, which is measurable during the process
- Quantifiable offline, which is measurable only after the process ended

An example of variables and groups for the constant voltage GMAW orbital process is shown in **Figure 1**, and some measurement techniques are described in the following sections.

2.1 Measurement of welding variables

The most common measurement variables in welding are related to the power supplied to the process by the unit of material length or area, and most are defined or measured by the power source. In conventional arc welding processes, these variables are *electric voltage*, *intensity of electric current* (also called simply *amperage* or *welding current*), and *wire feed speed* in processes with material addition. The parameters to control the waveform of voltage or electric current in the advanced arc welding process are important also. Other variables can be necessary to define and know, as *gas flow*, *pre-gas time* and *post-gas time*, *source impedance*, and time or position when the arc is open and closed (*arc status*). In laser welding processes, *laser power*, *pulse rate*, *focal distance*, and *spot size* are important too.

The modern welding power sources have one or various microcontrollers or microprocessors to control source operations, data acquisition, and

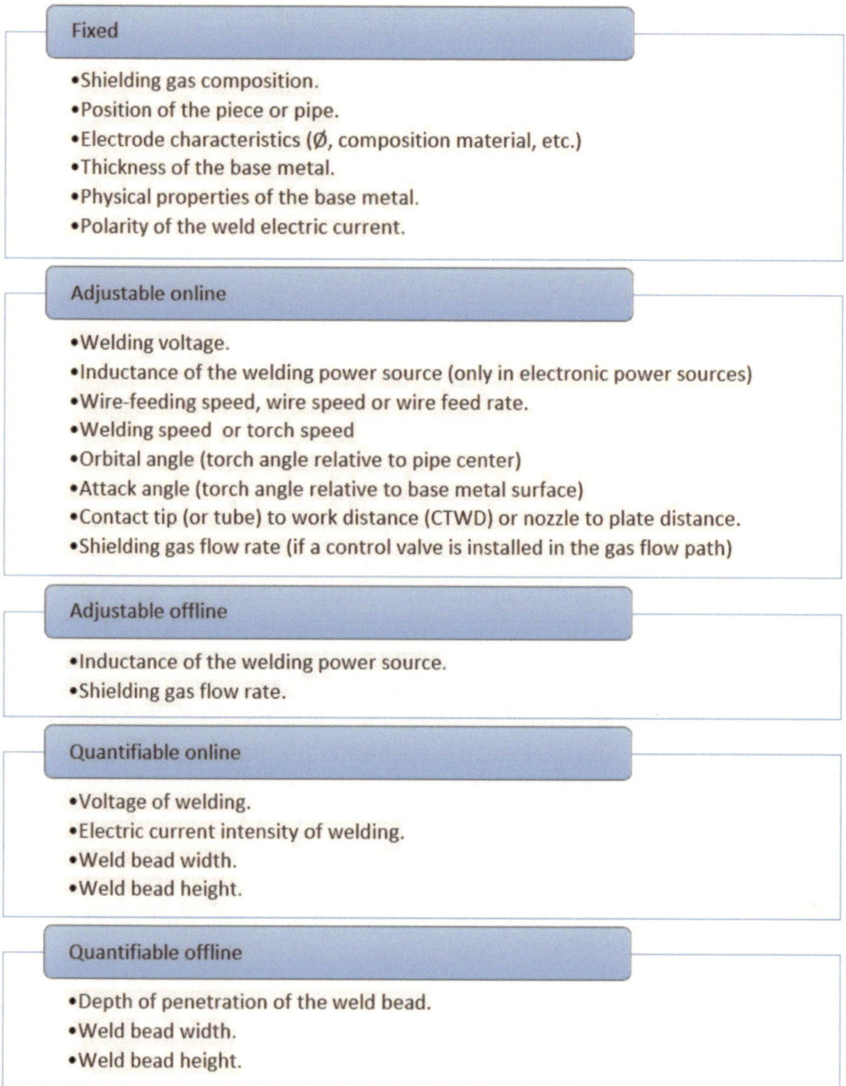

Figure 1.
Variables and classification groups to conventional constant voltage gas metal arc welding (GMAW) process (adapted from [8]).

communications. The communication interface usually implements a serial protocol or digital and analogical inputs and outputs to obtain or send information from an external supervisory control system (computer, programmable logic controller, and robotic system, among others). This interface can be used to synchronize work between power sources and robotic systems, other machines, control levels, or factory management. The welding variables, as voltage, electric current, and wire feed speed, can be measured and modified using this interface. Some commonly used standard network protocols are RS-232, RS-485, Modbus (RTU, TCP, or UDP), CAN Bus, DeviceNet, Field Bus, and Ethernet. Other companies define their protocol such as the Arclink developed by Lincoln Electric and SpeedNet by Fronius.

The communication protocol can be "open" or "proprietary." In the first case, the user can communicate his control system directly with the power source using the communication protocol description. In the second case, he needs to buy and use proprietary software or hardware. The implementation of protocols used in

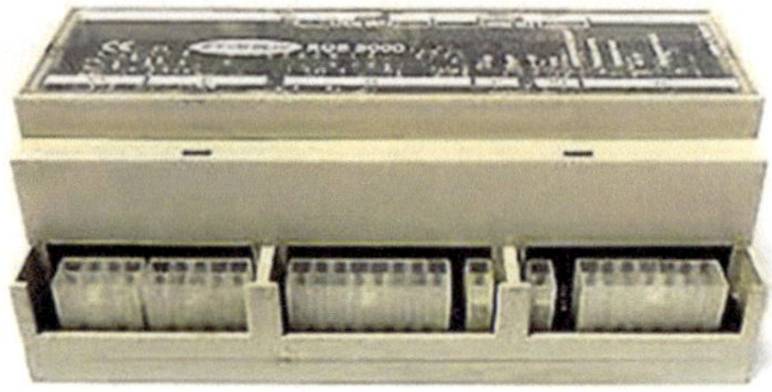

Figure 2.
Fronius interface to translate the information between serial protocol and data acquisition system with digital and analogical inputs and outputs.

modern power sources is serial[2], and information is obtained or sent in digital format.

The old welding power sources can have an interface with digital and analogical inputs and outputs, to let the monitoring and control of the source. These power sources need a dedicated data acquisition and control system to convert the analog values in digital information and vice versa. The user, using this system, can monitor and control the power source operation partially or fully. To design or select the system, the sampling time, resolution and range of analogical inputs and outputs, and range and type of digital inputs and outputs, based on the power source characteristics and application requirements, should be considered.

Some manufacturers offer hardware interfaces to translate the information from serial protocol to digital and analogical inputs and outputs. If you do not have conditions to use a serial protocol, these interfaces can help to get and send information to power sources. These systems can be more slow, inaccurate, and inefficient than serial protocol, because of the sequential conversions from analogical to digital (on the source), from digital to analogical (on the interface), and from analogical to digital again (on the acquisition system). The ROB 5000 of Fronius, shown in **Figure 2**, is an example [9].

2.2 Measurement of welding speed, welding angles, and orbital angle

Other variables that must be measured to control robotic welding processes are those that characterize the relative movement and position between the piece and torch. These variables can be obtained from the position control system of the robot, but many times an initial or absolute reference is necessary.

The relative movement between the piece and torch is defined by the *welding speed*. Sometimes, the torch is coupled to the robot manipulator and moves while the piece is fixed. Other systems fixed the torch and move the piece. More complex robotic systems move the torch and piece. In all cases, it is important to consider the relative speed between the torch and the piece to obtain the welding speed. This speed can be calculated using the torch and the piece speed obtained from the robotic system. On plane welding, the speed can be expressed in longitude per time unit, such as mm/s.

In orbital welding, the *orbital angle* is used to indicate the torch position in the polar coordinate system with the origin on the pipe axis. This measurement is very important because the welding conditions can be different for each orbital position. Its value must be obtained from the robotic system too. In this case, it is possible to express the welding speed using orbital angle per time unit, such as $^o/s$.

[2] The information bits are sent one by one through a communication channel.

The *forehand angle* or *attack angle* is the torch angle side view relative to the base metal surface, which is the angle in direction of torch travel. The *work angle* is defined by the torch angle end view relative to the base metal surface as shown in **Figure 3**. These angles can be fixed before starting the welding and staying constant or can be controlled by the robotic system. In the second case, the angles can be calculated or obtained from the robotic system also.

2.3 Measurement of contact tip-to-work distance

The *contact tip-to-work distance (CTWD)* can be obtained from the robot control system related to torch or piece position, or it can be measured with a laser distance sensor. Other variables, such as *electrical stick-out* or *electrode extension* and *arc length* (see **Figure 4**) need more complex procedures to obtain a measurement because they depend on the fusion rate. All these variables are expressed in longitude unit, such as *mm*.

The first method to obtain the *CTWD* is more economical but has low accuracy. The zero references of the robotic system are calibrated with the workpiece position, but the surface variations and thickness of the material can affect the accuracy of the value. In arc welding processes, small variations in CTWD can affect the electric resistance of the arc, and consequently, the welding current and the heat input can be significantly affected. For example, reduction of the arc length causes

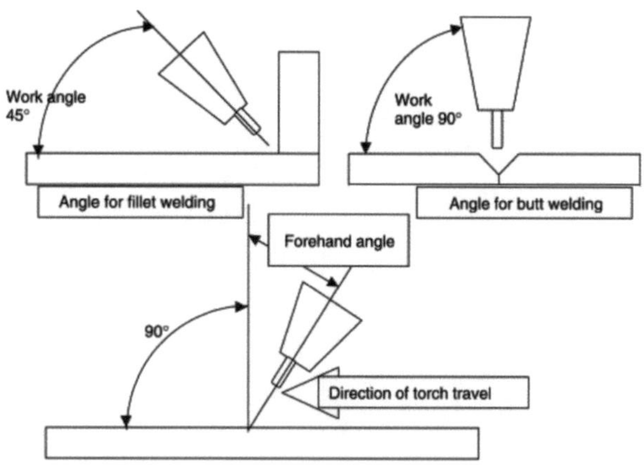

Figure 3.
Welding speed, torch angles, and orbital angle (adapted from [10]).

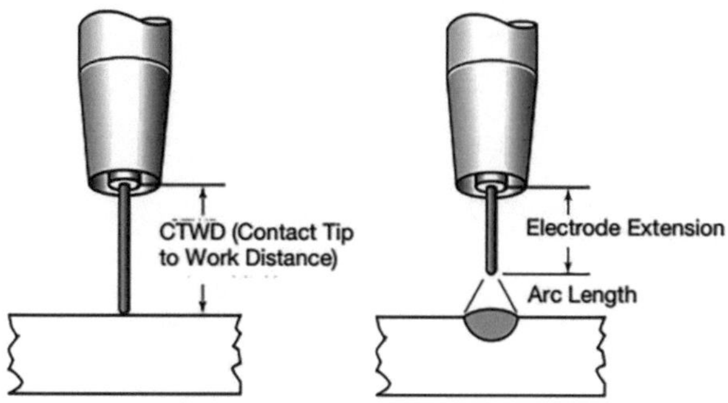

Figure 4.
Contact tip-to-work distance and electrical stick-out (or electrode extension) differences (adapted from [11]).

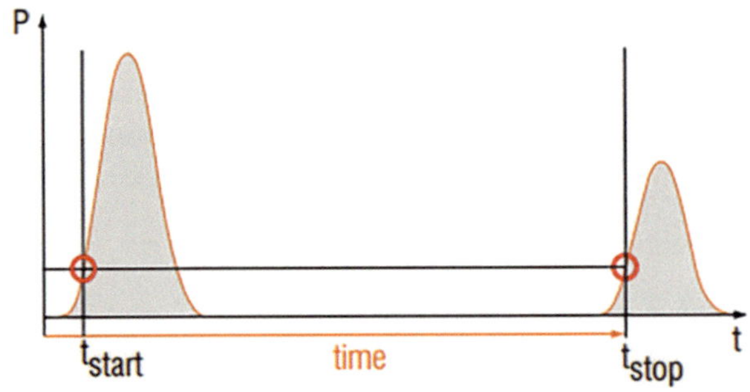

Figure 5.
Time of flight measurement principle (adapted from [12]).

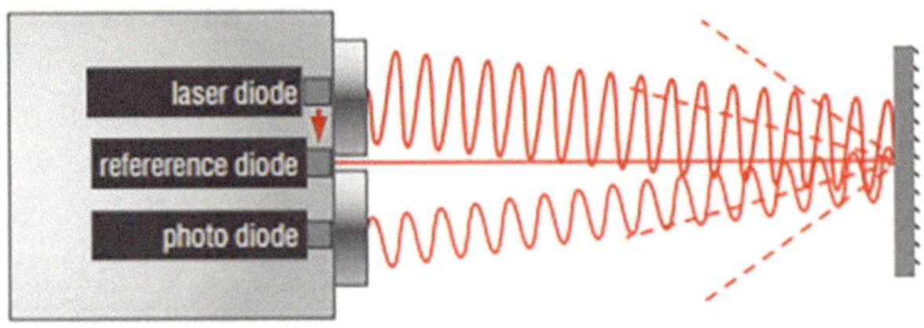

Figure 6.
Phase comparison measuring principle (adapted from [12]).

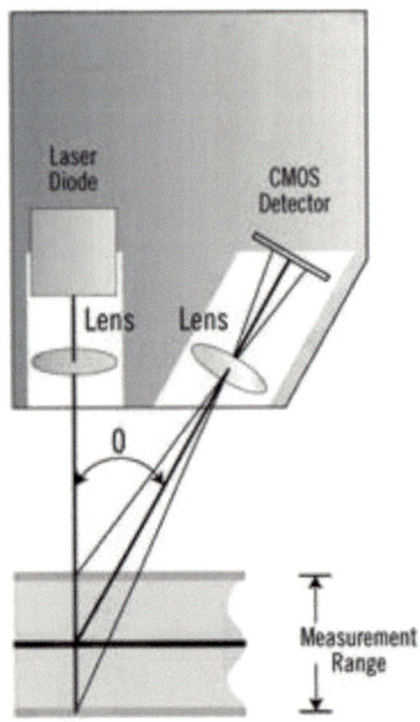

Figure 7.
Triangulation measuring principle (adapted from [13]).

the increase of heat input (and more current because it reduces the equivalent resistance of arc) which makes the wire electrode melt more quickly and thereby restore the original arc length.

A laser sensor can be more accurate, but the measurement point needs to be selected correctly. This sensor measures the distance to a point on the surface of the base metal of the piece, and it has three basic principles of operation: time of flight, phase comparison, or triangulation method.

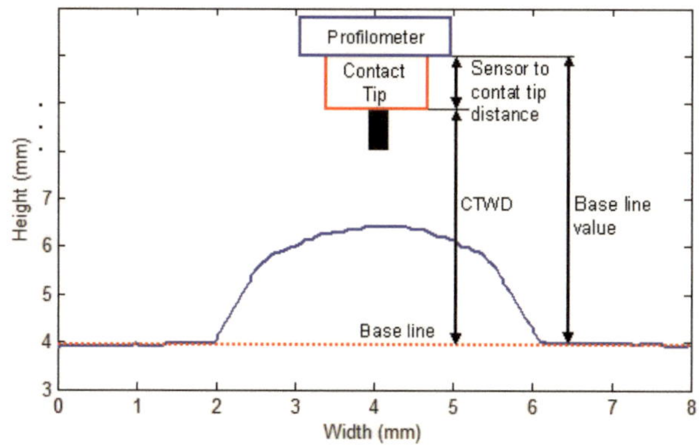

Figure 8.
Contact tip-to-work distance measurement [8].

In the time of flight measurement principle, shown in **Figure 5**, a laser diode produces short pulses which are projected onto the target. The light reflected from the target is recorded by the sensor element. The time of flight of the light pulse to the target and back determines the measured distance. The integrated electronics in the sensor compute the distance using the time of flight. Sensors using this principle are not sensitive to external light.

In the phase comparison measuring principle, a high-frequency modulated laser light with low amplitude is transmitted to the target as shown in **Figure 6**. Depending on the distance of the object, it changes the phase relationship between transmitted and received signals. Sensors using this principle operate with high accuracy for measurement distances up to 150 m.

In the triangulation method, shown in **Figure 7**, the laser beam is projected and reflected from a target surface to a collection lens. The lens focuses an image of the spot on a linear array camera. The camera views the measurement range from an angle at the center of the measurement range. The position of the spot image on the pixels of the camera is then processed to determine the distance to the target. The camera integrates the light falling on it, so long exposure times allow greater sensitivity to weak reflections.

The most used laser emitter distance sensors are in the wavelengths of red color (close to 658 *nm*), but in recent years, blue-violet lasers with a shorter wavelength than a red laser (close to 405 *nm*) have been used in welding processes and others that work with red-hot glowing metals. This shorter wavelength provides higher optical resolution and noise reduction. The blue-violet laser sensors enable more reliable measurements on these processes than red laser sensors.

Other research methods use the voltage and current feedback signals from the welding process, computing the minimum resistance during the short circuit period, and uses this value to estimate the *CTWD* after applying a correction factor for the duration of the short circuit. The effect of wire feed speed, actual CTWD, and shielding gas on the correction factor is determined experimentally [14].

The *CTWD* can be calculated with the same bead profile obtained from the laser profilometer. The distance between the torch contact tip and the sensor reference is fixed and known. The CTWD is the difference between the baseline value and this distance, as shown in **Figure 8**.

2.4 Measurement of welding joint and weld bead geometry

The quality of arc welding is commonly described by the geometry of the molten weld pool and the weld bead because the mechanical properties of the welding joint are reflected in these geometry characteristics. The geometry of the weld bead is a set of parameters defined in the design stage, and to achieve the required quality, it should be measured and controlled throughout the process. The parameters or variables which define the most important characteristics of the geometry of the weld bead (including the weld pool) are the *weld bead width*, the *weld bead reinforcement*, and *weld bead depth* or *weld bead penetration*, as shown in **Figure 9**.

The visual information of the molten weld pool is used by expert operators to control the welding process in manual welding. In automatic welding processes, this information can be used to improve the control behavior and achieve the desired quality in the welding joint. In arc welding processes, the geometry parameters are governed by many factors, such as *welding current*, *welding voltage*, *wire feed speed*, *welding speed*, and the *contact tip-to-work distance*. Then, for successful control of geometric variables, it is necessary to provide feedback to the control system.

For feedback implementation, it is important to know that the process' adverse environmental conditions, in the vicinity of the electric arc and the molten pool,

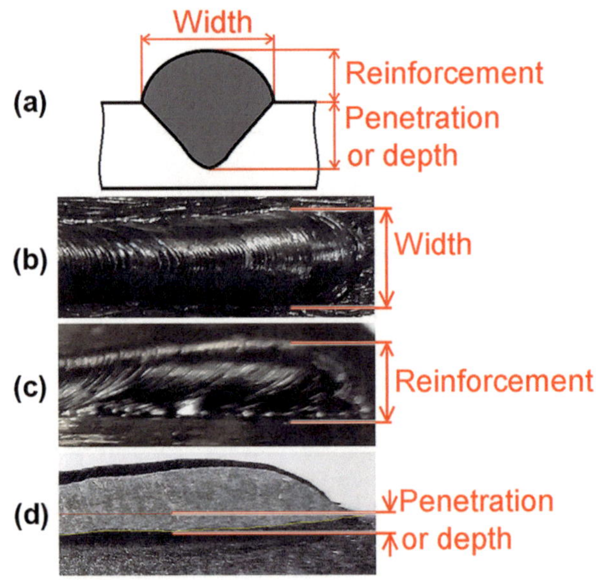

Figure 9.
Geometric dimensions of the weld bead: (a) cross-section, (b) top view, (c) side view, and (d) side view of a longitudinal cut [7].

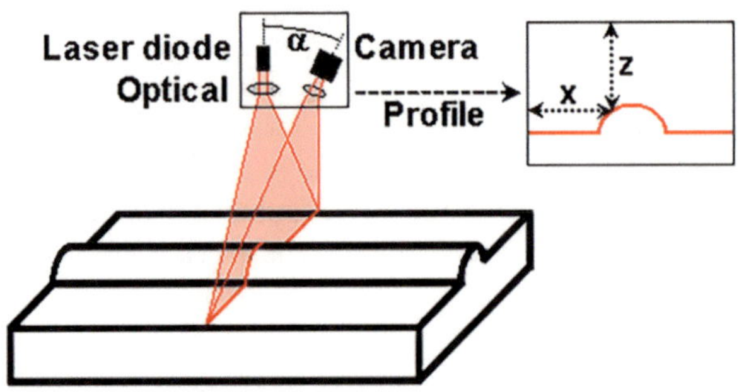

Figure 10.
Two-dimensional laser triangulation principle [7].

would damage the measuring instruments that require physical contact with the surface of the piece. These conditions make the measuring of the weld bead geometry a difficult task using conventional measuring principles. Instead, noncontact techniques have been developed and employed successfully.

For this purpose, vision sensing is a promising solution. One common method is formed by a laser beam, which draws one or more lines on the surface to be measured, the image is filtered to obtain only the wavelength emitted by the laser, and a camera or matrix image sensor captures the line created by the laser. Subsequently, using image processing algorithms and triangulation techniques, a profile of the piece with the required information is obtained.

This system, referred to as a laser scanner or profile sensor, is shown in **Figure 10**. In it, the optical system projects the diffusely reflected light of this laser line onto a highly sensitive sensor matrix. From this matrix image and the angle between the camera and the laser diode (α), the controller calculates the distance information (z-axis) and the position alongside the laser line (x-axis). These measured values are then projected in a two-dimensional coordinate system that is fixed for the sensor. To obtain three-dimensional measurement values, the sensor or workpiece can be moved in a controlled manner.

Another way to obtain a three-dimensional profile is by using a laser pattern with a dot matrix or line grill, as shown in **Figure 11**. The two-dimensional information is processed using triangulation techniques to obtain a three-dimensional profile. In [15], the weld pool surface deformation is obtained from the projection pattern using the deformation of the lines (see **Figure 11b**) or the distance between points (see **Figure 11c**). Other implementations of this method are shown in

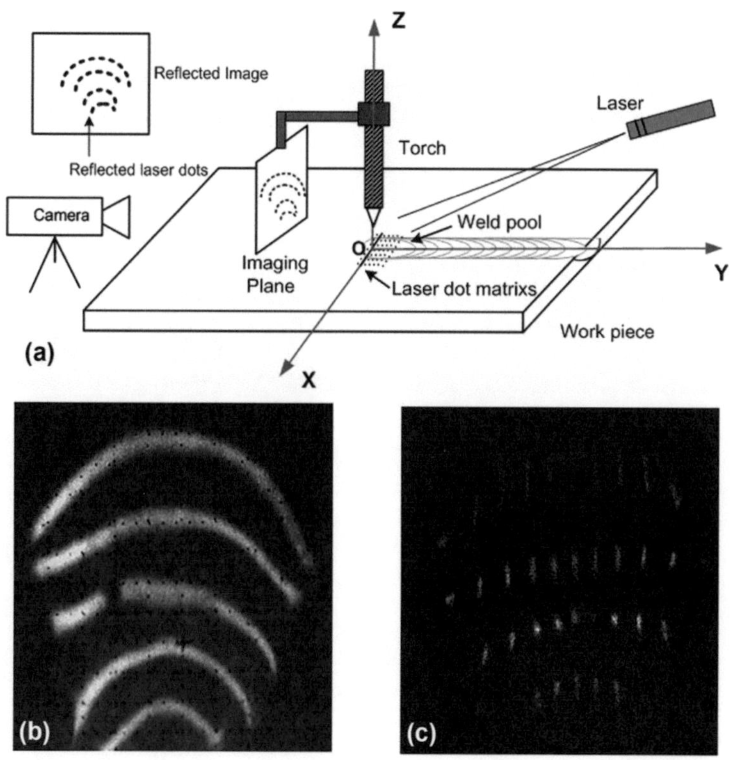

Figure 11.
Three-dimensional profile of the weld pool surface using structured light and triangulation method: (a) diagram of measure system, (b) reflected image using line laser pattern, and (c) reflected image using dot laser matrix (adapted from [15]).

[16–18]. In these methods, the uncertainty to recognize the real position is a problem. For this reason, a point of the pattern dot is intentionally missed to serve as a reference.

It can also use the same measuring principle to obtain a profile of the weld joint before joining the one. This allows the implementation of algorithms to define or adjust the trajectory to be followed by the torch (seam-tracking algorithms). Similarly, it is possible to estimate the amount of material required (deposition rate) for the formation of a bead with the desired dimensions [7].

The use of video cameras to measure the weld bead width and reinforcement is possible too, as is shown in [19–21], but these methods need optimal light conditions and are difficult to apply in the industrial environment.

These principles cannot be applied to penetration measurement, and this variable could be estimated from a different way. Due to the complexity of the measurement methods, we are going to dedicate another section to show some estimation methods of this magnitude.

2.5 Measurement of weld bead depth or penetration

Total penetration in welding processes is important to ensure weld quality. When the total penetration takes place, the melt weld pool crosses to the bottom side of the workpiece, as shown in **Figure 12**. The depth of penetration of the weld bead can be determined by nondestructive testing techniques such as ultrasound or X-ray. However, the portability and robustness of these traditional instruments are not a good option for the harsh conditions of the process and to develop an online measuring system. Because of that, many research works attempt to detect total or

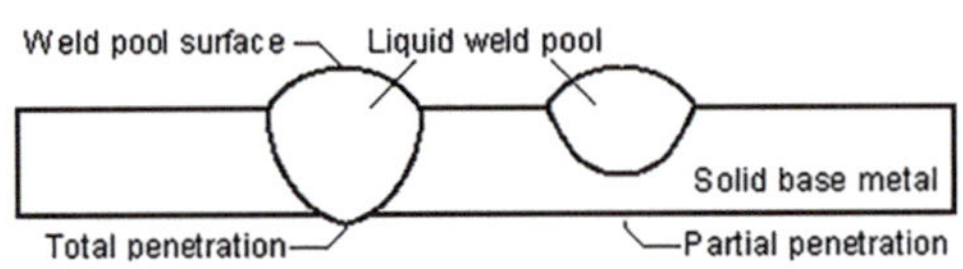

Figure 12.
Total and partial penetration in the weld bead [7].

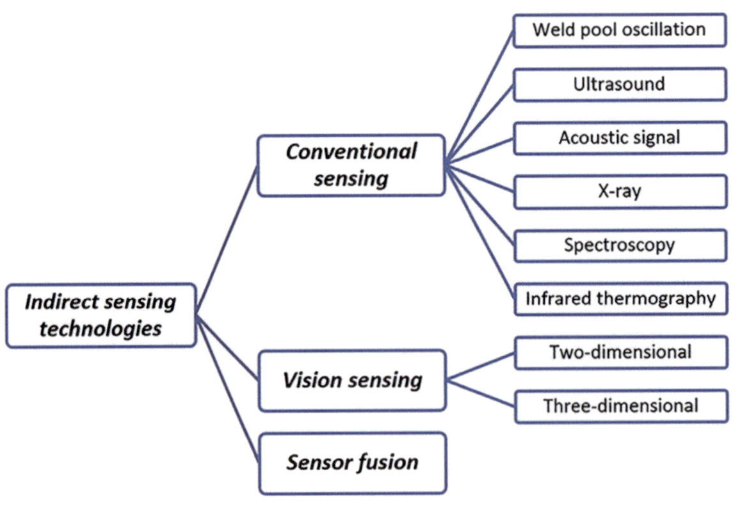

Figure 13.
Indirect sensing technologies used to monitor the bead and weld pool [7].

partial penetration with other methods, but few obtain a useful value for the control system algorithm.

Measurement of the backside width of the weld pool can be used to ensure total penetration, but this method is difficult because of the reduced space, limited movements, and restricted access, among other conditions, that make it impossible to place the sensors under the weld pool.

The front side of the weld pool offers information about the total penetration status. This information includes the *temperature of arc and weld pool*, *arc voltage*, *arc light*, *arc sound*, *geometry parameters*, *oscillation frequency*, and *resonance frequency*, among others. For obtaining this information, indirect sensing technologies are used. These measuring technologies can be classified as conventional, vision, and multi-sensor or sensor fusion technologies, as shown in **Figure 13**.

Conventional sensing technologies monitor parameters closely related to the weld bead and the weld pool geometry. It includes ultrasound, infrared thermography, weld pool oscillation, arc sound, and X-ray, among others. Vision sensing technologies obtain these features as skilled welders do. It can be divided into two-dimensional and three-dimensional sensing. These methods are applied to obtain the geometric shape of the weld pool and weld bead with good results. Sensor fusion integrates several sensing technologies in the same monitoring system. **Figure 14** shows the literature review statistics, obtained in [7], about the use of indirect monitoring technologies to estimate the weld bead geometry.

2.5.1 Modeling and estimating

Some of these measuring methods need a model to obtain the desired information from the process. To obtain a representative model, the estimators have been used successfully under specific conditions. But it is imperative to create a model that can be easily programmed and fed into the control system. The model must have satisfactory precision in the prediction of the depth of the weld bead and cover all of the positions used in the welding work. It is very useful if it also represents a wide range of thicknesses of the material, but this is not always possible.

The research about modeling the weld bead depth tries to relate this variable with the welding electric current intensity, welding voltage, wire feed speed, and welding speed. Documents analyzed in [7], show that the research is making mainly in horizontal or flat welding to obtain static models and the researchers are using artificial intelligence algorithms. The most used methods to estimate the weld bead geometry

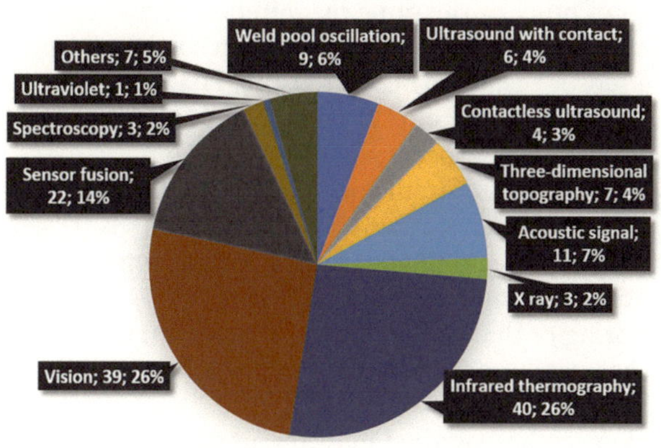

Figure 14.
Literature review statistics about indirect monitoring technologies used to obtain measurements of the weld bead geometry [7].

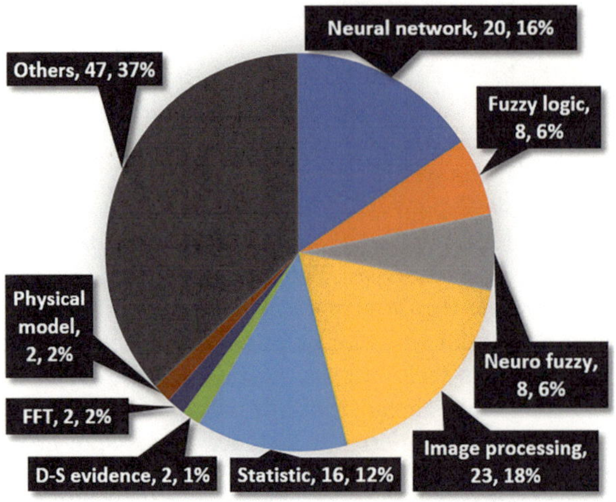

Figure 15.
Literature review statistics about analysis method used to estimate the weld bead geometry [7].

are artificial neural networks, fuzzy logic, and their combinations. This group represents 28% of the works, as shown in **Figure 15**. Image processing and statistical techniques such as multiple regression analysis, least squares, or factorial design are also frequently used. All these make up 58% of the total of the publications found.

In the welding process, the thermal energy is supplied through an electric current and stored in the material. Due to thermal inertia, dynamic models can be a better representation of the process. In these models, the historical values of the welding parameters are also selected as inputs. Dynamic behavior is essential to estimate the current and future state form to the past state. Despite this, [7] shows dynamic models in a minority. The main cause of the selection of the static model is the difficulty in obtaining a continuous data set of the weld bead depth. One way to obtain it is to perform a longitudinal cut and a macrographic analysis on the weld bead with an image processing algorithm as made in [22]. In the traditional cross-section cut, it is not possible to obtain enough information to make a dynamic model.

2.5.2 Sensor fusion

Measurements obtained by several sensors or measuring systems can be used to estimate the values of other or the same magnitudes. These techniques that combined different sensing technologies or information sources to obtain better sensing results are named sensor fusion or fusion sensory data. The sensor fusion is a multilevel process that needs a model to combine the information and describe the static or dynamic behavior of processes.

There are applications in different spheres such as aerial and ground navigation of mobile robots, systems for environmental monitoring, visual sensor networks, medicine, security, fault detection, and quality control, among others, as shown in [23]. This is a relatively young research area, but it is the third method used for the indirect monitoring of the welding process, as shown in **Figure 14**. In recent years, these methods have been studied to achieve effective welding and sense of the weld bead.

The sensor fusion can be classified in different ways as discussed in [24]. One of the most representative classifications for welding processes is *according to the relationships of the input data*, which defines how the information relates between the sensors. It can be complementary, competitive or redundant, and cooperative,

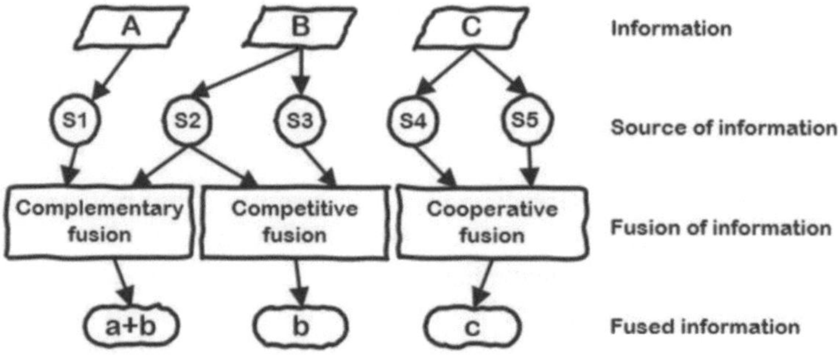

Figure 16.
Classification according to the relationships of the input data (adapted from [24]).

as shown in **Figure 16**, but it is very common to find more than one way of working on the same group of sensors.

Figure 16 shows four levels of the fusion process. The following list exemplifies welding process parameters:

- Information level represents the input magnitudes and zones that are measured by sensors, for example, the weld bead and weld pool dimensions.
- Source of information is the sensors installed on the process, for example, the video and thermographic cameras.
- The fusion of information is the algorithms used to obtain the fused information, for example, an image processing and neural network algorithms.
- Fused information is the final result, for example, the weld bead depth.

The *complementary fusion* is about fusing incomplete information that is obtained starting from different sources. This is the case in that several sensors are measuring different parts of an atmosphere or phenomenon, covering a bigger area, and allowing a more complete and more global vision of the process. For example, you can combine the weld pool thermographic information (A) obtained from an infrared camera ($S1$) and weld bead dimensions (B) obtained from a video camera ($S2$) to calculate the weld pool dimensions ($a + b$).

In *competitive or redundant fusion*, all the sensors are monitoring the same area, working redundantly and competitively. These sensors can have similar or different measurement principles. An example is the dimensional information (B) about the weld bead obtained from two video cameras ($S2$ and $S3$) used to calculate the weld bead dimensions (b).

Cooperative or coordinated fusion uses the information from independent several sensors to obtain new information, for example, the combination of the information of the weld bead (C) obtained from a vision system ($S4$) and pyrometer ($S5$) to estimate the weld bead penetration value (c).

A cooperative sensor fusion algorithm is used in [22] to obtain an estimator of the weld bead depth $\hat{D}$ for GMAW process. The developed algorithm tries to obtain information about the amount and spatial distribution of the energy supplied to the workpiece to estimate the depth of the weld bead using a thermographic camera and welding electric current measurements. The fusion algorithm is based on a perceptron neural network that combines the infrared features T of the weld molten pool, the welding current in the actual $i(nT)$ and previous sample $i(nT - T)$, and previous depth estimation $\hat{D}(nT - T)$. The symbol T is the sample time and n is the sample number. These previous values allow capturing the dynamic behavior of the process.

The artificial neural network has 8 neurons in the input layer, 12 neurons in the hidden layer, and 1 in the output layer. The activate function is the hyperbolic tangent sigmoid transfer function. The network training should be done with experimental measurements of the parameters of input and output and by using the backpropagation algorithm. A block diagram is shown in **Figure 17**.

The thermographic matrix, supplied by the thermographic camera, is processed with a moving average filter to obtain the thermographic peak T_p, base plane T_b, thermographic curve width T_w, thermographic area T_a, and thermographic volume T_v. The thermographic image is taken on the weld pool area as a physical reference, but the welding arc and the electrode are included. These features from a sample are shown in **Figure 18**.

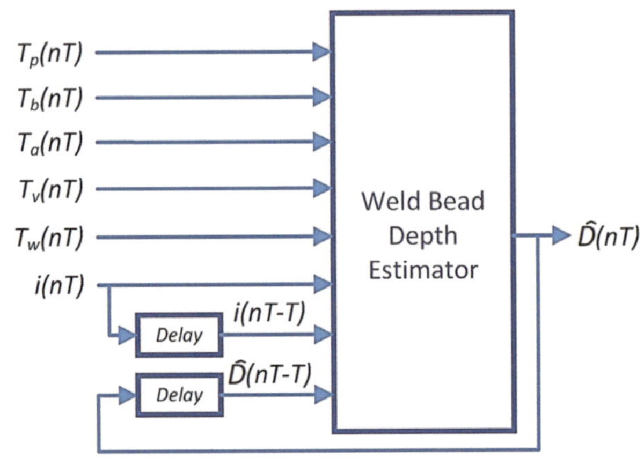

Figure 17.
Weld bead depth estimator block diagram, based on artificial neural network, developed in [22].

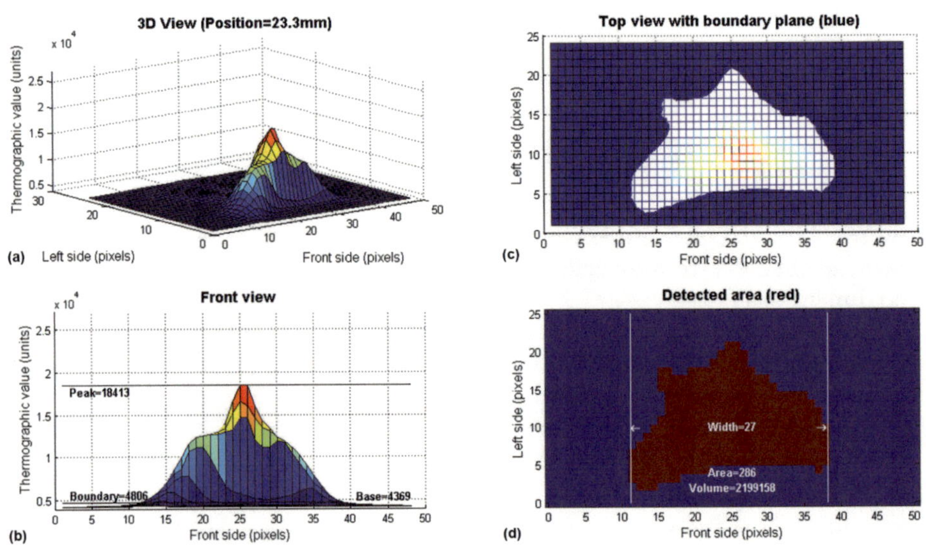

Figure 18.
Features extracted from an infrared image: (a) three-dimensional curve, (b) front view of the curve, (c) intersection between the boundary plane and three-dimensional curve, (d) and the calculation of the width, area and thermographic volume.

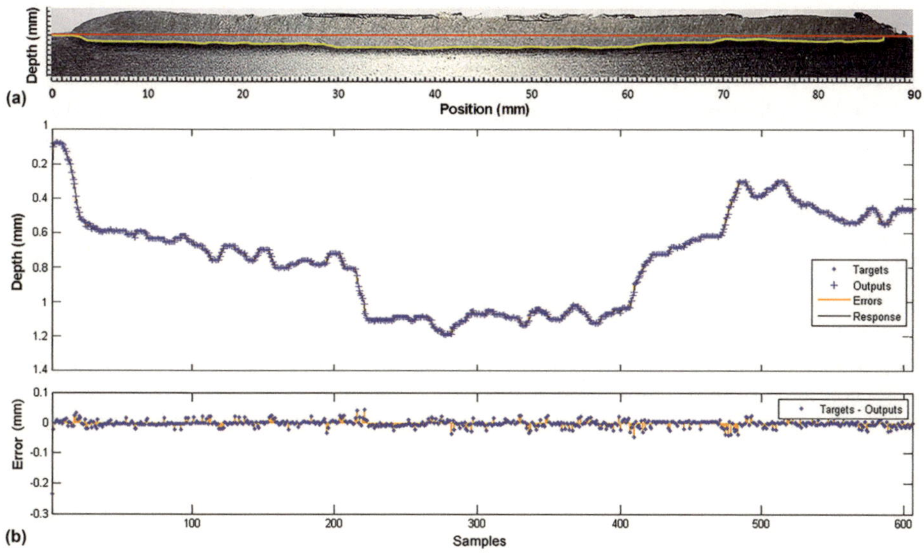

Figure 19.
Results obtained in the estimation of the weld bead depth in [22]: (a) weld bead depth longitudinal profile and (b) model response and experimental measurements. In the sample axis, each value corresponds to a position in the piece.

The base plane is calculated as the average of 10% of the values on the left and right sides, as shown in **Figure 18b**. The boundary plane is 10% above the base plane. The sum of active pixels in the intersection plane between the thermographic surface and the boundary plane is the area. The sum of the thermographic values within the intersection plane is the thermographic volume, shown in **Figure 18d**. This algorithm was optimized for implementation in embedded devices.

A contribution of the method is the minimization of errors when multiple inflection points are found because it does not use the second derivative for the calculation of thermographic width. Also, volume calculations are performed using the actual thermographic curve instead of the ideal Gaussian curve used in most research as an approximation value and using more complex equations. This approach uses only addition operations, simplifies calculations, and improves model accuracy.

The weld bead depth profile to network training was obtained using an image processing algorithm on the macrographic picture of the longitudinal cut of the piece. **Figure 19a** shows 610 measurements of weld bead depth in a yellow line and the base metal surface in a red line. The model has a fit of 0.99844, a performance or median square error (MSE) of 7.61×10^{-4}, and an estimation error less than 0.05 mm (less than 5% of full range). The error curves in **Figure 19b** show a model response that represents the behavior of the process with great accuracy.

It can be noticed that accurate sensing results were obtained based on multi-sensor information fusion technology, due to more weld pool information and effective information fusion techniques.

3. Evolution and comparison of the techniques and methods used for measurement and estimation of the geometry of weld bead

An analysis made in [7] about techniques and methods used for measurement and estimation of the geometry of the weld bead found publications and patents since 1971, but it is not until the 1980s that a greater interest in this subject was observed. Recently, in the last decade, the number of publications is increased. **Figure 20** shows this growing interest.

The technological development and cost reduction of sensors and the need for welding process control in an industrial environment (robotic welding) stimulated the number of scientific works about these topics.

An example is the development of infrared sensors that since the 1970s were available but in the 1980s were given classified contracts by the US Department of Defense to Honeywell and Texas Instruments to develop uncooled infrared sensor technology. In 1992, the US Government de-classified this technology for commercial products, allowing the sale of their devices to foreign countries, but kept close the manufacturing technologies. In the next decade, several countries developed uncooled imaging systems [25] with a drastic reduction of uncooled array cost.

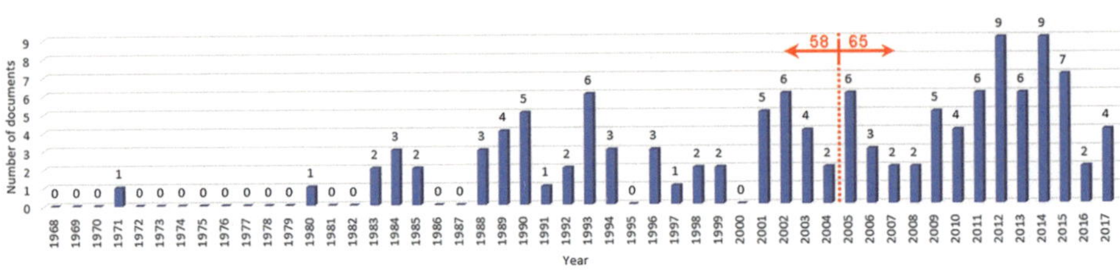

Figure 20.
Evolution of the number of publications about measurement and estimation of the weld bead geometry (adapted from [7]).

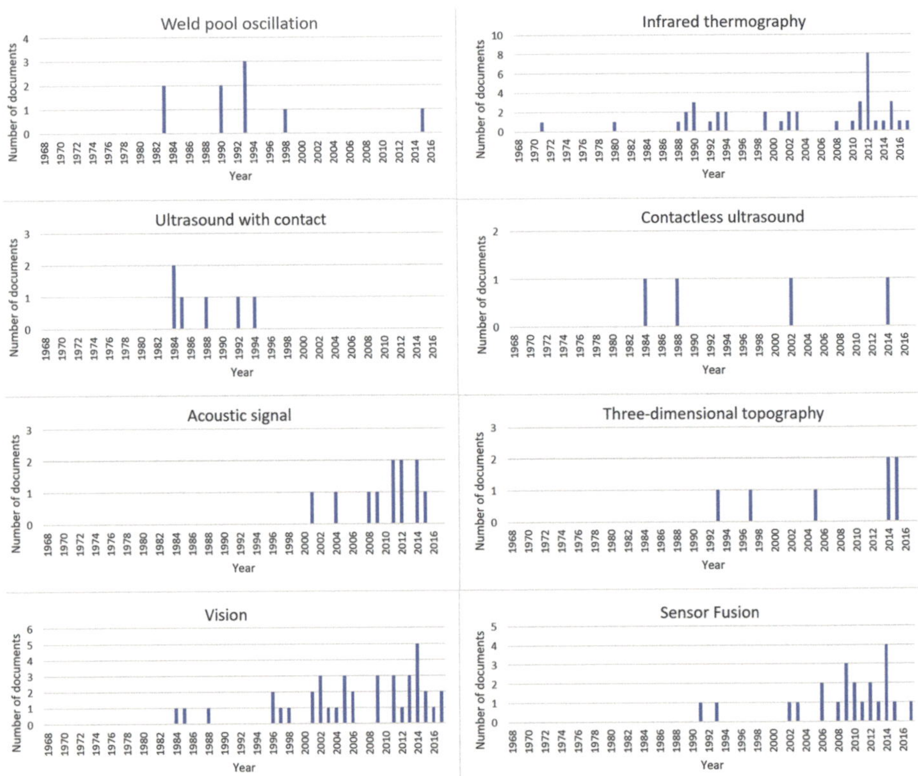

Figure 21.
Evolution of techniques used to measure the weld bead geometry [7].

Variable	Cost[a]/accuracy[b] of measuring method								
	From welding power source	From robotic system	Laser distance sensor	Laser scanner	Structural light[c]	Pyrometer[d]	Thermographic camera[d]	Acoustic signal[d]	Weld pool oscillation[c]
Welding voltage	L/H								
Welding current	L/H								
Wire feed speed	L/M								
Welding speed		L/H							
Welding angle		L/H							
Orbital angle		L/H							
Contact tip-to-work distance		L/M	M/H	H/H					
Welding joint dimensions				H/H	M/H				
Weld bead width			M/M	H/H	M/H				
Weld bead reinforcement			M/M	H/H	M/H				
Weld bead penetration						M/M	H/M	L/L	M/L

[a]L—Low cost; M—medium cost; and H—high cost.
[b]L—Low accuracy; M—medium accuracy; H—high accuracy.
[c]This method is used together with one or several video cameras.
[d]This method is used alone or together with welding power source measurements (e.g., arc variables).

Table 1.
Comparison between the most used methods to measure the variables of welding processes, considering the cost and accuracy of the method.

This evolution can be observed in **Figure 21** with a significant increase in scientific publications in this decade.

The estimation of geometry weld bead using acoustic signal and vision techniques shows more activity in the last 20 years because of the great development of cameras, audio systems, and digital signal processor devices with high speed and quality, little size, and low cost.

In **Table 1**, some commonly used methods to obtain measurements of welding processes online are compared. The comparison criteria are the cost of implementation and the accuracy of the method. Both criteria are evaluated as low, medium, or high. In the table, a blank cell indicates a method that is not used to measure a specific variable. Some methods are used in conjunction with other measurement methods to obtain or estimate the value of the variable. Various measurement methods can be combined using a sensor fusion technique.

The analysis techniques most used in the last century were regression models, least mean square algorithms, Kalman filter, and other statistical methods. In the last few years, the principal techniques used are the artificial neural network, fuzzy logic, and neuro-fuzzy. The intense development of image processing algorithms was observed in the last 20 years.

4. Conclusions

The correct selection of measuring techniques and the use of sensor fusion algorithms, combined with indirect measurement techniques, can help to reduce the cost of welding production and increase productivity by the detection or prediction of many welding defects or set point deviations. These measurements allow online adjusting of the welding power source and robot parameters in a closed control loop. Online estimation of variables that cannot be measured improves control systems and reduces the number of parts rejected in the final quality inspection.

To take advantage of a modern welding power source, it is important to equip the monitoring and control system with serial communication capabilities. The modeling of estimators is a critical step to obtain an accurate measuring system, and dynamic models have a better representation of the welding process than static models due to thermal inertia of the process. Vision and thermographic measuring techniques, image processing, and neural network algorithms, despite consuming more computing resources, are the most used to estimate the weld bead geometry, and excellent results have been observed.

The research on sensor fusion algorithms is grown. Following this trend, in this work a novel modeling method that uses arc welding measurements and thermographic information to create a dynamic model to estimate the weld bead penetration is presented. This new approach obtains information about the amount and spatial distribution of the energy in the workpiece and uses only addition operations, simplifies calculations, and improves model accuracy. A satisfactory solution was shown to be applied in welding automatic control using computers or embedded devices.

DOI: http://dx.doi.org/10.5772/intechopen.91771

Author details

Guillermo Alvarez Bestard
University of Brasilia, Brasilia, Brazil

*Address all correspondence to: guillermo@unb.br

References

[1] Pal K, Pal SK. Sensor based prediction of weld microstructure in pulsed MIG welding. International Journal of Microstructure and Materials Properties. 2015;**10**(5–6):402-434

[2] Alfaro SCA. Sensors for quality control in welding. In: Arc Welding. Vol. 17. London, UK: InTech; 2011. pp. 192-200

[3] Kolaík L, Kolaíková M, Kovanda K, Pantček M, Vondrouš P. Advanced functions of a modern power source for GMAW welding of steel. Acta Polytechnica. 2012;**52**(4):83-88

[4] Chen W, Chin BA. Monitoring joint penetration using infrared sensing techniques. Welding Journal. 1990; **69**(4):18-85

[5] Kerr HW, Hellina MC, Huissoon JP. Identifying welding pool dynamics for GMA fillet welds. Science and Technology of Welding and Joining. 1999;**4**:15-20

[6] Hong LY. Vision Based GTA Weld Pool Sensing and Control Using Neurofuzzy Logic. Technical Report. China: North University of China; 2000

[7] Bestard GA, Alfaro SCA. Measurement and estimation of the weld bead geometry in arc welding processes: The last 50 years of development. Journal of the Brazilian Society of Mechanical Sciences and Engineering. 2018;**40**(9):444

[8] Bestard GA. Sensor fusion and embedded devices to estimate and control the depth and width of the weld bead in real time [PhD thesis]. Brasilia, Brazil: Universidade de Brasília; 2017

[9] Fronius. Rob 4000/5000 Operating Instructions. Pettenbach, Áustria: Fronius; 2001

[10] Mathers G. The Welding of Aluminium and its Alloys. Cambridge, England: Woodhead; 2002

[11] Lincoln Electric. Gas Metal Arc Welding Product and Procedure Selection. Ohio, EUA: Lincoln Electric; 2017

[12] Micro-Epsilon. optoNCDT ILR Laser Distance Sensors. Raleigh, Nort Carolina, USA: Micro-Epsilon; 2017

[13] MTI Instruments. Laser Triangulation Principle. Albany, New York, USA: MTI Instruments; 2017

[14] Cuiuri D. Control of the short-circuit gas metal arc welding process using instantaneous current regulation [PhD thesis]. Wollongong, New Gales, Australia: University of Wollongong; 2000

[15] Zhang WJ, Zhang X, Zhang YM. Robust pattern recognition for measurement of three dimensional weld pool surface in GTAW. Journal of Intelligent Manufacturing. 2013;**26**:1-18

[16] Kovacevic R, Zhang YM. Sensing free surface of arc weld pool using specular reflection: Principle and analysis. Journal of Engineering Manufacture. 1996;**210**

[17] Kovacevic R, Zhang YM. Real-time image processing for monitoring of free weld pool surface. Journal of Manufacturing Science and Engineering. 1997;**119**(5):161-169

[18] Liu YK, Zhang YM. Model-based predictive control of weld penetration in gas tungsten arc welding. IEEE Transactions on Control Systems Technology. 2014;**22**(3):955-966

[19] Chen S-B, Jing W. Intelligentized methodology for arc welding dynamical process. Visual information acquiring,

knowledge modeling and intelligent control. In: Lecture Notes in Electrical Engineering. Berlin, German: Springer; 2009

[20] Font comas T, Diao C, Ding J, Williams S, Zhao Y. A passive imaging system for geometry measurement for the plasma arc welding process. IEEE Transactions on Industrial Electronics. 2017;**64**(9):7201-7209

[21] Pinto-Lopera J, Motta JMST, Alfaro SCA. Real-time measurement of width and height of weld beads in GMAW processes. Sensors. 2016;**16**(9): 1500

[22] Bestard GA, Sampaio RC, Vargas JAR, Absi Alfaro SC. Sensor fusion to estimate the depth and width of the weld bead in real time in GMAW processes. Sensors. 2018;**18**(4):962

[23] Bestard GA, Alfaro SCA. Sensor fusion: Theory review and applications. In: 23rd ABCM International Congress of Mechanical Engineering COBEM 2015, Rio de Janeiro, Brazil. 2015

[24] Sotela FC. Fusión de Datos Distribuida en Redes de Sensores Visuales Utilizando Sistemas Multi-Agente [PhD thesis]. Madrid, Spain: Universidad Carlos III de Madrid; 2010

[25] Rogalski A. Infrared detectors: Status and trends. Progress in Quantum Electronics. 2003;**27**(2–3):59-210

Section 2

Welding Processes

Chapter 5

Vacuum Brazing of Dissimilar Joints Mo-SS with Cu-Mn-Ni Brazing Filler Metal

Maksymova Svitlana

Abstract

The complexity of joining dissimilar materials, such as molybdenum-stainless steel, is due to the difference in thermal coefficients of linear expansion and low oxidation resistance of molybdenum. In this connection, brazing of this pair of materials should be performed in vacuum. The selection of chemical composition of brazing filler metal and its melting temperature range is very important. This work presents the results of metallographic and micro X-Ray spectral analysis investigations of dissimilar brazed joints of molybdenum-stainless steel and shows the features of the formation of brazed seams at application of brazing filler metals of Cu-Mn-Ni(Me) system. It is found that at brazed seam, crystallize reaction layers form the side of molybdenum. One layer is based on molybdenum, enriched in iron and silicon. The second layer is based on iron, enriched in silicon. At brazing temperature of 1100°C, base metal dispersion occurs, which can be avoided at temperature lowering to 1084°C. The structure of solid solution with a small amount of iron-enriched dispersed phase crystallizes in the central zone of brazed seams. The brazed joints produced with application of brazing filler metal based on Cu-Mn-Ni system are characterized by maximum values of shear strength.

Keywords: brazing, microstructure, brazing filler metal, molybdenum, stainless steel, micro X-ray spectral analysis, shear strength

1. Introduction

At present materials which are difficult to join by traditional welding methods are widely applied in different industries. These, primarily, are dissimilar materials, for joining which brazing is extensively used. This method of material joining has been known since ancient times, it is developing continuously, and the area of its application becomes wider. Producing permanent joints of a refractory material—molybdenum with stainless steel by brazing is highly important for many industries, related to structure operation at high temperatures. This is due to high melting temperature, high modulus of elasticity, relatively low density, and excellent specific strength of molybdenum at high temperatures. Recrystallization temperature and mechanical properties of molybdenum depend on many factors and, primarily, on the degree of its purity, method of its production, as it is sensitive to interstitial impurities. Recrystallization temperature of unalloyed molybdenum is in the range of 900–1000°C. It depends on metal purity, temperature, degree of deformation, and

IntechOpen

recrystallization duration [1]. Molybdenum becomes brittle after recrystallization, so that the melting temperature range of brazing filler metal is important at its selection.

Joining dissimilar materials is a more complex task than joining similar materials. The complexity of joining dissimilar materials is due to a significant difference in thermal coefficients of linear expansion (TCLE) and low oxidation resistance of molybdenum. So, at a temperature exceeding 500°C, the sublimation of MO_3 oxide begins on molybdenum surface. It becomes significant at the temperature of 600°C. At further temperature increase above 800°C, this oxide melts, leading to superactive oxidation of molybdenum in a standard atmosphere [1]. In this connection, it is better to conduct molybdenum brazing in a vacuum. Vacuum brazing has several advantages, compared to the traditional methods of brazing in an air atmosphere. In a vacuum furnace atmosphere, a practically complete absence of any substances is achieved. The most important feature of vacuum brazing is the possibility of conducting the process without the application of fluxes. In addition to eliminating the operation of flux washing, it allows producing joints with high strength, corrosion resistance, and vacuum tightness, which is very important for the fabrication of many structures.

In brazing dissimilar materials, an important task is a correct selection of the chemical composition of brazing filler metal, its solidus, and liquidus temperature. Vacuum brazing of stainless steels is usually performed in the temperature range of 1000–1200°C. This is the temperature range of quenching of most alloy steels, which allows you to combine brazing with heat treatment of the material and thereby achieve high strength brazed joints. The brazing filler metal should readily wet the materials being brazed, it should be sufficiently strong and, at the same time, ductile, should readily deform, and promote relaxation of stresses, arising during brazing and cooling to room temperature.

The authors of [2] presented the problems that arise when brazing dissimilar materials. We focused on the difference in the coefficients of thermal expansion of brazing filler metal and base metals, which can lead to the appearance of internal stresses, as well as to deformations of the base metal.

In various industries, brazing filler metal based on nickel and copper are widely used: VPR1, VPR4, BNi-1, BNi-2, BNi-3, BNi-4, BNi-5, BNi-7, BNi-8, etc. [3–10]. They are used for brazing steels of various grades, heat-resistant nickel alloys, and many other materials. As a rule, these brazing filler metals contain Si and B as depressants (**Table 1**), which provide an acceptable temperature range for melting and good wetting of the brazed metals.

The disadvantages of these brazing filler metal include active diffusion of boron, the formation of fusible boride, and silicide phases that are released in brazing joints (**Figure 1(a)**) and in the base metal during brazing (**Figure 1(b)**).

They relate to brittle intermetallic compounds that adversely affect the performance of brazed joints during prolonged use.

Brazed joints obtained using BNi-2, BNi-3, and BNi-4 brazing filler metal consist of three phases [9]: a nickel-based solid solution adjacent to the base metal and located in the center of the brazed joint of nickel borides and eutectic consisting of nickel silicides and borides.

Boron actively diffuses into stainless steel adjacent to the seam. In the process of brazing at high temperature, it forms intermetallic boride phases along grain boundaries. When these phases are in large numbers, they reduce the fatigue strength and corrosion resistance of steel. The brittle phases determine the brittleness of the joint as a whole, and crack development occurs along these phases. It is possible to increase the strength of brazed joints by forming a structure of a solid solution in the brazed seams, which effectively inhibits the development of cracks. Rabinkin has extensively studied the problems associated with the presence of

Grade of BFM	Chemical composition of the main elements, % (wt.)						
	Cr	B	Si	Fe	Mn	Cu	Ni
BNi-1	13.0–15.0	2.75–3.5	4.0–5.0	4.0–5.0	—	—	Base
BNi-1a	13.0–15.0	2.75–3.5	4.0–5.0	4.0–5.0	—	—	Base
BNi-2	6.0–8.0	2.75–3.5	4.0–5.0	2.5–3.5	—	—	Base
BNi-3	—	2.75–3.5	4.0–5.0	0.5	—	—	Base
BNi-4	—	1.50–2.20	3.0–4.0	1.5	—	—	Base
BNi-5	18.5–19.5	0.03	9.75–10.50	—	—	—	Base
BNi-8	—	—	6.0–8.0	—	21.5–24.5	4.0–5.0	Base
VPr1	—	0.1–0.3	1.5–2.0	0.1–1.5	—	base	27.0–30.0
VPr4*	—	0.15–0.25	—	1.0–1.5	27.0–30.0	base	28.0–30.0
VPr7	—	0.07–0.2	0.8–1.2	0.1–0.8	32–35	—	Base
VPr11**	14.0–16.0	2.0–3.0	4.0–5.0	3.5	—	—	Base

*4.0–6.0% Co; 0.1–0.2% P
**0.1–1.0% Al

Table 1.
Chemical composition of brazing filler metals.

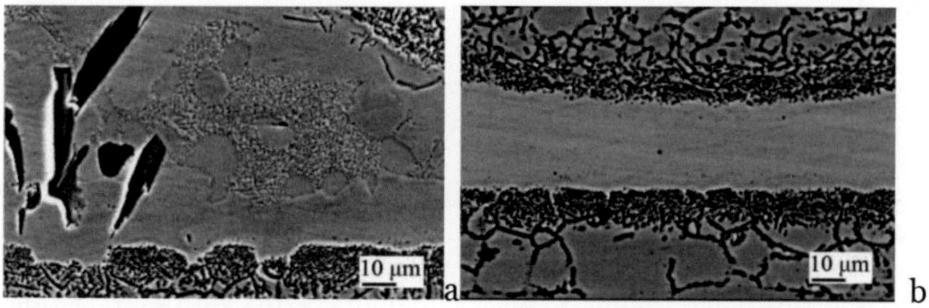

Figure 1.
Microstructure of brazed joints of 1Kh18N9 steel, produced using brazing filler metal Ni-7Cr-4.5Si-3Fe-3.2B: fillet section, (a) brazed seam (b) [10].

borides in brazed joints [7]. He suggests the use of prolonged heat treatment to dissolve borides and silicides, which complicates the process of obtaining joints and does not always allow you to get rid of them [7, 8] completely.

The use of silver brazing filler metals for brazing an alloy of Mo60%-Cu40% provides shear strength in the range of 170–220 MPa [11]. The application of brazing filler metals based on Ni-Cr-Si system containing boron [12] for brazing molybdenum (Mo60%-Cu40%) with stainless steel does allow achieving a high strength—200–230 MPa.

Promising are brazing filler metals with solid solution structure, which are characterized by acceptable melting temperature and high mechanical properties and act as a damper between two dissimilar metals, which promotes relaxation of stresses in brazed joints. The presence of solid solutions characterized alloys based on copper-nickel and copper-manganese systems [13].

The copper-manganese system has a minimum melting point (821°C) at a manganese concentration of 33.7 atom. %. With decreasing temperature, ordering processes occur in the alloys of this system, and the ordered phases Cu_5Mn and Cu_3Mn precipitate, increasing the strength of the solid solution. Analysis of nickel-manganese binary system is indicative of complete solubility of manganese in nickel in the liquid state at increased temperature but with temperature lowering precipitation of several phases takes place.

Proceeding from analysis of binary state diagrams of Cu-Ni, Cu-Mn, and Mn-Ni systems [13], Cu-Mn-Ni ternary system was selected as the base one [14, 15]. This system has a wide range of solid solutions. So, in the Cu-Mn-Ni alloy, additional alloying with silicon should improve the spreading of the surface of stainless steel, and alloying with iron should reduce erosion of stainless steel by brazing during the brazing process.

This work aims to study the features of the formation of the microstructure of brazed joints, the relationship between the structure, the initial composition of the brazing filler metal, and the strength of dissimilar brazed joints Mo-stainless steel obtained by vacuum brazing using brazing filler metal of the Cu-Mn-Ni system.

2. Experimental procedure

As the base metal, molybdenum, stainless steel 09Kh18N10, and brazing alloys based on copper-manganese system were applied. The brazing filler metal was applied in the cast form and was produced by melting in the laboratory installation in the shielding atmosphere of argon. The produced ingots were overturned and melted down (up to 5 times) in order to average the chemical composition and provide a uniform distribution of elements. The solidus and liquidus temperatures of cast brazing alloys were determined using the installation of high-temperature differential analysis in the shielding atmosphere of helium at constant heating and cooling rate (40°C/min).

Before brazing, the samples were machined and cleaned (degreased). The prepared samples were overlapped, and the brazing filler metal (**Table 1**) was placed on the surface of the base metal (near the gap) and loaded into a vacuum furnace with radiation heating to conduct capillary vacuum brazing with a rarefaction of the working space of 1×10^{-3} Pa.

Brazing filler metal (**Table 2**) in the cast state was placed at the gap (size of fixed brazing gap was 50 μm).

For metallographic examinations the overlapped joints were brazed, and the specimens were cut out perpendicular to the brazing seam; the microsections were manufactured according to the standard procedure and examined using the scanning electron microscope TescanMira 3 LMU.

No.	Base system of brazing filler metal alloying	Brazing temperature, °C/time, min
1	Cu-Mn-Ni-Fe-1.0Si	1050 /3
2	Cu-Mn-Ni-0.2 Si	1100 /5
3	Cu-Mn-Ni	1084 /3

Table 2.
Used brazing filler metals and brazing modes.

The distribution of chemical elements was examined using the method of a local micro X-ray spectrum analysis applying the energy dispersion spectrometer Oxford Instruments X-max (80 mm2) under the control of the software package INCA. The locality of micro X-ray spectrum measurements did not exceed 1 mm; the filming of microstructures was carried out in back-scattered electrons (BSE), which allowed examining the microsections without chemical etching.

For mechanical tests, the plane overlap joints of the 100 × 30 × 3 mm size (three samples for each brazing filler metal) were brazed and tested using the installation MTS-810.

3. Results and discussion

3.1 Microstructure of brazed joints of molybdenum with stainless steel at the application of brazing filler metal of Cu-Mn-Ni-Fe-1Si system

In vacuum brazing of dissimilar materials such as molybdenum-stainless steel by brazing filler metal of Cu-Mn-Ni-Fe-1Si [16] system, good wetting of both the materials is observed, namely, molybdenum and stainless steel. This ensures the formation of smooth and tight fillets (**Figure 2(a)** and **(b)**).

In the central zone (matrix) of the brazed seam, a copper-based solid solution (92.58% Cu) solidifies, which contains a small amount of iron—2.87%, in addition to brazing filler metal component elements (**Figure 3(a)** and **(b)**; **Table 3**).

The more detailed study of chemical inhomogeneity of brazed seam matrix by mapping showed that dispersed particles of 0.5–1 μm size, enriched in iron and silicon, precipitate in the copper-based solid solution (**Figure 4**).

Alloys of the copper-manganese-silicon system contain two phases: a solid solution based on copper and manganese silicides [15]. In the presence of iron in the alloy, silicides are compounds having a hexagonal lattice isomorphic to the lattices Mn_5Si_3 and Fe_5Si_3 [17].

In the peripheral zone of the brazed seam, which borders on molybdenum, two reaction layers are observed, which precipitate in the form of narrow continuous bands along the brazed seam. One of them, based on molybdenum (51.21%), is enriched in iron (31.71%) and silicon (5.88%) and is located closer to molybdenum (**Figure 3(b)**). The second one—based on iron (68.02%)—is also enriched in silicon but contains no molybdenum. It borders on the copper-based solid solution. The width of these reaction layers is variable but does not exceed 5 μm (each). Their common feature is an increased concentration of silicon from 4.83 to 5.88% (**Table 3**). In some areas brazing filler metal penetrates along the grain boundaries of the stainless steel to a maximum depth down to 20 μm (**Figure 3(a)**).

It is evident that during brazing, the liquid brazing filler metal is saturated by steel component elements. Diffusion processes take place at the cooling of

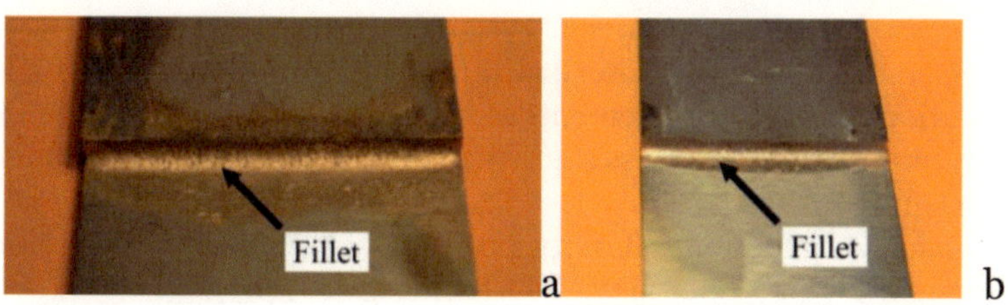

Figure 2.
The appearance of the brazed sample of molybdenum-stainless steel, produced using Cu-Mn-Ni-Fe-1Si brazing filler metal: direct (a); reverse (b) fillet.

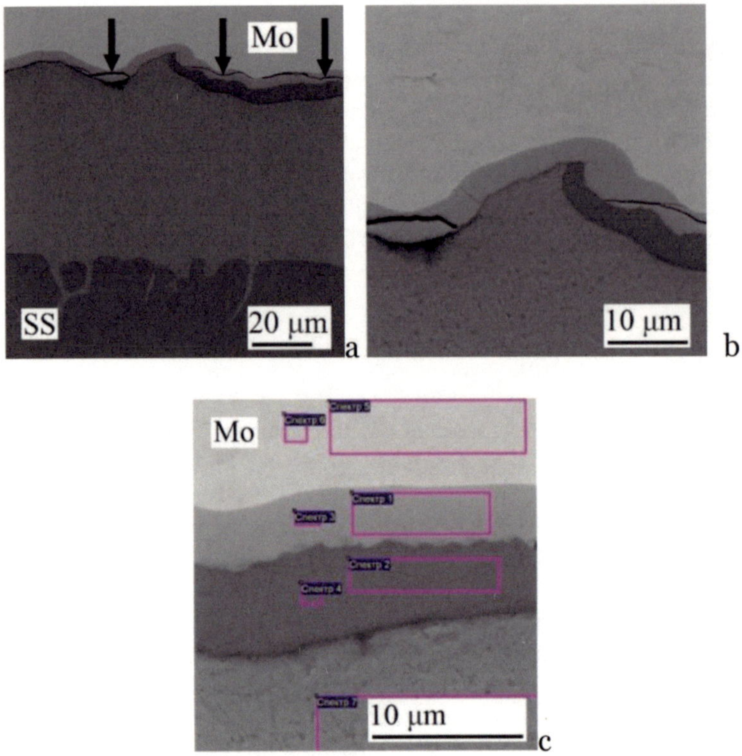

Figure 3.
Microstructure (a, b) and studied regions (c) of the brazed joint at the application of Cu-Mn-Ni-Fe-1Si brazing filler metal.

Spectrum No.	Chemical elements, wt. %							
	O	Si	Cr	Mn	Fe	Ni	Cu	Mo
1	—	5.88	7.08	0.77	31.71	2.54	0.81	51.21
2	—	4.83	16.79	2.23	68.02	4.46	3.67	—
3	—	5.65	6.56	0.86	32.07	2.27	0.00	52.59
4	—	4.92	16.50	2.48	66.79	4.63	4.68	—
5	1.74	—	—	—	—	—	—	98.26
6	1.85	—	—	—	—	—	—	98.15
7	—	—	0.30	3.04	2.87	1.21	92.58	—

Table 3.
Chemical composition of individual phases of brazed joint.

brazed joints under nonequilibrium conditions and in the presence of a gradient of concentrations of chemical elements of base metal and brazing filler metal. Silicon and iron from the stainless steel diffuse into the brazing filler metal (**Figure 5(a)** and **(b)**).

In connection with limited solubility of the latter, two reaction layers form along the molybdenum-brazing filler metal interface (**Figure 5(b)**).

Longitudinal cracks (**Figure 4(a)**) are observed in the molybdenum-based reaction layer (51.21–52.59%), enriched in iron 31.71–32.07% (**Table 3**—Spectrum 1, 3), and on molybdenum-brazing filler metal interface (along the seam).

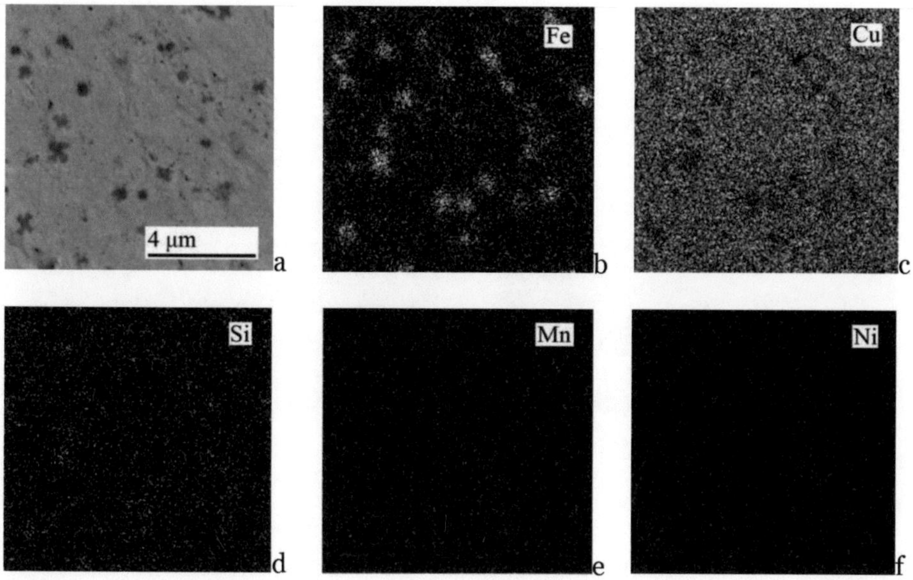

Figure 4.
The electronic image of the microstructure of the brazed seam matrix (a) and the qualitative distribution of iron (b), copper (c), silicon (d), manganese (e), and nickel (f).

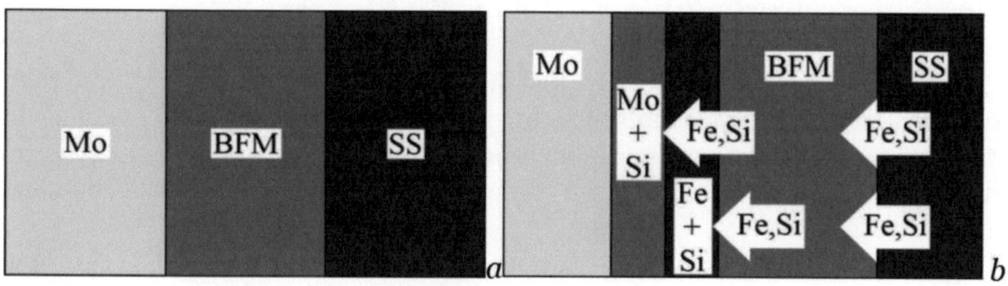

Figure 5.
Schematic image of the formation of the brazed joint: before (a) and after brazing (b).

3.2 Microstructure of brazed joints of molybdenum with stainless steel at the application of brazing filler metal of Cu-Mn-Ni-0.2Si

Lowering of silicon concentration in brazing filler metal from 1 to 0.2 wt. % does not eliminate the formation of reaction layers on the interface of brazing filler metal—molybdenum (**Figure 6(a)** and **(b)**). At brazing temperature of 1100°C (τ = 5 min), reaction layers also form along the brazed seam at the interface with molybdenum.

The results of micro X-ray spectral analysis showed that silicon concentration in the reaction layer based on molybdenum (63.41%) does not exceed 0.92% (**Figure 6**), which is significantly lower than in the previous sample (in brazing with Cu-Mn-Ni-Fe-1Si brazing filler metal). The quantity of the other component elements, namely, iron, chromium, nickel, and manganese, is on the same level as in the previous sample (**Table 3**—Spectrum 1). The obtained investigation results show that microcracks form in some regions of the reactive layer (**Figure 6 (a)**). They are located normal to the reaction layer and base metal plates. Cracks are absent in the central zone of the brazed seam, consisting of a solid solution based on copper and dispersed inclusions.

According to binary phase diagrams, consisting of metallic systems, the molybdenum-iron system has considerable regions of solubility at high temperatures.

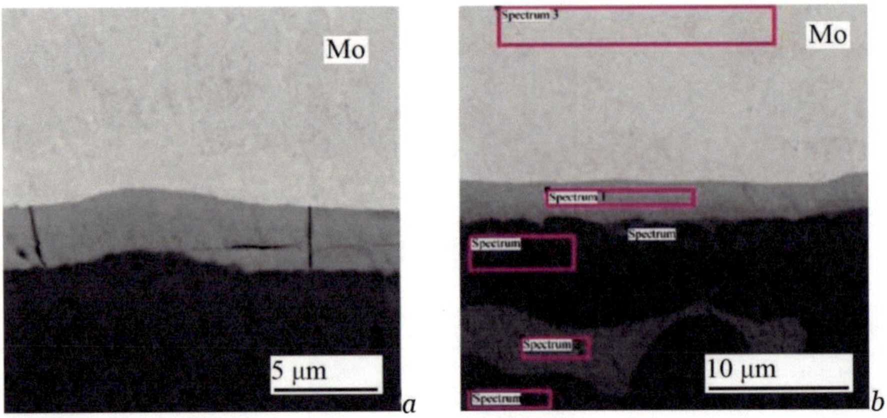

Figure 6.
Microstructure (a) and studied regions, in which chemical composition (b) was determined in the brazed joint of molybdenum-stainless steel (Cu-Mn-Ni-0.2Si brazing filler metal).

However, with temperature lowering, these regions are quickly reduced, and at room temperature, the mutual solubility is practically absent. Some intermetallic phases form between the considered areas [13].

3.3 Microstructure of brazed joints of molybdenum with stainless steel at the application of brazing filler metal of Cu-Mn-Ni system

Brazing filler metal of Cu-Mn-Ni system, not containing silicon [18], was used, to prevent cracking in brazed joints. In brazing with this brazing filler metal of dissimilar joints of stainless steel-molybdenum (Tb = 1100°C, τ =3 min), the structure of direct fillet differs from that of the reverse one by its morphology and chemical composition (**Figure 7(a)** and **(b)**).

This is due to the features of the sample assembly before brazing. The cast brazing filler metal is placed at the gap on the plane of the plate to be brazed. During brazing, the brazing filler metal melts, and the liquid phase flows into the capillary gap, wets the solid surface being brazed, and is saturated by elements of base metal (steel). Brazed joint forms as a result of thermal and physicochemical interaction of the brazing filler metal and base metal [19]. The interaction of liquid copper brazing filler metal and solid base metal at brazing temperature results in the dispersion of the latter (stainless steel).

Micro X-ray spectral analysis showed that the direct fillet consists of copper matrix-solid solution, containing 4.31 wt. % iron (**Figure 7(a)** and **(b)**). In the copper matrix of reverse fillet, the weight fraction of iron practically does not change (4.19 wt. %), but a considerable number of dispersed particles based on iron (67.83–67.89 wt. %, **Figure 7 (c)**, **Table 4**) appears. They also contain nickel (8.80–9.31%), chromium (17.81%), and a small amount of the other component elements of the brazing filler metal and base metal (**Figure 7 (c)**, **Table 4**).

These results confirm the identity of the chemical composition of stainless steel and dispersed particles located in the solid solution.

From the side of the reverse fillet, brazed seam forms similarly. Round particles of different size based on iron (44.15%) are located in the brazed seam matrix—in the solid solution, and take up a large area of the braze seam [19]. Their composition includes all the elements of brazed metals and brazing filler metal: chromium, nickel, manganese, copper, silicon, and molybdenum (**Figure 8 (a)**, **Table 5**).

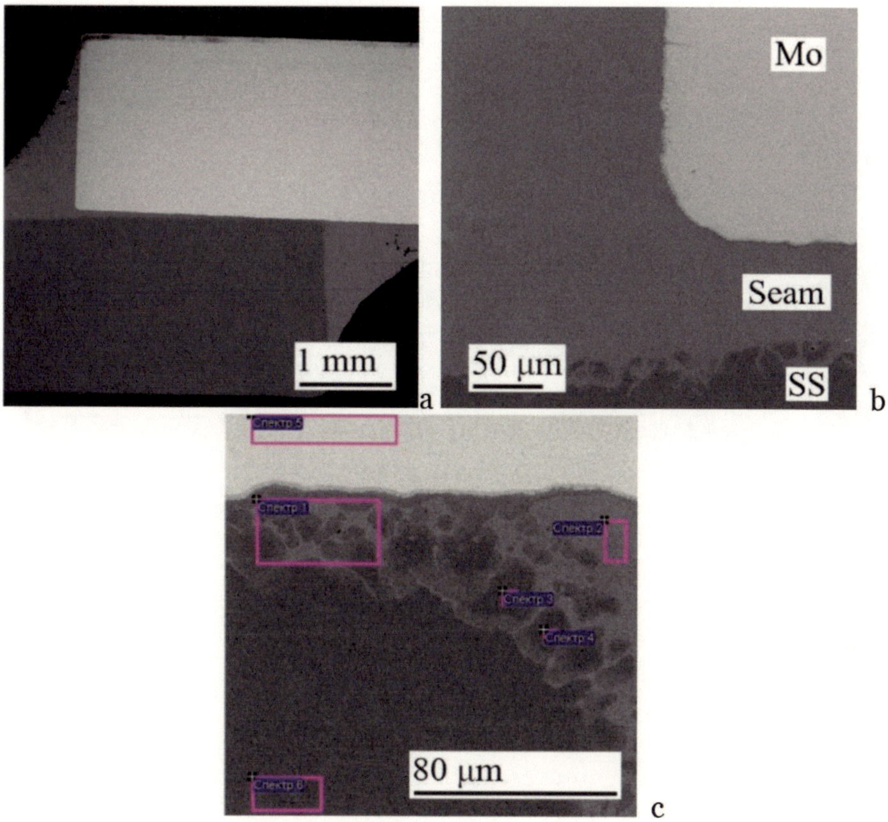

Figure 7.
The appearance of the brazed sample (a), the microstructure of direct fillet regions (b), and studied areas of reverse fillet region (c).

Spectrum No.	Chemical elements wt. %							
	Si	Ti	Cr	Mn	Fe	Ni	Cu	Mo
1	0.22	0.10	5.73	3.30	22.47	11.92	55.00	1.24
2	0.15	0.00	0.84	5.30	4.19	8.59	80.51	0.44
3	0.63	0.46	17.81	1.88	67.83	9.31	1.80	0.27
4	0.67	0.26	17.81	1.93	67.89	8.80	2.25	0.39
5	0.10	0.00	0.00	0.00	0.00	0.00	0.00	99.90
6	0.70	0.39	18.44	1.94	68.69	8.92	0.91	0.00

Table 4.
Chemical composition of the fillet area.

Such stainless steel elements, as iron and chromium, are found in the copper-based solid solution, but in much smaller amounts of 3.27–3.45 and 0.62–0.64%, respectively.

Two reaction layers in the form of continuous bands (about 2.4 μm width) are also observed along the brazed seam from molybdenum side. One is based on molybdenum and contains an increased concentration (wt. %) of iron, 22.27%; chromium, 7.29%; and silicon, 0.84% (**Figure 8(a)** and **(b)**; **Table 5**). The second one is based on iron.

Obtained data of X-ray spectral analysis show, that similar formation of brazed joints proceeds at the application of brazing filler metals of Cu-Mn-Ni-Fe-1Si and

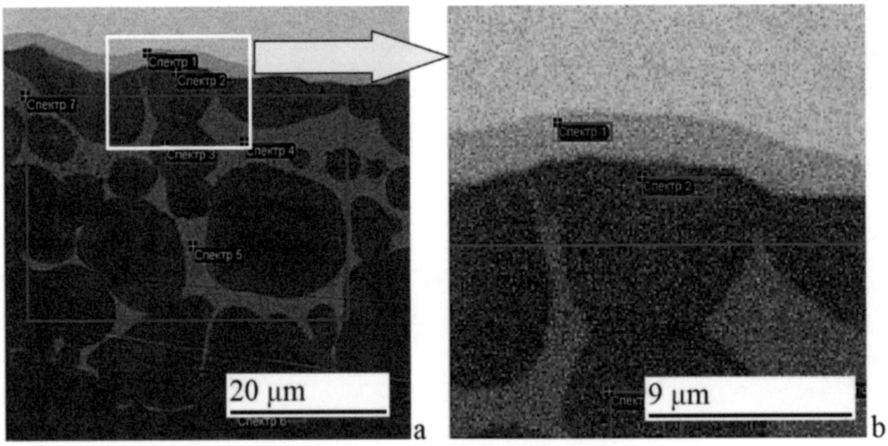

Figure 8.
The microstructure of brazed seam of Mo-SS joint produced at 1100°C brazing temperature.

Spectrum No.	Chemical elements, wt. %							
	Si	Ti	Cr	Mn	Fe	Ni	Cu	Mo
1	0.84	0.00	7.29	0.99	22.27	4.11	0.71	63.78
2	0.29	0.09	6.77	3.38	24.82	8.25	48.08	8.32
3	0.46	0.10	11.25	3.55	44.15	16.77	21.57	2.15
4	0.00	0.08	0.64	5.29	3.27	3.79	86.66	0.27
5	0.08	0.00	0.62	5.14	3.45	3.61	86.86	0.24
6	0.84	0.68	18.33	1.95	68.13	8.59	1.14	0.34
7	0.29	0.16	8.34	2.98	30.10	7.84	49.10	1.19

Table 5.
Composition of brazed seam.

Cu-Mn-Ni systems. In both the variants, reaction layers form on molybdenum-brazing filler metal interface. In the first case (Cu-Mn-Ni-Fe-1Si), silicon and iron are present in the brazing filler metal. In the second variant (Cu-Mn-Ni), these elements diffuse from the base metal into brazed seam metal, leading to its saturation with component elements of stainless steel and formation of reaction layers on the interface. The difference between the chemical compositions of these layers consists in that silicon concentration is significantly lower in the second variant, compared to the first one.

The results of the conducted studies show that lowering of the temperature of brazing the dissimilar joints to 1084°C allows avoiding base metal dispersion and ensures the formation of tight homogeneous brazed seams (**Figure 9(a)** and **(b)**).

In individual areas of the seams, the brazing filler metal penetrates along the boundaries of stainless steel grains to the depth of 15–20 μm.

It should be noted that the reaction layers at the interface of molybdenum-brazing filler metal form in a similar way (**Figure 9 (c)**, **Table 6**), as in brazing other samples, described above. However, their width decreases: from 2.4 μm to 1.7 for the molybdenum-based layer and to 1.8 μm—for the iron-based layer.

Results of local micro X-ray spectral analysis show that the maximum concentration of silicon in the layer near molybdenum does not exceed 0.78%. Brazed seam matrix, similar to the previous samples, is represented by copper-based solid solution (**Table 6**—Spectrum 4; **Figure 9(c)**) with inclusions of dispersed particles. It contains

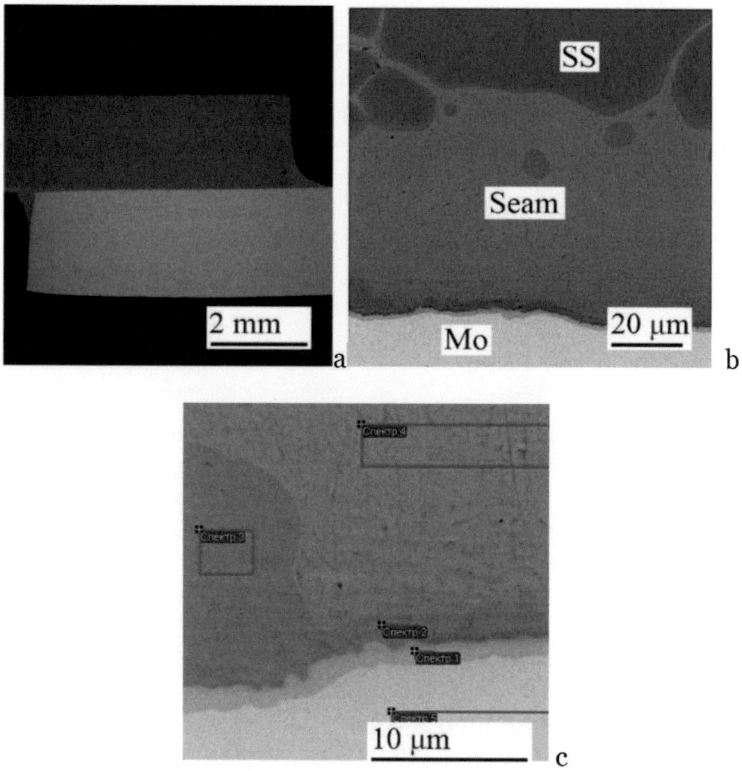

Figure 9.
The appearance (a), microstructure (b), and studied areas (c) of brazed seam of Mo-SS *joint, produced at 1084°c brazing temperature.*

the same concentration of component chemical elements, as does the solid solution at brazing temperature of 1100°C [19]. A small area of the microsection field of view (about 1%) is made up by particles (of 2–10 μm size) enriched in iron (35.45–39.54%) and other component elements of stainless steel (**Table 6**, **Figure 9(c)**).

In keeping with state diagrams of binary metal systems, iron and copper have limited solubility at elevated temperature, but with temperature lowering their solubility decreases, and at 20°C it is practically absent. The iron-nickel binary system is characterized by the existence of a continuous series of solid solutions between γ-iron and nickel at elevated temperature. Temperature lowering leads to the formation of several intermediate ordered phases (Fe_3Ni, FeNi, $FeNi_3$). Copper-nickel system is characterized by the formation of a continuous series of solid solutions [13, 20]. Thus, it can be assumed that individual particles of intermediate

Spectrum No.	Chemical elements, wt. %						
	Si	Cr	Mn	Fe	Ni	Cu	Mo
1	0.78	7.17	1.09	26.15	9.81	2.65	52.34
2	0.23	7.40	3.72	36.66	28.62	18.86	4.52
3	0.26	7.36	4.00	35.45	29.33	18.18	5.42
4	0.07	1.25	6.17	6.86	12.60	72.67	0.38
5	—	—	—	—	—	—	100.00

Table 6.
Composition of brazed joint at 1084°C brazing temperature.

phases form against the background of copper-based solid solution at brazed seam solidification. At the same time, it should be noted that at brazing, the brazed seam metal solidification proceeds under nonequilibrium conditions (in the capillary gap) in the presence of a concentration gradient on the interface, leading to saturation of brazed seam metal with the component elements of the brazed metal [21]. Diffusion processes in the brazed seam result in the formation of particles based on the iron-nickel-copper system. In the brazed seam—copper-based solid solution, iron concentration is significantly lower and is equal to 6–7%.

4. Mechanical properties of dissimilar brazed joints of molybdenum-stainless steel

Metallographic and micro X-ray spectral investigations of brazed joints were followed by mechanical shear testing (at room temperature) of overlap flat samples of dissimilar joints of molybdenum-stainless steel (**Figure 10**).

Brazed overlap samples were tested by axial tension. To ensure the conditions of combining the load axis and the plane of the brazed seam to the sections of the samples that are placed in the grips of the testing machine, stainless steel plates were welded. They help to reduce the eccentricity of the sample during mechanical tests.

The mechanical properties of brazed joints depend on the microstructure of the brazed joints [6, 16]. The obtained results of mechanical tests of flat overlap samples of dissimilar joints of molybdenum-stainless steel are in good agreement with previous structural studies. Brazing filler metal of the copper-manganese-nickel system containing silicon contributes to the formation of phases enriched in silicon (silicides). They stand out at the interface between the base metal-brazing filler metal [22] and reduce shear strength. The use of brazing filler metal with a solid solution structure without silicon allows obtaining higher values of shear strength.

Performed testing showed that application of brazing filler metal based on Cu-Mn-Ni-Fe-Si system, containing up to 1% silicon, cannot ensure the shear strength above 110 MPa (**Figure 11**).

Lowering of silicon concentration to 0.2% in brazing filler metal №2 ensured an increase of shear strength.

At application of brazing filler metals based on Cu-Mn-Ni-Fe-Si and Cu-Mn-Ni-Si systems, samples fail in the brazing seam. Samples produced using brazing filler metal based on Cu-Mn-Ni system fail in the brazing seam, near-seam zone (mixed nature of fracture), and the base metal—molybdenum (**Figure 12**).

In case of fracture in the seam, the shear strength was on the level of 200–210 (average value of 205 MPa). At fracture through molybdenum, the maximum shear strength was 300 MPa. In some cases, a mixed nature of fracture was observed—partially in the seam and partially in the base metal (**Figure 12(c)**).

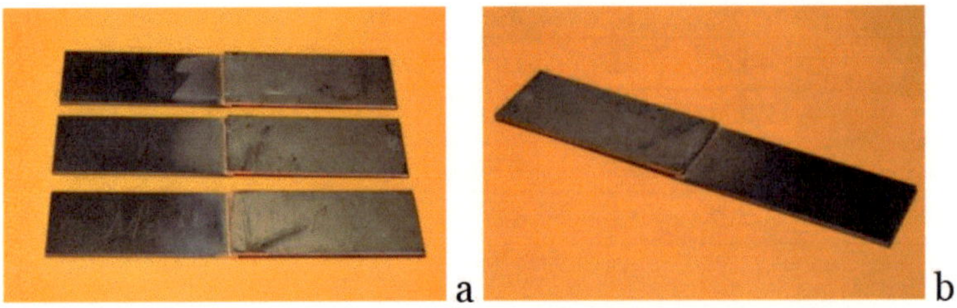

Figure 10.
Appearance of brazed samples of molybdenum-stainless steel.

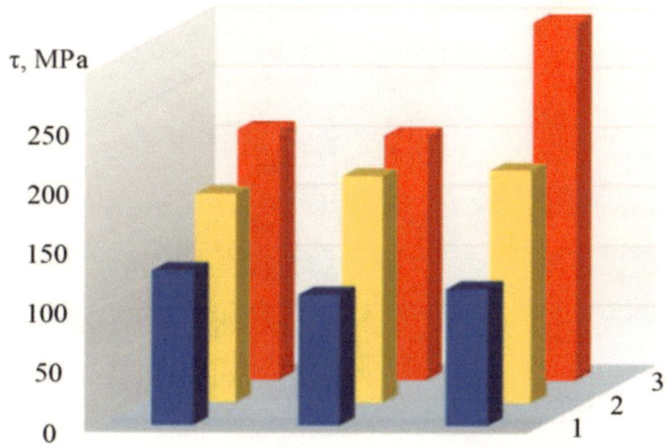

Figure 11.
Shear strength of brazed dissimilar overlap joints of molybdenum-stainless steel.

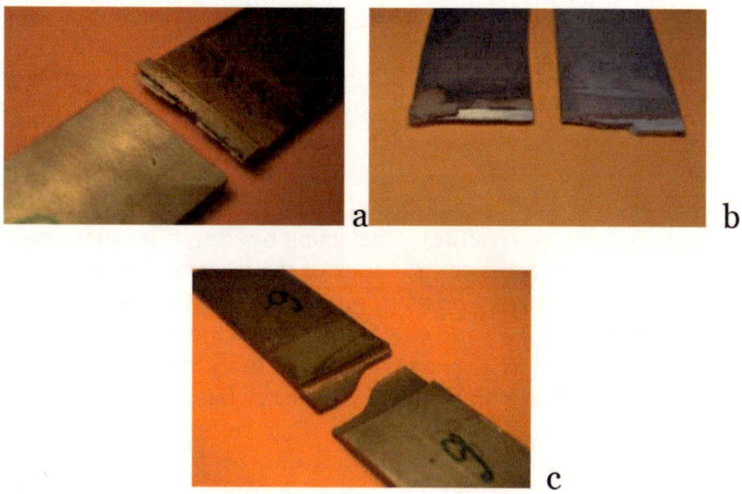

Figure 12.
Brazed samples after mechanical testing: a fracture in the brazed seam (a), near-seam zone (b), and base metal—molybdenum (c).

When designing telescopic tubular joints from dissimilar materials, molybdenum-stainless steel, it is necessary to take into account the difference in thermal expansion coefficients. The inner tube must be made of molybdenum (with a low coefficient of thermal expansion), and the outer one is made of stainless steel. This design provides a high-quality formation of brazed joints and seams. When brazing some lap joints, the main indicator of quality is tightness (vacuum density), and the strength of the joints is ensured by the large length of the overlap (when using brazing filler metal with a solid solution structure).

The brazed molybdenum-stainless steel telescopic tubular joints with dense seam by high temperature vacuum brazing using system brazing filler metal Cu-Mn-Ni-Fe-Si [23] were manufactured (**Figure 13**).

Checking the brazed telescopic tubular joint Mo-SS for vacuum density gave a positive result. Testing the brazed telescopic tubular dissimilar joints Mo-SS for vacuum density gave a positive result.

In some cases, nonstandard tubular joints of dissimilar metals Mo-stainless steel are used for mechanical testing. In this case, some parameters influence the stability

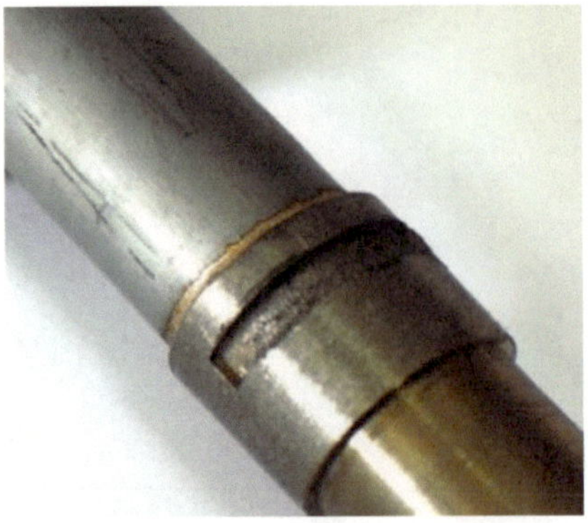

Figure 13.
The brazed telescopic tubular joint Mo-SS.

of the results obtained: assembly accuracy, the size of the gaps, test equipment, and many others. To improve the quality of brazed lapped tubular joints, a threaded fit is used [24].

5. Conclusions

X-ray microspectral studies established that during vacuum brazing of dissimilar joints, molybdenum-stainless steel using brazing filler metal based on the Cu-Mn-Ni-Fe-1Si system in the central zone of the brazed seam forms a copper-based solid solution structure. The mapping showed that dispersed particles of 0.5–1 μm size, enriched in iron and silicon, precipitate in the copper-based solid solution. The peripheral zone of the seam (on the molybdenum side) is formed by reactive layers, based on iron (width 5.3 μm) and based on molybdenum (width 3.2 μm), which stand out in the form of continuous strips along the soldered seam. At a silicon concentration in the brazing filler metal (1%), these zones are enriched in the latter, which leads to the formation of silicides at the interface and cracking.

It is shown that when using brazing filler metal based on the Cu-Mn-Ni system (Tn = 1100°C), the base metal is dispersed with the release of particles based on iron in a solid solution based on copper and reactive layer about 2.4 μm.

Lowering of brazing temperature to 1084°C allows avoiding base metal dispersion. It ensures the formation of brazed seams with the homogeneous solid solution structure based on copper and single dispersed inclusions of the phase enriched in iron. It was found that the width of the reactive layer decreases to 1.7–1.80 microns, but the concentration ratio of chemical elements remains the same as at a temperature of 1100°C. In some areas, brazing filler metal penetration along the grain boundaries of the base metal of stainless steel is observed.

Mechanical testing of brazed joints of molybdenum-stainless steel proved that brazed joints produced with application of brazing filler metal based on Cu-Mn-Ni system are characterized by maximum values of shear strength. Brazed samples fail both in the seam and in the near-seam zone (strength level of 200–210 MPa) and the base metal (molybdenum) at 300 MPa.

Developed technological process of brazing such dissimilar materials as molybdenum and stainless steel can be applied, when producing individual brazed components of dissimilar materials in the nuclear and aerospace industry and at the development of the fusion reactor.

Author details

Maksymova Svitlana
E.O. Paton Electric Welding Institute of the NAS of Ukraine, Kyiv, Ukraine

*Address all correspondence to: maksymova@paton.kiev.ua

References

[1] Titts T, Wilson J. Refractory Metals and Alloys. Metallurgia: Moscow; 1968. p. 352

[2] Paton BE, editor. Technology of Welding, Brazing and Cutting. Encyclopedia. V. 111-4. Moscow: Mashinostroenie; 2006. p. 768

[3] Petrunin IE. Brazing Reference Book. Moscow: Mashinostroenie; 2003. p. 480

[4] Lashko SV, Lashko NF. Brazing of Metals. Moscow: Mashinostroenie; 1998. p. 376

[5] Ermolaev GV, Kvasnitskii VV, Kvasnitskii VF, Maksymova SV, Khorunov VF, Chigarev VV. Brazing of Materials. Textbook. Mykolaiv: National University of Shipbuilding; 2015. p. 340

[6] Schwartz MM. Brazing. Ohio: Materials Park, ASM International; 1987. p. 455

[7] Rabinkin A. Advances in Brazing: Science, Technology and Applications. Cambridge, UK: Woodhead Publishing Limited; 2013. pp. 121-159

[8] Khorunov V, Maksymova S. Selection of brazing filler metals for brazing thin-walled heat-exchangers. The Paton Welding Journal. 2008;**3**:17-21

[9] Khorunov V, Maksimova S. Production and application of rapidly quenched brazing alloys. Welding International. 2006;**20**(5):405-409. DOI: 10.1533/wint.2006.3639

[10] Maksymova SV. Amorphous brazing filler metals for brazing stainless steel and titanium and brazed joint structure. Adgeziya Rasplavov i Paika Materialov. 2007;**40**:70-81

[11] Vacuum active brazing process of molybdenum-copper alloys and stainless steel [Electronic resource]. Patent №102489813A КНР: IPC B23K 1/008, B23K 1/019, appl. 20.12.11; publ. 28.05.14. 2013. Available from: https://patents.google.com/patent/CN102489813A/en?oq=102489813

[12] Vacuum brazing solder and process of Mo-Cu alloy and stainless steel [Electronic resource]. Patent №102554509 China: IPC B23K 1/008, B23K 35/30, appl. 24.02.12; publ. 23.10.13. 2013. Available from: https://patents.google.com/patent/CN102554509A/en. oq=102554509

[13] Massalski TB et al. Binary Alloy Phase Diagrams. Bulletin of Alloy Phase Diagrams. 2nd ed. Ohio: Materials Park, ASM International; 1990

[14] Bao CJ. Studying ternary alloys. Cu–Mn–Ni. University News. Non-ferrous metallurgy. 1958;**5**:107-115

[15] Drits ME, Bochvar NP, Gusej IS, et al. Double and multicomponent cupper-based system. Moscow: Nauka. 1979. p. 248

[16] Khorunov VF, Maksymova SV, Levadniy LP. Brazing filler metal for steel brazing. Patent USSR №. 1656801, МКИ B23K 35/30 от 15.02.91

[17] Goldschmidt XD. Implanted alloys. In: Chebotarev NT, editor. Moscow: Mir. 1971. pp. 56-68

[18] Maksymova S, Voronov V, Kovalchuk P. Brazing filler metal for brazing dissimilar joints Mo-SS. Patent UA. №119396 C2, publ. 10-06-2019, Newsletter. №11

[19] Maksymova SV, Voronov VV, Kovalchuk PV, Larionov AV. Influence of temperature of the brazing on structure of brazed heterogeneous molybdenum–stainless steel joints. Metallofizika i Noveishie Tekhnologii. 2017;**39**(9):1227-1237. DOI: 10.15407/mfint.39.09.1227

[20] Lyakishev NP. Constitutional Diagrams of Binary Metallic Systems. Reference Book in 3 Volume. Vol. 2. Moscow: Mashinebuilding; 1997. p. 1024

[21] Radzievsky VN, Tkachenko GG. High-Temperature Vacuum Brazing in Compressor Engineering. Kiev: Ecotechnology; 2009. p. 400

[22] Maksymova SV, Voronov VV, Kovalchuk PV, Larionov AV. Producing dissimilar joints of molybdenum–stainless steel using vacuum brazing. The Paton Welding Journal. 2017;**2**: 13-18. DOI: 10.15407/as2017.02.03

[23] Maksymova SV, Khorunov VF. The experience of brazing molybdenum with stainless steel. Avtomaticheskaya Svarka. 1997;**4**(529):60-61

[24] Ivannikov AA, Sevryukov ON, Vertkov AV, Penyaz MA, Misnikov VE, Kochnov YO, et al. High-temperature brazing of molybdenum with steel 12X18H10T with rapidly quenched nickel brazing filler metal STEMET 1301. Welding Production. 2016;**4**:22-28

Chapter 6

Overview of Selected Issues Related to Soldering

Karel Dušek, David Bušek and Petr Veselý

Abstract

The formation of defects and imperfections in the soldering process can have many causes, which primarily include a poorly setup technological process, inappropriate or inappropriately used materials and their combinations, the effect of the surroundings and design errors. This chapter lists some examples of errors that can occur in soldering, while review is devoted to selected defects: non-wettability of the solder pads, dewetting, wrong solder mask design, warpage, head-in-pillow, cracks in the joints, pad cratering, black pad, solder beading, tombstoning, dendrites, voids, flux spattering from the solder paste, popcorning and whiskers.

Keywords: electronic assembly, soldering, reflow soldering, soldering defects, reliability issues

1. Introduction

A significant change in soldering technology was the transition to lead-free soldering. With this change and the increasing miniaturisation of electrical equipment comes the associated issue of developing suitable alloys, production technology, etc. Over time, soldering has become a very complex process, with many factors affecting the final quality of the solder joints, and thus the product.

The formation of defects and imperfections in the soldering process can have many causes, which primarily include a poorly setup technological process, inappropriate or inappropriately used materials and their combinations, the effect of the surroundings and design errors. This chapter lists some examples of errors that can occur in soldering.

2. Non-wettability

One of the problems is the non-wettability of the soldering pads. During the soldering process, the soldering pads are not sufficiently wetted with the solder alloy (the solder does not create a connection with the entire surface of the soldered area). An example of this defect is shown in **Figure 1**.

This defect may be caused by the soldering pads being too oxidised, the flux not being active enough, insufficient PCB warming, soldering pad contamination as well as, for instance, a poor solder mask coating [1].

Figure 1.
Problems with the non-wettability of soldering pads.

3. Dewetting

Dewetting usually happens in the case of a prolonged heating, leading to the formation of an intermetallic that results in a change in the composition of the solder alloy as is the case, for example, when soldering a tin-lead solder on copper substrate, where the tin is sucked out of the solder alloy, which is involved in the formation of an intermetallic layer, thus increasing the proportion of lead in the solder alloy, which has poor wetting. Dewetting may also arise due to the dissolution of the precious metals in the solder alloy or the influence of a poor soldering surface under the surface finish. Photos of dewetted surfaces are shown in **Figure 2**.

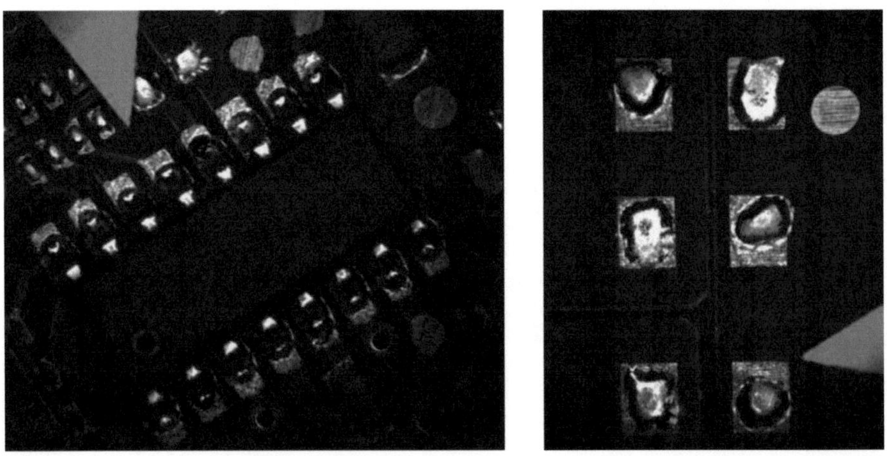

Figure 2.
Photos of soldering pads dewetting.

4. Wrong solder mask design

Another problem that can occur with the solder mask is due to wrong PCB design, especially if the solder mask is used to cover pads for unused BGA package leads (see **Figures 3** and **4**). In this case the BGA balls become deformed and the risk of bridging is increased.

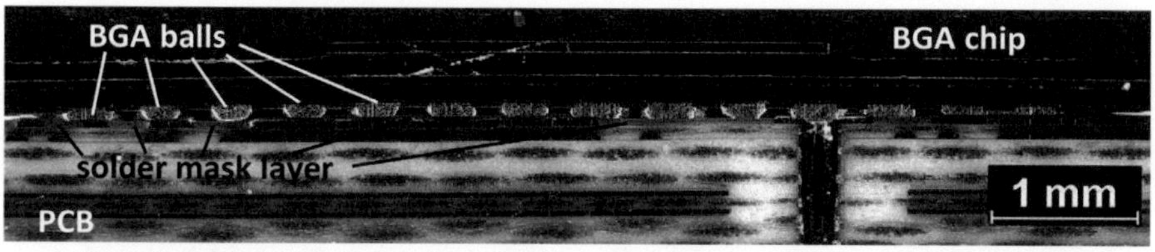

Figure 3.
Section—deformation of BGA balls due to inappropriate solder mask design.

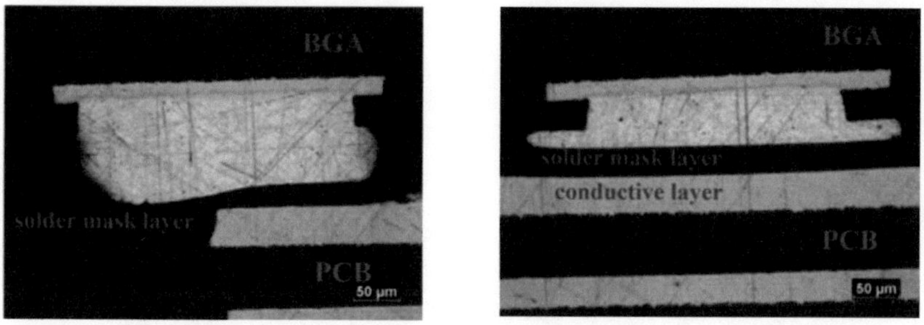

Figure 4.
Detail of the sections from ***Figure 3****, showing the deformation of BGA balls due to inappropriate solder mask design.*

This is mainly caused by the designer's unfamiliarity with correct PCB design. A soldering pad on the PCB must be present for each BGA terminal, even if the terminal is not used.

5. Warpage effect: head-in-pillow effect

In the case of differing thermal expansion, coefficients between the PCB and the component package, during heating/cooling, there is mutual deflection/deformation called the "warpage effect". This effect may result in correctable errors (bridges, open joints), but also in unrepairable errors—cracks (on components, inside the PCB) [2]. This effect, along with some other defects, including the "Head-in-Pillow" effect, is shown in **Figure 5**.

The Head-in-Pillow effect is a characteristic in the case of soldering of BGA components. During heating the BGA terminal/balls lift up from the soldering pad coated with solder paste. In the reflow zone, the soldering paste and the BGA solder alloy balls reflow independently. During cooling this sag is counteracted, nevertheless

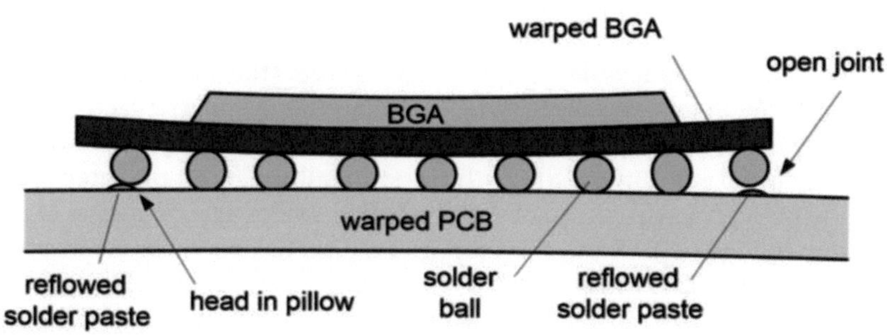

Figure 5.
Schematic representation of the warpage effect.

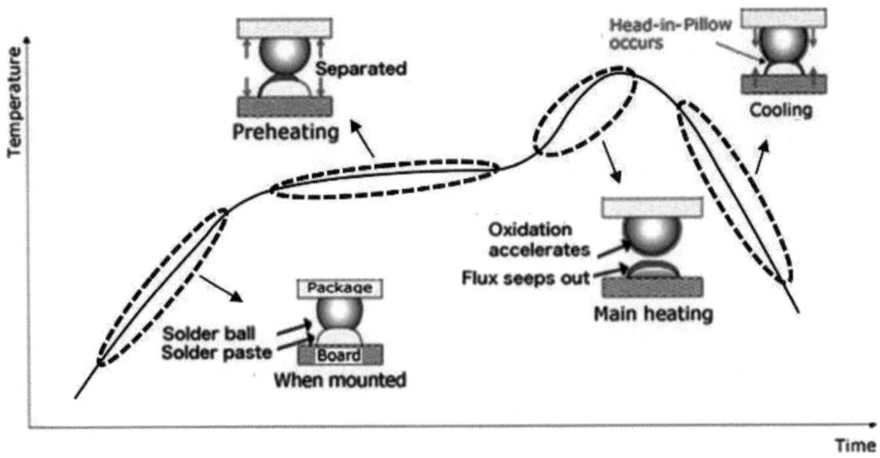

Figure 6.
Head-in-pillow effect in connection with the temperature profile.

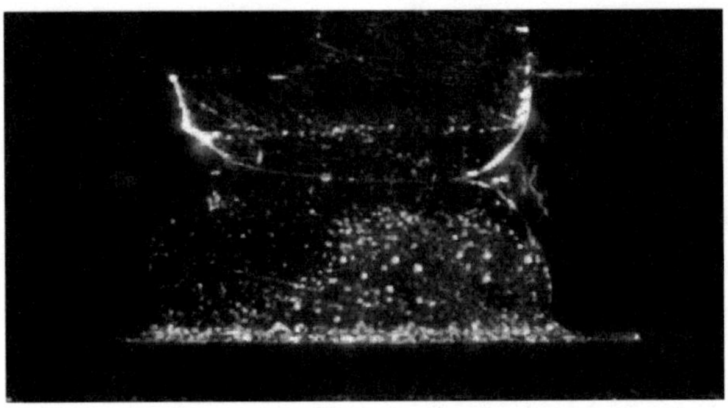

Figure 7.
Photo of a microsection of the head-in-pillow effect.

the reflowed solder alloy on the solder pad no longer joins with the reflowed BGA solder alloy ball. The Head-in-Pillow effect in connection with the temperature profile is shown in **Figure 6**, together with a photo of a microsection in **Figure 7**.

6. Cracks: pad cratering

Cracks occur not only in the soldered joints, but also in the components and PCB. Due to mechanical, thermal and combined stress, the material is under tension, which results in cracks forming in the weakest spot, together with a release of the tension [3].

The location of the cracks depends on many factors (the materials used, the type of soldering technology, the package, the PCB material, geometric factors, etc.) [4, 5]. **Figure 8** shows an overview of crack failures occurring in a PCB assembly.

An example of a crack in the site of the intermetallic alloy on the interface between the BGA package's terminal and the ball of solder alloy is shown in the microsection in **Figure 9**. Due to their minimum dimensions, these errors are difficult to detect using X-ray inspection.

Another example of a crack is pad cratering where a crack is formed within the PCB under the soldering pad (see **Figure 10**). This defect does not manifest during a functional electrical test, and it is not even possible to detect it with the non-destructive methods for the output check.

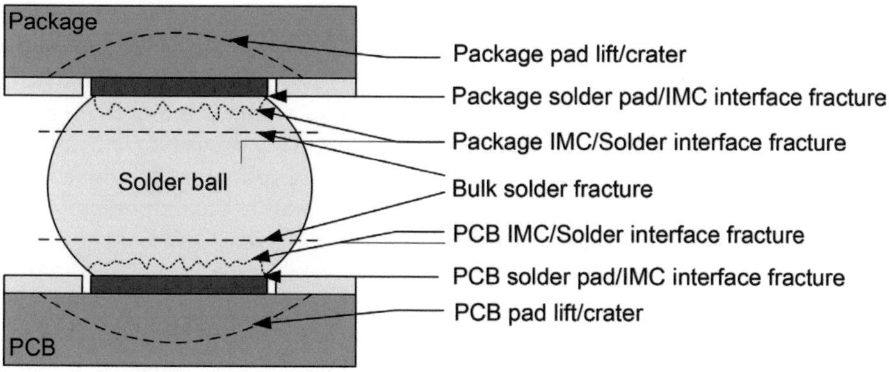

Figure 8.
Overview of crack failures occurring in a PCB assembly.

Figure 9.
A longitudinal crack between the solder joint and the BGA package's terminal.

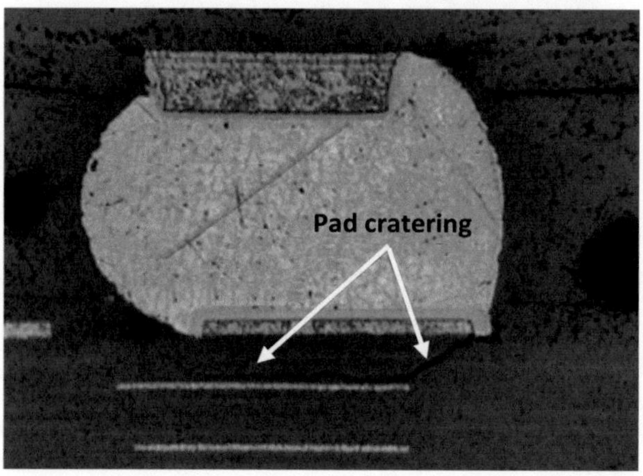

Figure 10.
Microsection of a BGA terminal soldered to a PCB with a pad cratering defect.

A pad cratering defect can be compounded by the use of lead-free solder alloys, where the higher temperatures used for lead-free solder alloys cause greater tension in the materials used [6, 7]. In addition, the lead-free solder alloys are significantly stiffer than tin-lead eutectic solder alloys; therefore, they transfer greater stress under the package pads during a mechanical stress [8].

Static or cyclic mechanical stress or thermal cycling can lead to the cracks spreading and the consequent failure of the device [9]. Stress can affect a mounted PCB even without external influences; this is called residual stress (tension that remains in the material even though the cause of stress has been removed). The source of the residual stress is primarily the manufacturing process in which many stress factors affect the components [10]. This primarily concerns the soldering process, where the elevated temperature leads to the fixation of components that often have different coefficients of thermal expansion.

7. Black pad effect

The black pad effect is a characteristic for Ni/Au surface finishes, where Ni contains higher amount of phosphor. During the Ni/Au surface layer creation, the nickel layer is covered with a thin layer of gold; this plating process may lead to corrosion of the nickel surface. The final Au coating can provide good wetting for the solder alloy, even though it has an oxidised Ni-P layer under it. Another cause of the black pad effect is the solder pads reacting with a lead-free alloy with a higher tin content at a higher temperature (longer reaction time). This reaction produces a thicker layer (rich in phosphorus) on the interface, which has a defective structure (microfractures, microvoids). Due to the black pad effect, the soldered joint is considerably weakened mechanically and ultimately will break the conductive connection between the component's terminal and the soldering pad. This fault is very difficult to detect; thus it may occur on devices that have passed output control tests and have already been sent to market (**Figure 11**).

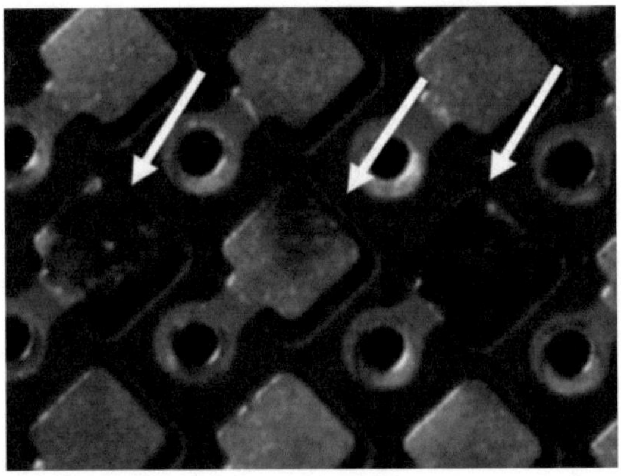

Figure 11.
Photo of black pad effect after stripping off the immersion gold layer.

8. Solder beading

The presence of balls of solder alloy next to the package (solder beading) is an error that is easily detectable by optical inspection methods. An example of balls next to the component package is shown in **Figure 12a**. The device's reliability is compromised in the event that the conductive ball becomes free, which can be a potential risk of a fault in the device, for example, it can cause an accidental short circuit. The occurrence of the balls is due to inaccurate deposition of solder paste (either due to solder paste misprint or due to excessive amount of the deposited paste), where particles of the solder alloy get under the package when attaching the component [11]. An example of inaccurate application of solder paste is shown in

Figure 12.
(a) Solder alloy balls occurring next to the component's package. (b) Inaccurate deposition of solder paste with regard to the solder pads.

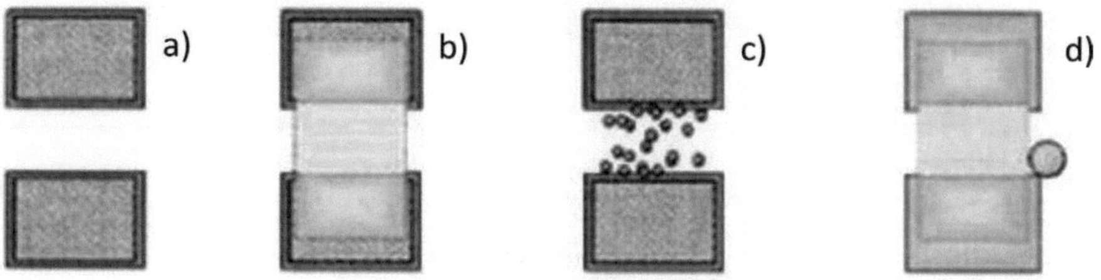

Figure 13.
Schematic representation of the principle of balls forming next to the component's package: (a) solder paste deposited on solder pads (b) mounting the component (c) during reflow particles of solder paste partially agglomerate (d) a ball forms next to the component's package.

Figure 12b. The solder alloy particles partially agglomerate under the component package during reflow, and the component's package squeezes them out of its side. This phenomenon is illustrated schematically in **Figure 13**. If the error occurs regularly during mass production for a specific group of components, it is necessary to modify the dimensions of apertures in the stencil.

9. Tombstone effect

The tombstone effect (or Manhattan effect, drawbridging or the Grabstein effect) is a phenomenon which is characterised by one of the sides of a small SMD component (typically in a 0805, 0603, 0402 and 0201 package) lifting up during reflow. A photo of the tombstone effect is shown in **Figure 14**.

Tombstone effect is caused by an imbalance of wetting forces during the reflow process [12]. This can be caused by unequal amount of solder paste applied to the connecting pads, differently sized soldering pads, eccentrically mounted component, different wettability of soldering pads, a different time of solder melting on each side of the component, etc. or by upward push by solvent vapours from flux during an asymmetric reflow process [13]. The effect of the wetting force, or upward push force from solvent vapours, is depicted in **Figure 15**.

The frequency of the tombstone effect is also influenced by the reflow technology used. The feedback from industrial practice says that this effect is more frequent in vapour-phase soldering technology. This has been experimentally verified and presented in the publication dealing with tombstone effect [14, 15]. Currently work is being done on a more detailed explanation of the cause of the higher occurrence of the tombstone effect in vapour-phase soldering.

Figure 14.
Photo of the tombstone effect.

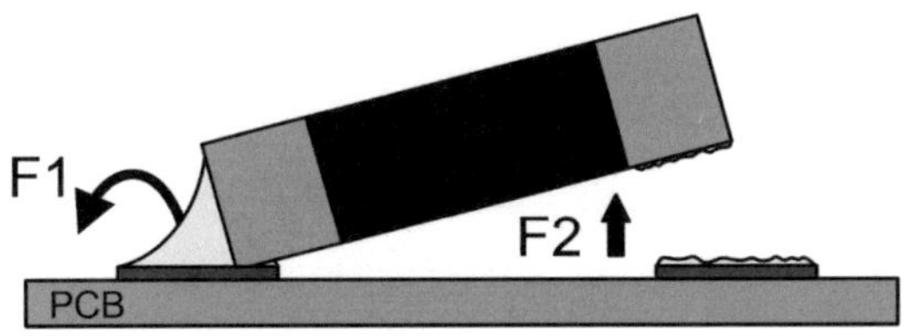

Figure 15.
Schematic representation of the wetting force (F1); upward push force from solvent vapours (F2).

10. Dendrites

Dendrites grow due to electrochemical migration when metal ions go into the electrolytic solution at the anode, plating out at the cathode and creating needle- or tree-like formations on the PCB substrate (see photos in **Figure 16**) [16]. Electrochemical migration can be defined as the movement of a metal ion in an electrolytic solution between two neighbouring conductors with different electrical potential.

The time required to create dendritic bridges, which cause short circuits between two conductive paths, is determined by several factors including relative humidity,

Figure 16.
Photo of dendrite.

temperature, conductor material, conductor spacing, voltage difference, contamination amount, contamination type, etc. [17–19].

The growth of dendrites has become a more serious issue, mainly due to the constant miniaturisation of electronic components (decreasing distance between the cathode and the anode) and the use of newer fluxes (especially no-clean fluxes). Manufacturers use no-clean fluxes to remove the washing process from production and reduce production costs. Despite this, these fluxes, under certain conditions (high humidity, significant temperature changes), are a good basis for creating an electrolytic solution. To prevent the growth of dendrites, it is necessary to thoroughly wash (even no-clean fluxes were used) the PCB after soldering.

11. Voids

Voids are non-conductive cavities within the soldered joint, and their excessive presence poses a significant reliability risk for the manufactured product, especially in power electronics, where higher currents are present [20]. Voids are formed during the soldering process, their presence in the soldered joint causes the displacement of electrical and thermal paths, the resistance is higher and the temperature stress is non-uniform. This subsequently causes cracks and lowers the mechanical shock tolerance of the whole PCB.

With the advent of lead-free soldering (new material base, different temperature profiles, different types of fluxes, higher surface tension of solder, etc.), a higher incidence of voids was detected, leading to the lower reliability of soldered joints. Increased attention is therefore paid to the voiding issue.

Voids can be classified into several categories [21]. There are macro- and microvoids, shrinkage voids, voids in microvias, Kirkendall voids and pinhole voids (see **Figure 17**).

11.1 Macrovoids

Macrovoids are the most commonly occurring type of voids. Macrovoids are formed by the evaporation of gases from fluxes and soldering pastes during the reflow process. Macrovoids may occur anywhere in the solder joint, and their diameter is around 100–300 μm.

The factors that affect the size of the macrovoids are the solder paste's properties (particle size, composition, melting temperature and oxide content), flux (viscosity,

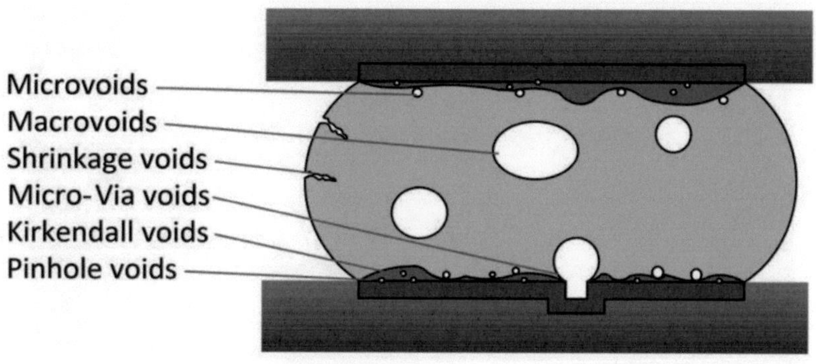

Figure 17.
Soldered joint and the location of different types of voids. Intermetallic layer contains Kirkendall voids; pinholes are located on the boundary between the intermetallic layer and the PCB.

surface tension, activator, solvent, etc.), components (geometry, shape, terminal oxidation, etc.) and the process (thickness, the shape and the parameters of the solder paste's printed layer, temperature profile) [22–26]. An example of macrovoids in the solder joints of a soldered BGA package on the PCB is given in **Figure 18**.

11.2 Planar microvoids

Microvoids are characterised by their small diameter, less than 25 μm. Their occurrence is typical for Ni/Au, OSP and Ag finishes [27]. There are planar (one plane) on the solder pad/solder alloy interface; due to this, the joint's mechanical strength is significantly reduced. The cause of these voids' formation is not yet fully clarified. An example of planar microvoids on the interface of the solder pad/solder alloy is shown in **Figure 19**.

11.3 Shrinkage voids

Shrinkage voids are voids with rough tree-like, branching edges pointing from the joint's surface towards the solder joints' core. They are characteristics for SAC solder

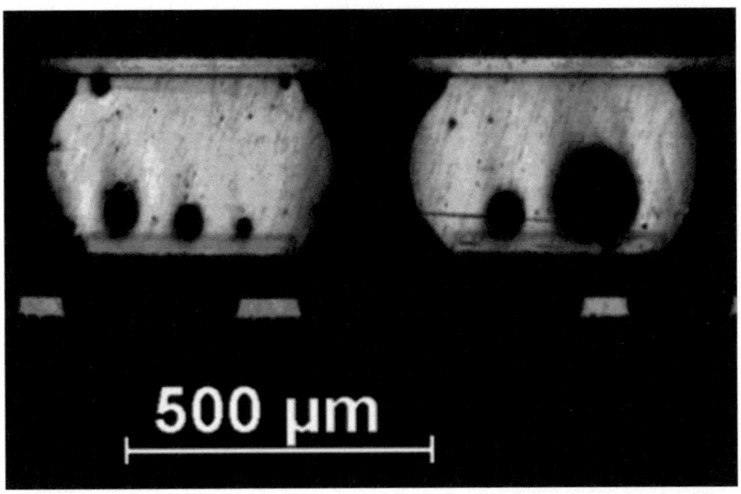

Figure 18.
Macrovoids within the BGA solder joints.

Figure 19.
Planar microvoids.

alloys. The cause of these voids' formation is the solder solidification process, when the solder shrinks during slow cooling. Their occurrence can be avoided by controlled cooling. It has not yet been proven that these voids have a negative effect on the reliability of soldered joints. An example of shrinkage voids is shown in **Figure 20**.

11.4 Microvia voids

This type of void arises from gases escaping from the microvia (plated holes with a diameter smaller or equal to 150 μm, used for interconnection of conductive paths between the individual PCB layers) during the soldering process. The solder has inadequate wettability and is not able to penetrate inside a microvia. These voids can greatly affect the reliability of a solder joint; therefore, it is recommended to take them into account during design (**Figure 21**).

11.5 Kirkendall (IMC) voids

Kirkendall voids arise in the intermetallic layer between the solder and the solder pad. They are created when joining two metals with different diffusion coefficients. These voids are most commonly located on the interface between a tin solder and

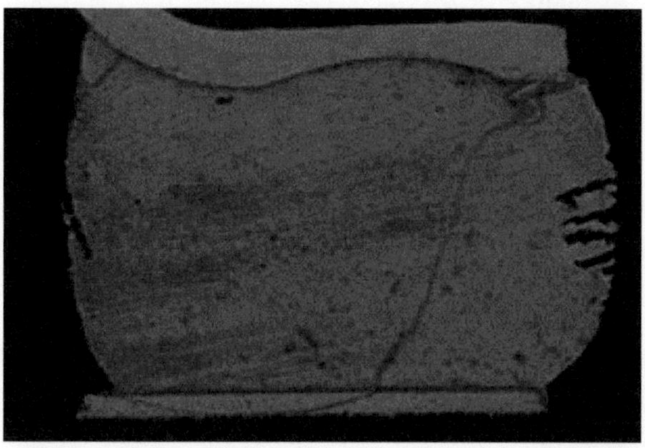

Figure 20.
Shrinkage voids.

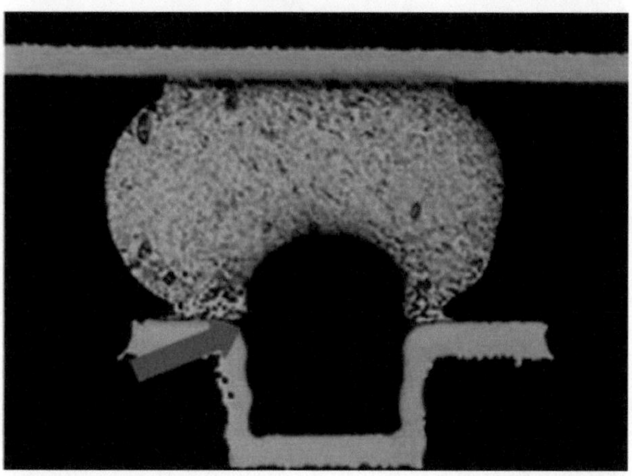

Figure 21.
Microvia void.

a copper solder pad. Tin has a lower diffusion coefficient and thus penetrates more slowly into the copper than copper into tin.

11.6 Pinhole voids

Pinhole voids arise when a gas leaks from metal soldering pads during the soldering process. These voids have very small dimensions, on average around 1–3 μm. Their appearance is due to substances absorbed by substrates from previous processes. They mostly occur in the case of galvanic copper surface finish.

12. Flux spattering from the soldering paste

During the reflow process, the flux may spatter from the soldering paste. If the PCB is cleaned after soldering, flux spattering need not necessarily be considered a problem. If PCB cleaning is not included in the process, then flux residues can not only be the cause of future corrosion, but flux spatter can leave stains on the test surfaces intended for the subsequent electrical testing methods.

If a flux stain on the test surface gets into sites where the contact test tip is in contact with the test surface, then, due to the non-conducting nature of the flux, the electrical test evaluates the product as a reject, even if the product is fully functional. In addition, these stains are, in most cases, clear and therefore difficult to detect. An example of a photo of the test pads on a PCB is given in **Figure 22** along with an image from an electron microscope, where a dark stain is evident; the source of the stain was flux spattering.

There are various sources of non-conducting stains on the test surfaces, for example, the solder mask, organic coatings, inappropriate materials used when maintaining technological equipment, etc. Diagnosis of the cause can be very problematic.

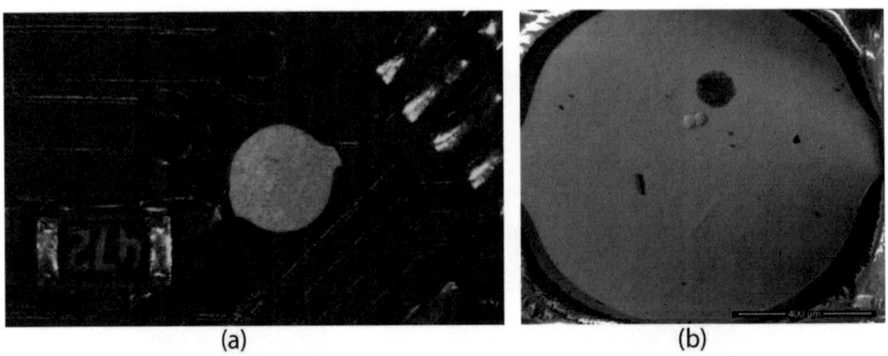

Figure 22.
Test surfaces for electrical testing methods: (a) photo from an optical microscope and (b) photo from an electron microscope—a dark stain is obvious, and the source of the stain was flux spattering.

13. Popcorn effect

During the soldering process, heating the mounted PCB, some components may experience absorbed moisture turning into steam, which may result in damage (delamination, cracking the component's package). The components absorb moisture during their manufacture and storage. It is therefore necessary to dry components that are sensitive to moisture absorption before they are mounted and

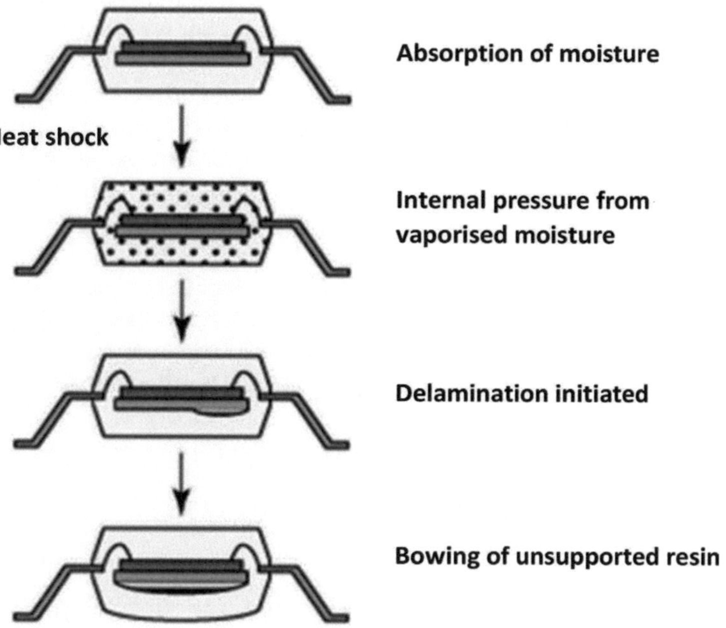

Figure 23.
Schematic representation of the popcorn effect.

soldered. The popcorn effect is schematically shown in **Figure 23**. With the advent of lead-free soldering, this effect occurs more frequently due to the use of higher soldering temperatures. In some cases, when the package cracks from the PCB side, this defect is very difficult to detect.

14. Whiskers

Whiskers are electrically conductive and relatively mechanically resistant crystals growing on the surface of some metals. They may grow on the surface of the solder alloys with a high tin content (majority of lead-free solder alloys). It most

Figure 24.
Photos of whiskers.

often concerns surfaces where tin (mainly galvanic) was used as a final finish [28, 29]. Whiskers can be straight, curved or kinked. Photos of whiskers are given in **Figure 24**.

Whiskers represent a very serious threat to the proper function and reliability of electronic equipment, mainly from the perspective of creating a short circuit (permanent, short-term or possible arcing). Therefore, materials with the potential risk of the growth of whiskers are not used in those industrial sectors that have higher demands on reliability and where they could cause large-scale damage or pose a threat to human life.

Whisker diameters range from 1 to 10 μm. Their length is typically in the order of micrometres, but in extreme cases they can reach lengths of up to 1 cm. The rate of whisker growth is determined by many factors; the literature [11] gives the approximate growth rate of whiskers in the range of 5 mm/year up to 1 cm/year. The following factors affect the whiskers' growth:

- Compressive stress
- External—caused by mechanical stress
- Internal—caused by the size of the grains inside the solder alloy, the type of surface finish of the solder pads, a different thermal expansion coefficient of the substrate and the solder and scratches on the surface
- The crystalline structure (grain shape and orientation) of the solder alloy and the presence of intermetallic compounds
- Temperature and humidity [30]

Due to their size and growth on shiny surfaces, whiskers are very difficult to detect by classical optical methods. The experimental results of whisker growth are unpredictable and unrepeatable, and so far the exact mechanism of their growth is not known.

Author details

Karel Dušek*, David Bušek and Petr Veselý
Department of Electrotechnology, Faculty of Electrical Engineering, Czech Technical University in Prague, Prague, Czech Republic

*Address all correspondence to: karel.dusek@fel.cvut.cz

References

[1] Dušek K, Plaček M, Bušek D, Dvořáková K, Rudajevová A. Study of influence of thermal capacity and flux activity on the solderability. In: Proceedings of the 2014 37th International Spring Seminar on Electronics Technology (ISSE), IEEE. 2014. pp. 185-188

[2] Bušek D, Dušek K, Plaček M, Urbánek J, Horník J, Holec J. Determination of BGA solder joint detachment cause-warpage effect. In: 2015 38th International Spring Seminar on Electronics Technology (ISSE), IEEE. 2015. pp. 306-309

[3] Otáhal A, Adamek M, Jansa V, Szendiuch I. Investigation of the mechanical properties of lead-free solder materials. Key Engineering Materials. 2014;**592-593**:453-456. DOI: 10.4028/www.scientific.net/KEM.592-593.453

[4] Dušek K, Bušek D, Beran T, Rudajevova A. Comparison of shear strength of soldered SMD resistors for various solder alloys. In: 2015 38th International Spring Seminar on Electronics Technology (ISSE), IEEE. 2015. pp. 237-240

[5] Garami T, Krammer O, Harsányi G, Martinek P. Method for validating CT length measurement of cracks inside solder joints. Soldering & Surface Mount Technology. 2016;**28**:13-17

[6] Long G, Embree T, Mukadam M, Parupalli S, Vasudevan V. Lead free assembly impacts on laminate material properties and pad crater failures. In: IPC APEX/EXPO Conference. 2007

[7] Dušek K, Rudajevová A. Influence of latent heat released from solder joints II: PCB deformation during reflow and pad cratering defects. Journal of Materials Science: Materials in Electronics. 2017;**28**:1070-1077

[8] Ma H. Effects of temperature and strain rate on the mechanical properties of lead-free solders. Journal of Materials Science. 2010;**45**:2351-2358

[9] Pietriková A, Durisin J, Ďurišin J. VPS and reliability of solder joint. In: 2009 15th International Symposium for Design and Technology of Electronics Packages (SIITME). 2009. pp. 395-398. DOI: 10.1109/SIITME.2009.5407338

[10] Rudajevova A, Dušek K. Residual strain in PCBs with Cu-plated holes. Journal of Electronic Materials. 2017;**46**:6984-6991

[11] Pietriková A, Ďurišin J, Mach P. Diagnostika a optimalizácia pou žitia ekologických materiálov pre vodivé spájanie v elektronike. 1. vydanie. Košice: Fakulta elektrotechniky a informatiky Technickej university v Košiciach; 2010

[12] Zero-defect printing shifts blame for poor-quality soldering. 31 July 2002. Zetech. Dataweek n.d. Available from: http://www.dataweek.co.za/news.aspx?pklnewsid=7525 [Accessed: 20 March 2016]

[13] How Reduce Tombstoning of Small Chip Components—Tombstoning explained.pdf n.d. Available from: http://metallicresources.com/documents/Tombstoning%20explained.pdf [Accessed: 20 March 2016]

[14] Dusek K, Straka V, Brejcha M, Beshajova Pelikanova I. Influence of type of reflow technology and type of surface finish on tomb stone effect. In: 2013 36th International Spring Seminar on Electronics Technology (ISSE), IEEE. 2013. pp. 132-135

[15] Dušek K, Bušek D, Plaček M, Géczy A, Krammer O, Illés B. Influence

of vapor phase soldering fluid Galden on wetting forces (tombstone effect). Journal of Materials Processing Technology. 2018;**251**:20-25

[16] Ready WJ, Turbini LJ. The effect of flux chemistry, applied voltage, conductor spacing, and temperature on conductive anodic filament formation. Journal of Electronic Materials. 2002;**31**:1208-1224

[17] Bumiller E, Hillman C. A review of models for time-to-failure due to metallic migration mechanisms. In: White Paper. DfR Solutions; 2006

[18] Steiner F, Hirman M, Rendl K, Wirth V. Optimization of soldering process to reduce contamination and related consequences. In: 2018 41st International Spring Seminar on Electronics Technology (ISSE). 2018. pp. 1-6. DOI: 10.1109/ISSE.2018.8443691

[19] Medgyes B, Horváth B, Illés B, Shinohara T, Tahara A, Harsányi G, et al. Microstructure and elemental composition of electrochemically formed dendrites on lead-free micro-alloyed low Ag solder alloys used in electronics. Corrosion Science. 2015;**92**:43-47. DOI: 10.1016/j.corsci.2014.11.004

[20] Diehm R, Nowottnick M, Pape U. Reduction of voids in solder joints an alternative to vacuum soldering. In: Proceedings of the IPC APEX EXPO. San Diego, CA; Vol. 28. 2012. p. 8

[21] Aspandiar Raiyo F. Voids in solder joints. Journal of SMT Article. 2006;**19-4**:406-415

[22] Primavera AA, Sturm R, Prasad S, Srihari K. Factors that affect void formation in BGA assembly. Journal of Surface Mount Technology. 1999;**12**:19-26

[23] Bušek D, Dušek K, Růžička D, Plaček M, Mach P, Urbánek J, et al. Flux effect on void quantity and size in soldered joints. Microelectronics Reliability. 2016;**60**:135-140

[24] Steiner F, Wirth V, Hirman M. Relationship of soldering profile, voids formation and strength of soldered joints. 2019 42nd International Spring Seminar on Electronics Technology (ISSE). 2019:1-6. DOI: 10.1109/ISSE.2019.8810303

[25] Hirman M, Steiner F. Optimization of solder paste quantity considering the properties of solder joints. Soldering & Surface Mount Technology. 2017;**29**:15-22

[26] Skácel J, Otáhal A, Szendiuch I. Influence of electric current at solidification of solder. In: 2019 42nd International Spring Seminar on Electronics Technology (ISSE). 2019. pp. 1-5. DOI: 10.1109/ISSE.2019.8810308

[27] Aspandiar R. Planar microvoids. In: Intel Lead Free Symposium, Scottsdale, AZ. 2006

[28] Bušek D, Vávra J, Dušek K. Whisker growth and its dependence on substrate type and applied stress. In: 2016 39th International Spring Seminar on Electronics Technology, IEEE. 2016. pp. 263-266

[29] Horváth B, Illés B, Shinohara T, Harsányi G. Whisker growth on annealed and recrystallized tin platings. Thin Solid Films. 2012;**520**:5733-5740

[30] Illés B, Horváth B, Géczy A, Krammer O, Dušek K. Corrosion-induced tin whisker growth in electronic devices: A review. Soldering & Surface Mount Technology. 2017;**29**:59-68. DOI: 10.1108/SSMT-10-2016-0023

Chapter 7

Hard Alloys with High Content of WC and TiC—Deposited by Arc Spraying Process

Stefan Lucian Toma, Radu Armand Haraga, Daniela Lucia Chicet, Viorel Paleu and Costica Bejinariu

Abstract

Obtained by different spraying technologies: in atmospheric plasma spray, High Velocity Oxygen Fuel (HVOF) or laser cladding, the layers of hard alloys with a high content of WC and TiC find their industrial applications due to their high hardness and resistance to wear. Recognized as being a process associated with welding, the arc spraying process is a method applied industrially both in obtaining new surfaces and for reconditioning worn ones. This chapter presents the technology for obtaining ultra-hard layers based on WC and TiC - by the arc spraying process, using a classic spray device equipped with a conical nozzle system and tubular wire additional material containing ultra-hard compounds (WC, TiC). To study both the quality of deposits and the influence of thermal spray process parameters on the properties of deposits with WC and TiC content, we approached various investigative techniques, such as optical scanning microscopy (SEM), X-ray diffraction, and determination of adhesion, porosity, Vickers micro-hardness and wear resistance.

Keywords: arc spray process, ultra-hard alloy, wear, WC, TiC

1. Introduction

There is a multitude of technological processes that allow the modification of the physico-chemical and mechanical properties of the surfaces in order to increase the performance in operation, the service life, or the esthetic aspect, [1–3].

The improving the surface properties, by thermal spraying, is carrying out by depositing new materials on the existing surface in order to give it new properties. Therefore, „thermal spray" is a general term that groups a set of processes, by which a material is melted under the action of a heat source, then is projected using a carrier gas on the surface of a part [4–6].

Coating via thermal spraying consists of the acceleration of fine particles from a certain metallic material, in a molten or semi-molten state, onto a specially prepared surface. The sprayed material is called coating material (CM) and the surface on which the deposition is made is called substrate (S), [7–10]. The layer produced by thermal spraying becomes a constituent part of the base material and gives it specific properties: hardness, corrosion resistance, wear resistance or new functional properties (chemical, electrical, magnetic).

1.1 Background

Depending on the melting method of the coating material, the thermal spraying processes can be classified into four main categories, as presented in **Figure 1**.

The thermal spraying processes are realized with the help of installations and equipment capable to develop the necessary heat for the CM melting, to achieve the dispersion of the formed droplets into fine particles and to transfer to them kinetic energy.

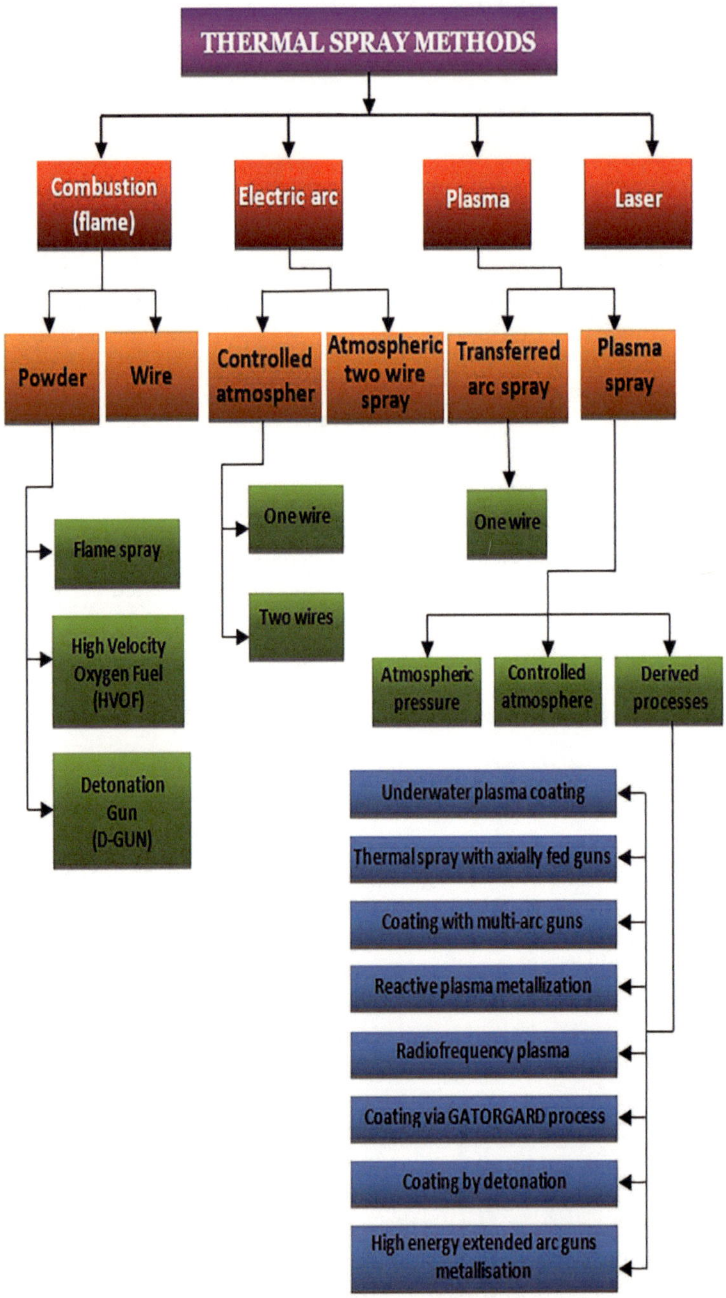

Figure 1.
Classification of thermal spray methods, [11].

The substrate material can be of any nature: ferrous, non-ferrous, plastic, ceramic, textile, glass, wood, but the coating material is limited by its capacity to be transformed into a liquid state, or to be processed into of powder, wire, cord or wand, [12–13]. Usually, spraying processes allow the deposition of a diverse range of materials: pure metals and metal alloys, cermets, ceramics and under certain conditions polymers, [2, 14, 15].

The thermal spraying methods, recognized in the specialized technical literature as synergetic and versatile [16], have as objectives:

- production of new layers, with physical, chemical, mechanical and technological properties different from those of the substrate;
- restoring the geometry of some parts with a high wear degree;
- obtaining pulverulent products used in different industrial sectors (sintering, spraying).

Recognized as the cheapest process for obtaining spray coatings, two-wire arc spraying finds numerous industrial applications that aim to obtain both surfaces with new properties and the restoration of used surfaces. [11–13].

The spraying method consists of making an electric arc between two consumable metal wires, followed by the atomization of the molten material with the help of a compressed air jet and the spraying of those particles on the substrate surface - **Figure 2**, [17]. Although the functional principle constructive of electric arc spraying devices is apparently simple, it must to allow the correlation between the wire feed speed, the intensity of the electric current in the circuit and the pressure of the compressed air, in the purpose of obtaining qualitative coatings and maximum yield, [11].

The electric arc coating devices also called spraying guns, are powered by direct current generators which have a rigid characteristic, develop voltages between 25 and 45 V and high currents with values in the range of 100–500 A, [18]. They are characterized by a modulated design, composed of:

- the wire entrainment device, also called the wire feed mechanism - according with SR EN 657/1995 standard;
- melting-spraying module, known as "spray head" - according with SR EN 657/1995 standard.

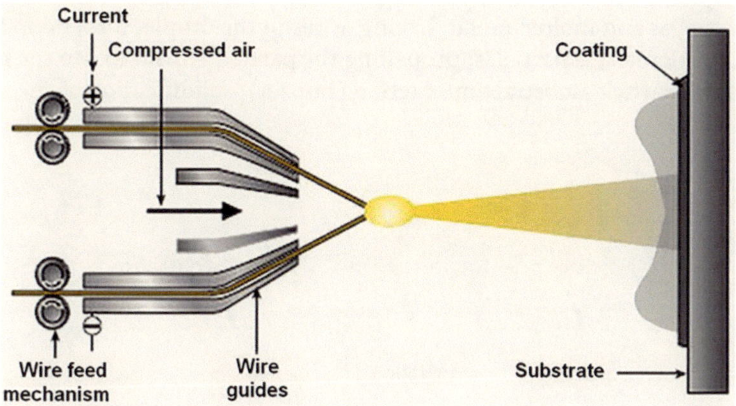

Figure 2.
Schematic representation of electric arc coating method, [17].

The wire feed mechanism is positioned directly on the gun or outside of it and has the role of directing the wire from the coils towards the area of electric arc formation.

The spray head includes the wire guides and the nozzle through which the compressed air passes - symbolically called carrier gas, [2]. During the device functioning, the nozzle is placed in the lower part of the melting zone, before the contact point of the wire electrodes, usually called "arc point". The role of the spray head is to direct the entrainment gas in the area where the electric arc is formed, in order to produce the division (atomization) of the droplet of molten filler material into particles, which it propels on the surface of the substrate, [19].

The dispersion of the molten droplets determines the interrupting of the circuit respectively of the electric arc. The electric arc priming is carrying out by advancing the wires in the melting area, where the medium is strongly ionized. The phenomenon has a periodic character, being composed of melting sequences of the input material followed by interruptions. In this case, the electric arc is a short-circuit arc, intermediate between the usual arc and the "breaking" arc produced at the interruption of a circuit, [20, 21].

1.2 State of art

The formation of coatings by thermal spraying in electric arc is carried out according to a well-precisely mechanism, presented in **Figure 3**, composed of the following stages, [18]:

- heating the coating material, used as wire;
- forming a drop of molten material;
- transformation of the coating material into fine particles - atomization;
- displacement of the melted particles towards the surface of the substrate;
- the impact of the particles with the surface of the substrate;
- coating formation by successive particles depositions.

Each stage is accompanied by a series of physical phenomena and interactions, such as: sequential melting of the coating material by Joule - Lentz effect; interaction between gas and molten metal, having as result the droplet dispersion ("atomization"), in micron size particles; propelling the particles formed onto the surface of the substrate; particle-substrate interaction (impact); solidification of the sprayed particles, [22].

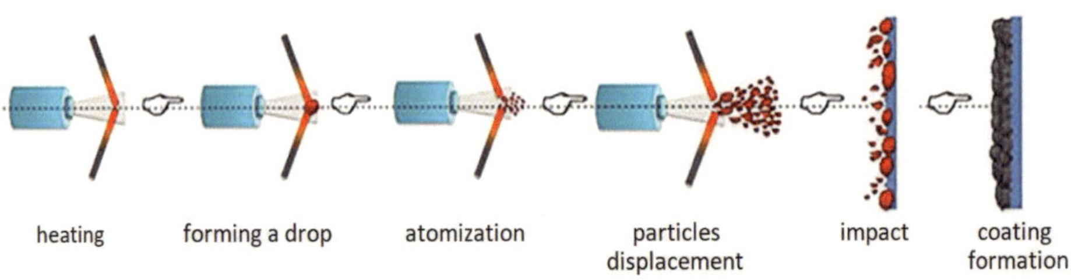

Figure 3.
Coatings formation mechanism by electric arc thermal spraying.

The interaction between the molten metal and the uninsulated jet of gaseous fluid, under pressure, at ambient temperature, has as effect the heat transfer by forced convection between droplets (particles) and jet, fact which can determine the apparition of rapid solidification phenomenon in the droplet (particle).

It is known that the properties of the coatings produced by electric arc thermal spraying are closely related to the velocity and the temperature of the particles before the impact [23, 24]. The two parameters that characterize the particles from a thermodynamic point of view are mainly influenced by the intensity of the electric arc and the pressure of the carrier gas flow. The intensity of the electric current determines the melting/overheating temperature of the filler material. The carrier gas pressure, in close connection with its velocity, influences the displacement velocity of the particles towards the substrate surface [25, 26].

The studies carried out by Zhao et al. [27], on the flow phenomenon of an un-insulated gas jet which pass through a nozzle, demonstrates the fact that the velocity and the temperature of the gaseous fluid decrease continuously along the jet. Due to the interaction between the fluid and the sprayed particles, it can be said that the molten particles inside the spray jet have a transient behavior, characterized by the continuous decrease of speed and temperature.

A large category of ductile metal materials, such as: aluminum, zinc, copper, bronze, steels as well as numerous wire drawing alloys can be sprayed by this technique, [28]. The exploitation of high temperature of the electric arc directed the scientific research towards the extension of the input materials range of use, by removing the technological barrier imposed by electrical ductility and conductivity. Thus, appeared the tubular wires, which have the exterior formed by a metal mantle characterized by high conductivity, and the interior is filled with powders of fragile materials, [29]. Recently, have been developed wires manufacturing technologies by crimping, in which the ductile material is folded and inside the folds are injected powders of hard materials. [30]. The widening of the spectrum of use of the input materials determined that the arc spraying process to be competing with the other spraying techniques (plasma, flame) in the technology of restoring of large surfaces.

The advantages of the arc spraying process are [2, 18]: efficiency - from an energy point of view (power used between 5 ÷ 10 kW), high productivity (15 ÷ 45 kg/h), no need to preheat the substrate, uses cheap equipment compared to other spraying processes.

The main disadvantages of this spraying process are related to the high porosity of the deposits (over 18%) and to carrying out deposits resistant to abrasive wear (WC, TiC) only by using high-performance equipment, equipped with command and control system, which are expensive, [31].

The purpose of our research is to present the technology of obtaining coatings of alloys resistant to abrasive wear - containing ultra-hard chemical compounds (WC and TiC), arc thermal the process using a classic spray device provided with a system of conical nozzles and tubular wire with a containing ultra-hard compound (WC and TiC), as additional material.

The objectives of our study are the following:

- the realization of a system of conical nozzles able to melt the additional material (which contains ultra-hard compounds: WC and TiC) and to transfer, to the formed particles, the high speed necessary to obtain dense coatings,
- the characterization of the ultra-hard alloys coatings obtained with the aforementioned spraying device equipped with the conical nozzle system,
- establishing the optimal process parameters.

2. Preparation and characterization of deposits

In order to carry out our own research on the possibility of obtaining ultra-hard coatings based on WC and TiC by thermal spraying, we made a spraying head system, which we have adapted to a classic drive mechanism, [32]. The schematic overview of the spraying device is presented in **Figure 4**.

The designed spray device consists of two distinct modules:

- wire feed module;
- spraying module.

The wire feed mechanism - realized by IOR Bucharest, Romania, consists of a direct current electric motor, two-step speed reducer (vertical stage) and electronic control system. It develops at the output a power P_{ie} of 0.063 kW and a speed n_{max} of 10 rpm. The overview of the wire feed mechanism is shown in **Figure 5**.

The spray module (usually called "spray head") is designed from a system of concentric nozzles - see **Figure 6**, convergent at the electric arc level, capable to ensure a convergent – divergent geometry, of the compressed air jet.

The components elements of the nozzle system are made of insulating materials such as: textolite, high density polypropylene and polyurethane of Moldotan type.

The concentric nozzle system, together with the module body, forms two compressed air circuits [32, 33], as presented in **Figure** 7:

- the main circuit - formed in the space between the module body, the conical nozzle with grooves and the conical insulator;
- the secondary circuit - formed in the space between the cover, the front nozzle, the conical nozzle with grooves and the body of the module.

The compressed air which passes through the *main circuit* determines the detachment of the liquid droplets from the wires surface and the dispersion of the droplet formed into fine particles. This circuit influence thus the size and the speed of the sprayed particles, [32]. The compressed air that passes through the secondary circuit has the role of constraining the electric arc, determining the increase of the current density and implicitly of the particle temperature. It is directed by the inner surface of the constraint frontal nozzle - see **Figure 8**, to the forming area of the electric arc. The two compressed air circuits are supplied from two different sources, with different pressures.

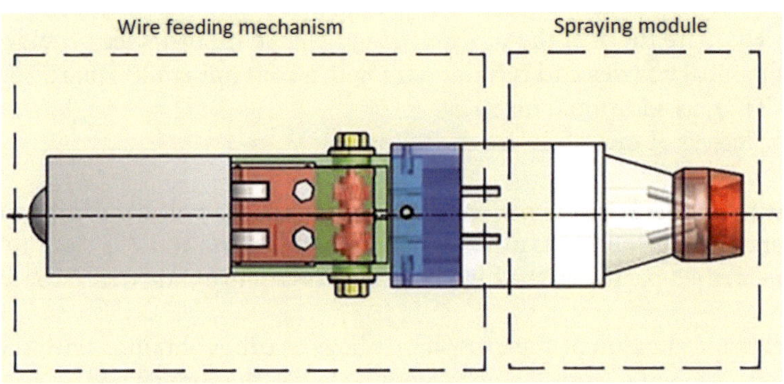

Figure 4.
Spray device - schematic overview.

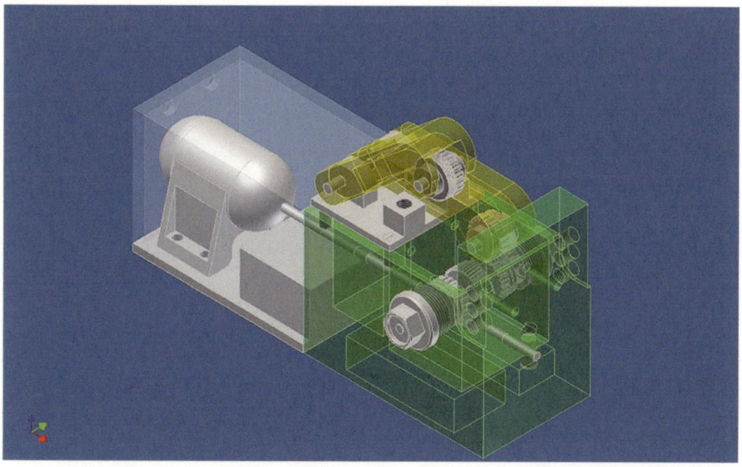

Figure 5.
Wire feed mechanism assembly.

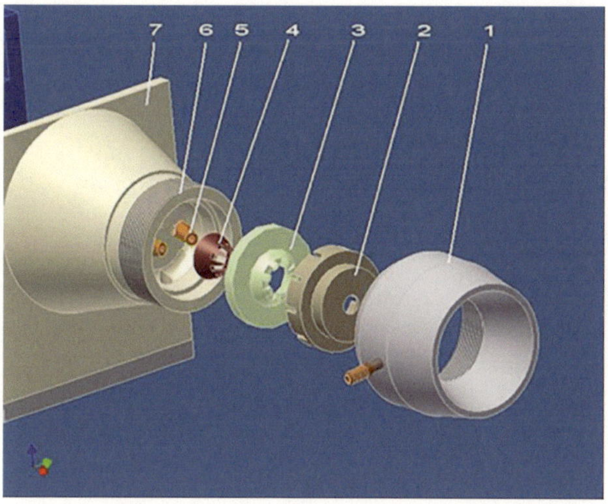

Figure 6.
Spray head: Isometric view - exploded:1- front cover; 2-front nozzle; 3-conical nozzle with grooves; 4- conical insulator; 5-wire guides; 6-thread; 7-module body.

2.1 Technological workflow

In our research substrates made of low alloy steel C15 –EN10083, with dimensions of 40 mm x 40 mm x 10 mm were covered by arc spraying process using as additional material 97MXC - in the form of cored wires, product of the company Praxair-Tafa, USA. The chemical composition of the substrate, as well as that of the additional material is presented in **Table 1**.

In order to produce the ultra-hard coatings with WC and TiC content, we used an electric arc spraying installation, provided with a compressed air compressor, which ensures pressures of 8 bar and a flow of up to 700mc/min, a direct current source RSC 400 type and the spray device presented in **Figure 4**.

The stages of the technological flow are schematically presented in **Figure 9**, [11].

The substrate surface activation included a series of preparatory operations for the metallization stage, which aimed both to clean the surface from oxides, oils and

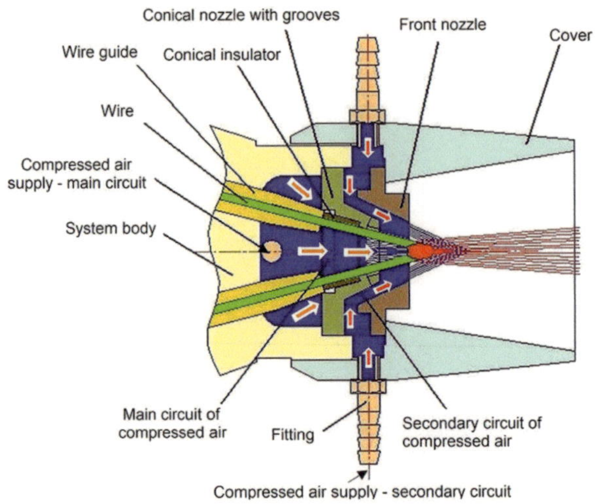

Figure 7.
Operating diagram of the spray head, [32].

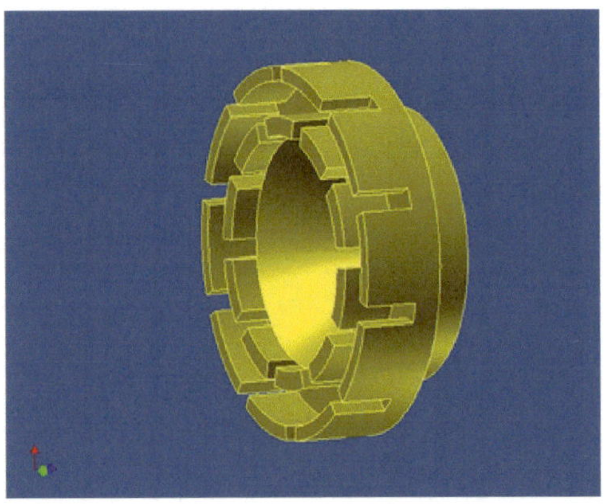

Figure 8.
The frontal constraint nozzle, [32].

Materials	Elements (wt%)										
	C	Si	Cr	Ni	Mn	B	WC	TiC	Fe	P	S
C15	0,14	0,15	0,3	0,3	0,43	—	—	—	—	0,04	0,04
97MXC	—	1,25	14,0	4,5	0,55	1,87	26,0	6,0	balance	—	—

Table 1.
Chemical composition of the materials.

greases, and to increase its roughness, in order to ensure good adhesion of the coating to the substrate, [32, 34]. The surface cleaning operation was performed in two successive phases: chemical cleaning and mechanical cleaning.

Chemical cleaning consisted of several steps: washing the substrate surface in a jet of liquid solution with a concentration of 10% (5% caustic soda, 5% soda salt), at a temperature of 50°C; rinsing in a stream of hot water at a temperature of 90°C; degreasing in methylene tetrachloride; wiping the substrate surface until dry with a textile material, [31].

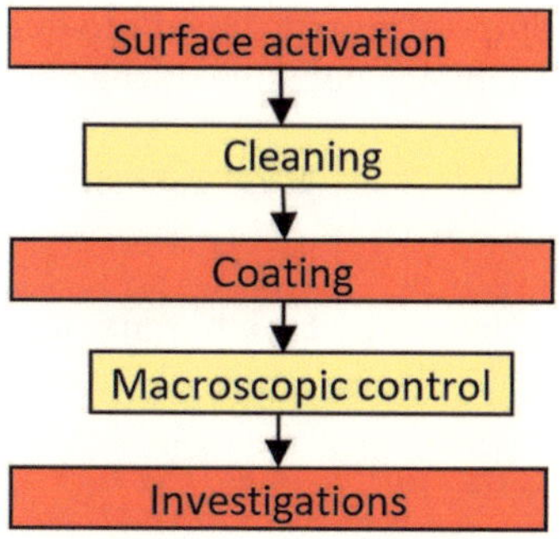

Figure 9.
Technological workflow.

Parameters	Value
Current intensity (A)	200/220/250
Voltage (U)	32
Air pressure in the primary circuit (bar)	5.5/6.0/6.5
Air pressure in the secondary circuit (bar)	3.0
Movement speed of the gun (m/s)	0.14
Coating thickness (mm)	0.5–0.7
Stand-off distance (SOD) (mm)	110

Table 2.
Thermal spray parameters.

The subsequent phase, respectively the mechanical cleaning, consisted in sand-blasting the surface of the substrate with abrasive particles. This process required pressurizing the abrasive medium with the help of compressed air and directing the flow of abrasive particles on the surface of the substrate [35]. Hard, abrasive and sharp particles of aluminum oxide with an average diameter of 536 ± 124 μm, sprayed with a pressure of 4barr at a distance of 30 mm, were used for blasting. The surface roughness obtained was between 46 and 62 μm. However, sandblasting always carries great risk, respectively it leaves the surface of the substrate contaminated with abrasive particles entrapped in the material, [36–38]. These residues have negative effects on the mechanical properties of the coatings, such as: they reduce the adhesion of the layer to the substrate, reduce the fatigue resistance properties of the substrate, limits the diffusion between coating and substrate and reduces the contact surface between particles [39–41]. To prevent the mentioned aspects, the substrate surface was ultrasonically cleaned by immersing the specimens in an ethanol bath (C_2H_6O) for 10 minutes, followed by drying them under a jet of filtered compressed air.

In our research, two process parameters, respectively: the compressed air pressure passing on the primary circuit and the intensity of the electric current, varied on three levels. For a good analysis of the effect of these variations, the rest of the technological parameters were kept constant. **Table 2** shows the parameters of the electric arc thermal spray process.

2.2 The characterization of ultra-hard 97MXC coatings

This subchapter presents the following investigations performed for the analysis of the WC/TiC ultra-hard coatings obtained by arc spraying process:

- elementary chemical composition - necessary to determine the presence and the proportions of the chemical elements of the analyzed coatings;
- structural characterization - necessary for the study of microstructure, identification of phases and constituents;
- mechanical characterization - which highlights the mechanical properties of 97MXC coatings: porosity, adhesion, microhardness, wear behavior.

2.2.1 The characterization of ultra-hard 97MXC coatings

A complete characterization of a deposit requires the most accurate knowledge of its chemical composition, the concentration of various alloying elements or impurities. The chemical composition of the 97MXC deposits was determined by semi-quantitative elementary chemical analysis of EDX type (Energy Dispersive X-ray Spectroscopy). The analysis system used (EDAX-AMETEK, Holland, 2008) is attached to an electron microscope (QUANTA 200 3D, FEI, Holland, 2008), being a microanalysis detector that records the energy of X-rays emitted from the surface of the specimen when scanned with an electron beam. This laboratory investigation allowed us to highlight the types of existing chemical elements and the proportions in which they are present.

To determine the chemical composition of the sprayed hard alloys, each sample - measuring 10 mm x 10 mm x 5 mm, was investigated by spot analysis, at 10 different points located on the cross-section. Before being investigated, the samples were metallographically prepared, by sanding on abrasive paper and polishing to remove impurities and oxides formed on the surface of the deposits.

Table 3 presents the mass percentages of the chemical elements identified in the composition of the ultra-hard deposit - depending on the intensity of the electric current used in the thermal spraying process.

As an example, **Figure 10** presents the EDX spectrum and the mapping (distribution of chemical elements on the scanned surface) of the chemical elements present in the 97MXC layer deposited by arc spray process, at current I = 220A and pressure p = 5.5 bar.

In all the cases (of coatings at different intensities with 97MXC material) on the scanned areas, the following chemical elements were identified following the EDX elemental chemical analysis: Fe, W, Cr, C, O Ni and Ti. Correlating these results with the data in **Table 3**, it had been able to conclude that the concentration of the alloying elements does not show changes due to the increase of the electric current intensity, respectively of the increase of the electric arc temperature.

2.2.2 The XRD analysis of the 97MXC coatings

The XRD analysis (XRD - X-ray diffraction) is a fast-analytical technique, used primarily for phase identification of a crystalline material. This type of analysis

Current intensity, (A)	Chemical element, (% weight)						
	Fe	W	Cr	C	O	Ni	Ti
200	48.67	13.84	13.08	11.43	5.63	4.21	3.14
220	48.89	13.64	12.30	12.10	8.65	3.85	2.57
250	46.42	14.21	12.84	12.42	7.32	3.91	2.88

Table 3.
Chemical composition of the ultra-hard coatings (p = 5,5 bar).

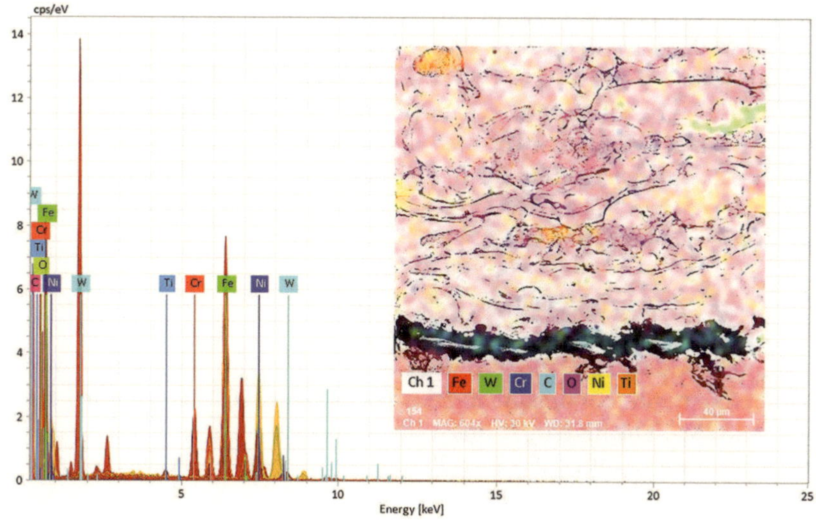

Figure 10.
EDX spectrum and chemical elements distribution map in case of 97MXC coating (I = 220A, p = 5.5 bar).

is very important because it highlights the existing phases and constituents in the electric arc coated layer. Knowing the phases and constituents can help to understand the coatings behavior during the different types of tests they have been subjected to. The testss were performed using the X'PERT PRO MRD diffractometer (Panalytical, Holland, 2008) with the following working configuration: Cu anode with λ = 1.54 Å, open eulerian cradle sample support, 2θ = 20–90°, [42].

Figure 11 shows the X-ray diffraction patterns of the 97MXC coatings obtained at a pressure of 6.0 bar, at a spray distance of 110 mm and different intensities of the electric current: 200 A - sample Ia, 220 A - sample Ib and 250 A - sample Ic.

From the XRD patterns it is observed that the three coatings contain the Fe-Cr alloy, complex carbides of the FeW_3C, Fe_3W_3C and Fe_6W_6C type and fractions of

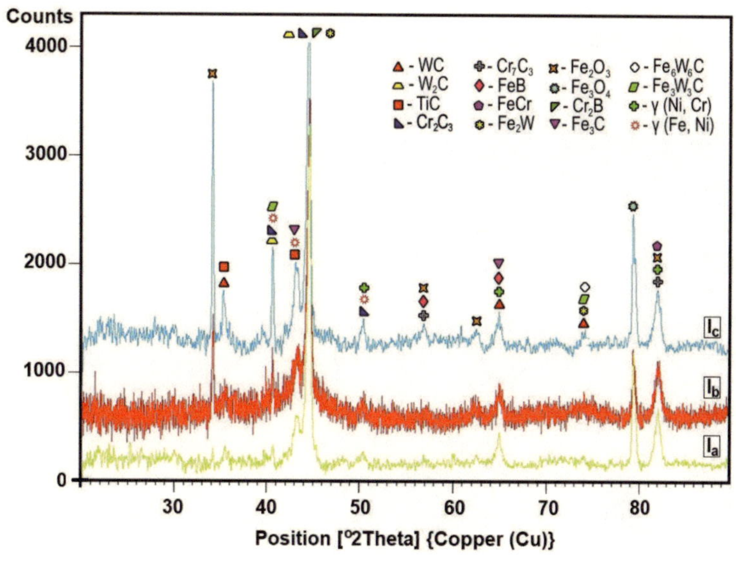

Figure 11.
X-ray diffraction patterns of 97MXC coatings.

FeB, Cr_2B, Fe_2O_3 and Fe_3O_4. Peaks of WC and W_2C are present in all three coatings, formed as a result of decomposition during the thermal spraying process, similar to the results reported by He et al. [43]. In additional to the eutectic phases of WC, W_2C and TiC, alloyed solid solutions of γ(Fe, Ni) and γ(Ni, Cr) were also identified. It is noted that the intensity of the W_2C peak increases with increasing current intensity. It is also suggested that the high temperature of the electric arc favors the decomposition of the WC carbides into single elements, respectively the formation of C-poor compounds such as W_2C.

2.2.3 Structural characterization of the 97MXC coatings

The morphology and metallographic structure of 96MXC deposits were investigated on micrographs obtained using the FEI - Quanta 200 3D scanning electron microscope (SEM). The investigations were performed on the cross section. The samples, measuring 10 mm x 10 mm x 10 mm, were obtained by cutting, after which there were incorporated into epoxy resin, sanded and polished. For the metallographic characterization of the compounds present in the analyzed coatings structures, the surfaces were chemically attacked with Vilella reagent (a solution with 10 ml of HF, 5 ml of HNO_3, and 85 ml of H_2O) for 10 min. After preparation, the samples were analyzed with the help of the SEM microscope.

Figure 12 presents two representative SE (secondary electron) images of the 97MXC deposits, obtained by arc spray process (I - 220 A), at different values of the compressed air pressure passing through the primary circuit: $p_{primary\ air}$ = 5.5 bar; and $p_{primary\ air}$ = 6.5 bar.

The deposits from **Figure 12** present a heterogeneous microstructure formed by flattened lamellas (usually called splats), oriented parallel to the substrate, polygonal formations, unmelted spherical particles and pores - being specific microstructure to the deposits obtained by arc spray process. In the cross section of the samples are observed several variations of the structural elements brightness (various shades of gray), an aspect that suggests the inhomogeneity of the chemical composition.

XRD patterns, EDX analyses and SE images allow the identification of the various phases of the coatings. Thus, the matrix formed by light gray metal splats corresponds to FeCr and FeW phases, the dark gray flattened splats correspond to hard Fe_2B phases, at the limit of the splats some interstitial oxides appear and the areas with dark contrast correspond to the pores.

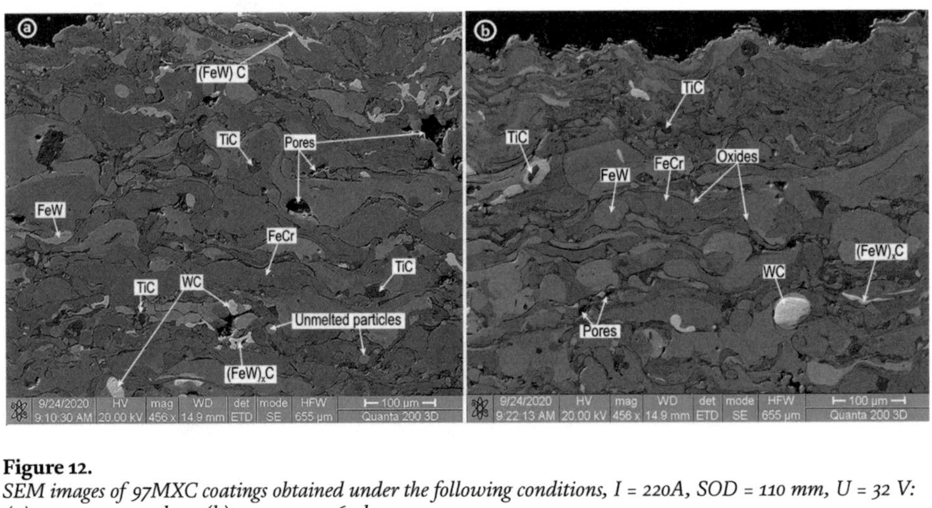

Figure 12.
SEM images of 97MXC coatings obtained under the following conditions, I = 220A, SOD = 110 mm, U = 32 V: (a) $p_{primary\ air}$ *= 5.5 bar; (b)* $p_{primary\ air}$ *= 6.5 bar.*

In the micrographs taken on the cross section, light-colored and bright polygonal formations appear, which correspond to the W-rich hard phase (WC) embedded in a matrix composed of eta type carbide $(FeW)_xC$ of light gray color. There are also present some dark gray polygonal formations which corresponds to titanium carbide (TiC). W-rich alloyed areas have a heterogeneous distribution, and inside them are formed eutectic phases of WC and W_2C type - similar to the results reported by Tillmann et al., [44].

It can be observed that at low values of the compressed air pressure passing through the primary circuit - see **Figure 12a**, in the deposit are obtained particles of larger dimensions, compared to the particle sizes presented in **Figure 12b**, which have a flattened shape. This aspect is explained by the fact that by increasing the pressure of the compressed air, the speed of the gas jet and implicitly the impact speed of the particles increase. The presence of a small quantity of unmelted particles inside the coating suggests that the temperature of the particles at the impact moment together with the spray distance were optimally chosen.

These aspects are confirmed by the investigations performed at the interface between the W-rich eutectic carbides (WC) and the metallic matrix based on Fe - see **Figure 13**. The deposits obtained at high values of electric current intensity present, around the polygonal eutectic carbides, at the interface between the W-rich phase and the Fe-based matrix, some transition areas (the luminous phase to darkness), presented in **Figure 13a,c**. The EDX analyses carried out in the dark area indicate the presence of a zone rich alloyed in Fe.

We found that the width of the transition area decreases until it disappears with the decrease of the current intensity that supplies the electric arc during the

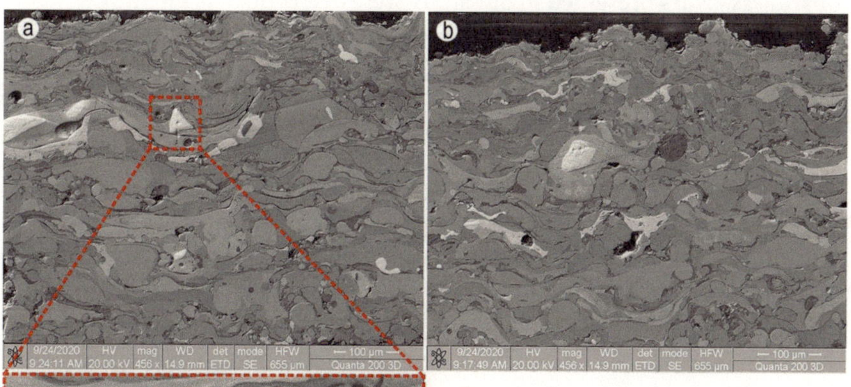

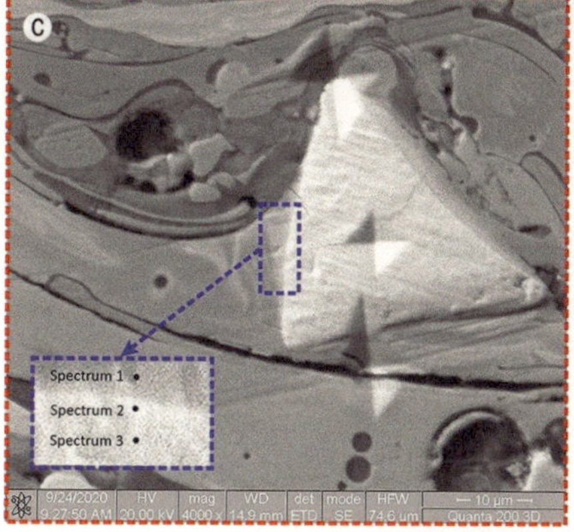

d

Spot analyses	Spectrum 1	Spectrum 2	Spectrum 3
Element distribution determined by EDX spectrum			
C [%wt]	9.21	15.52	9.82
Fe [%wt]	67.47	36.72	4.32
W [%wt]	23.32	47.76	85.68

Figure 13.
Cross-section images taken by SEM microscopy showing the embedment polygonal carbide WC at $p_{air\ secundary}$ *= 6.5 bar: (a) I = 250A; (b) I = 200A; (c) detail figure a; (d) EDX spot analysis.*

spraying process. As an example, in **Figure 13b** is presented polygonal WC carbide - where the absence of the transition zone is observed.

It can be observed that the W content decreases and the Fe amount increases in the transition area, as the distance to the eutectic polygonal carbide increases (see **Figure 13d**). These aspects suggest the fact that, by increasing the temperature of the sprayed particles and due to the increase of the electric arc intensity, the appearance of a transition zone between the polygonal eutectic carbides' WC type and the metallic matrix is favored. It can be suggested that high particle temperatures permit a better integration of the unmelted polygonal eutectic carbides into the matrix of Fe.

2.2.4 Microhardness of the 97MXC coatings

The hardness is the capacity of a piece to oppose the tendency to destroy the surface coatings by another piece, which acts on it with localized pressures on very small areas. In determining the hardness of materials, account shall be taken of the size of the traces produced by a penetrating piece, characterized by a certain shape and size and of the force acting on it. The hardness of a material is appreciated by the value of some conventional characteristics, obtained after some non-destructive tests.

Because the porous structure of the sprayed coatings does not permit the exactly determination of the hardness by conventional methods, in order to carry out investigations regarding the micro-hardness of the 97MXC deposits, we considered that the most suitable method is the Vickers method. The microhardness represents the Vickers hardness of some elements from the metallographic structure (phases, structural constituents, inclusions, etc.) and of some very thin coatings. The Vickers microhardness values presented in our study were determined using the CV - 400DAT digital microdurimeter, produced by CV Instruments, with a 100 g load, for 10s.

In order to establish the microhardness of the 97MXC coatings, we performed 5 determinations in points located at a minimum distance of 0,5 mm from each other, arranged on the transverse direction of the coating - according to the norm SR EN ISO 14923/2004.

The average values of the microhardness of the 97MXC deposits, produced in different experimental conditions are presented in **Figure 14**.

As the data presented in **Figure 14** show, the micro-hardness of 97MXC coatings is relatively high. This aspect is due both to the phases, rich alloyed in W or Ti and to the chemical compounds based on Fe_2Cr or of the eta type carbide $(FeW)_xC$ - which

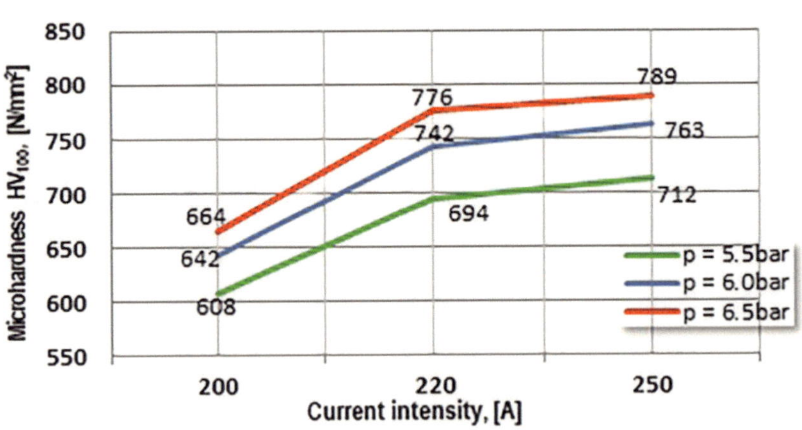

Figure 14.
HV_{100} microhardness of 97MXC coatings.

microhardness is relatively high. However, it is observed that the micro hardness of the 97MXC coatings varies in limits of up to 125 units with the pressure of the compressed air passing through the primary circuit and with the intensity of the spray current. Thus, for low values of the electric current intensity, the microhardness of the deposits is relatively low of HV_{100} = 664 ± 42 N/mm^2 compared to the microhardness of the depositions obtained at values of the electric current intensity higher than HV_{100} = 789 ± 32 N/mm^2. It can be affirmed that the high degree of homogeneity of depositions obtained at high values of the electric current intensity determines the increase of the microhardness of 97MXC deposits.

Microhardness investigations were also performed on the W, Ti and Cr hard phases. In **Figure 13b** and **c** are presented traces of the Vickers penetrator produced on the W-rich phase, in the area of the Fe phase highly alloyed with W and C (at interface) - **Figure 13b**, as well as in the low alloyed Fe phase - **Figure 13b**.

For the light-colored phases, the average microhardness value was 2842 N/mm^2, and inside the Fe-rich phase, the average microhardness value was 473 N/mm^2. At the interface level, the average microhardness value was 1387 N/mm^2. The TiC–rich phases had an average microhardness of 845 N/mm^2, the Cr-rich phases had an average microhardness of 636 N/mm^2 and those of interstitial oxides (the dark phase, positioned between the flattened particles) had an average microhardness of 258 N/mm^2. It can be suggested the fact that the transition phase, formed at the interface level between the W-rich eutectic phase and the metal matrix, rich alloyed in Fe, obtained the increasing the electric arc intensity, contains complex compounds based on Fe, W and C of $(FeW)_xC$ type, whose hardness is relatively high.

2.2.5 The wear behavior of the 97MCX ultra hard coatings

The wear behavior of the 97MCX coatings was evaluated by sliding wear tests conducted on an Amsler type system.

A general view of the AMSLER machine is given in **Figure 15**. The AMSLER machine was equipped with a data acquisition system based on tensometric strain gauges system [45], calibrated by deadweights method. The interface of the data acquisition system was realized in LabVIEW program [46]. The coated parallelepiped sample on rotating steel disc testing arrangement is presented detail.

Each coated sample was tested twice: at 20 N and 40 N normal load. The speed of the disk was kept constant, N = 100 rpm. The testing time was 3600 s. No lubricant was used. Before each test, samples were cleaned with acetone.

Figure 15.
General view of testing machine.

The tests were performed on three samples of 97MXC coatings obtained by arc spraying process, at pressure p = 6.5 bar and at different values of electric current intensity– see **Table 4**. The turning disk used in tribological tests was made of AISI52100 steel, hardness 64 HRC. The roughness of the tested samples was measured on Taylor-Hobson profilometer. The values of the roughness on longitudinal and transversal direction of tested samples are given in **Table 4**.

Figures 16 and **17** show the time variations of the friction coefficient and the friction torque at loads of 20 N, respectively 40 N.

The mean values of the coefficient of friction (CoF), as well as the weight loss of the samples after performing the tests at loads of 20 N and 40 N are shown in **Table 5**.

Sample	P_1 I = 200A	P_2 I = 220A	P_3 I = 250A	Disk AISI 52100
Longitudinal roughness, Ra [μm]	14.91	14.35	14.26	0.83
Transversal roughness, Ra [μm]	14.42	14.66	14.78	1.28

Table 4.
Roughness of the tested samples.

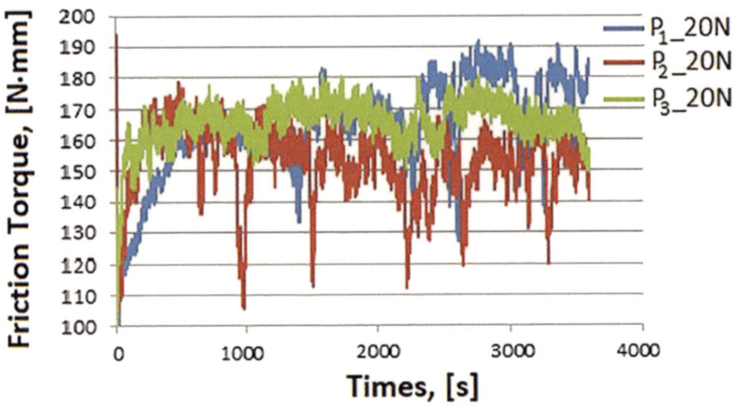

Figure 16.
Friction torque versus time at 20 N.

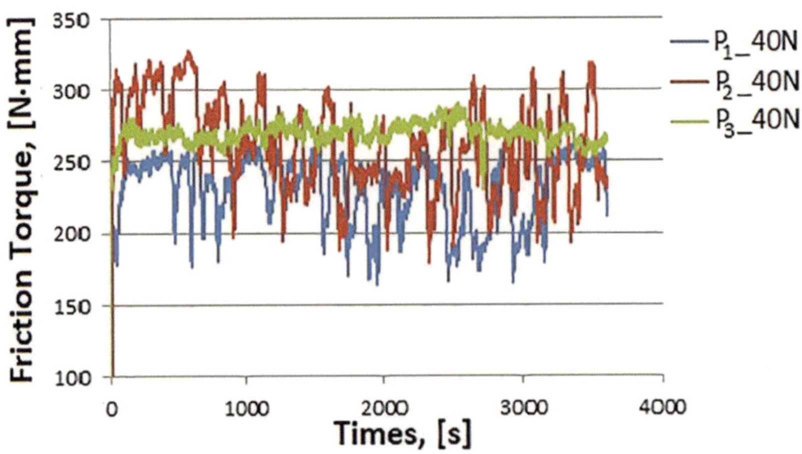

Figure 17.
Friction torque versus time at 40 N.

	Mean friction coefficients CoF					
Sample	**P_1**		**P_2**		**P_3**	
Load [N]	20	40	20	40	20	40
CoF	0.217	0.207	0.215	0.226	0.242	0.237
Wear, [%wgt]	0.000	0.028	0.000	0.004	0.000	0.011
Initial mass, g	28.976	28.976	47.874	47.855	47.226	47.226
Final mass, g	28.976	28.968	47.874	47.853	47.226	47.221

Table 5.
Mean CoF and wear in wt%.

Analyzing **Figures 16** and **17**, one can be observed that the friction torque in tribological system was more constant for P_3 sample, especially at high load - **Figure 17**). Correlated with the wear rate and general friction coefficients (CoF) results from **Table 5**, it can be concluded that the P_3 coating assure an almost constant CoF during an hour of continuous testing, while the friction torque of P_1 and P_2 manifested between large limits.

The data presented in **Table 5** show that the CoF values do not vary much by increasing the applied load, respectively from 20 N to 40 N. The average CoF values of the samples were as follows: CoF ≈ 0.21 for P1; CoF ≈ 0.22 for P2; CoF ≈ 0.24 for P_3. The CoF varied according to the quality of the tested surfaces of samples, but also with possible increase in temperature over the pad on disk contacts. A higher and constant CoF was obtained for the P_3 sample, especially at high load – see **Figure 17**.

The superior friction behavior of the P_3 against P_1 and P_2 coatings, given by the stability of the friction torque and the low wear rate, can be explained by the presence inside the deposit of the transition area between the W-rich hard phase and the $(FeW)_xC$ eta type carbide matrix, which allows fixing and maintaining the hard phase in conditions of advanced wear for a long time.

2.2.6 Analysis of 97MXC coatings adhesion

The adherence of the deposits obtained by thermal spraying is defined as being the force necessary detaching the layer from the substrate. The studies carried out by Haraga et al. [47] have demonstrated that the adhesion of the coatings is predominantly mechanical and is due to the solidification of the sprayed liquid particles or to the deformation of the semi-viscous particles on the substrate asperities. The adhesion of the deposits was determined by the traction test - in accordance with EN 582.

Figure 18 presents the adhesion variation of the deposits with the pressure of the compressed air passing through the primary circuit. It is noticed that at low values of the compressed air pressure the layer adhesion to substrate has low values. It can observe that for the same value of the compressed air pressure which passes through the primary circuit adhesion values of the deposit vary with the current intensity.

Thus, for the same pressure value, the adhesion of the 97MXC coatings obtained at I = 250A is superior to the one of the other coatings obtained at I = 200A and I = 220A, in the same technological conditions. Knowing that by the increasing of the current intensity of the electric arc temperature increases, we can suggest that, at high values of the current intensity, the drop formed in the electric arc is atomized into small particles with reduced inertia. Studies carried out by Toma et al. [33], have demonstrated that the high pressure of the compressed air favors the

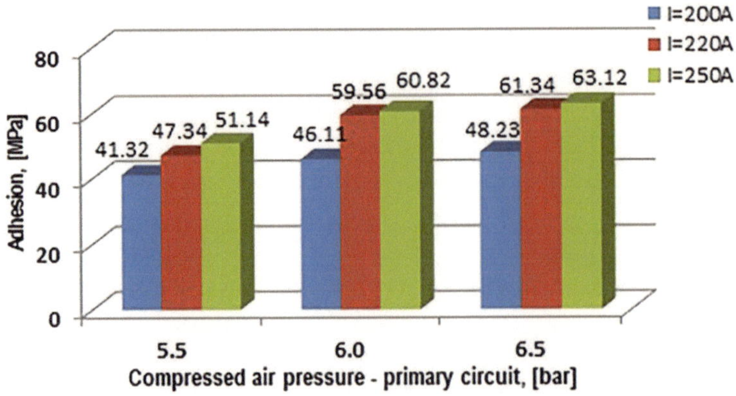

Figure 18.
Variation in the adhesion of 97MXC deposits with the compressed air pressure passing through the primary circuit.

increase of the impact velocity of the particles with the substrate surface, respectively their better fixation in the surface roughness. This aspect determines the increasing the adhesion of the 97MXC deposits with the increase of the compressed air pressure.

2.2.7 Porosity measurements of the 97MXC coatings

Porosity is a defining physical characteristic of the deposits obtained by thermal spray, due to the presence of interlamellar voids and interconnection channels between them. It is expressed by the degree of porosity of the deposit, measured in percentages, [48]. In the case of the 97MXC coatings, the porosity was investigated by image analysis of the transversal section of the specimens, using the IQ Materials software (Japan).

In **Figure 19** it is presented the variation of the average porosity of the 97MXC deposits with the compressed air pressure passing through the primary circuit - for different values of the current intensity that supplies the electric arc.

As expected, the variation of the average porosity of the 97MXC coatings presents a decreasing tendency both by the increasing of the compressed air pressure value and by the increasing of the current intensity value. Thus, it is observed that

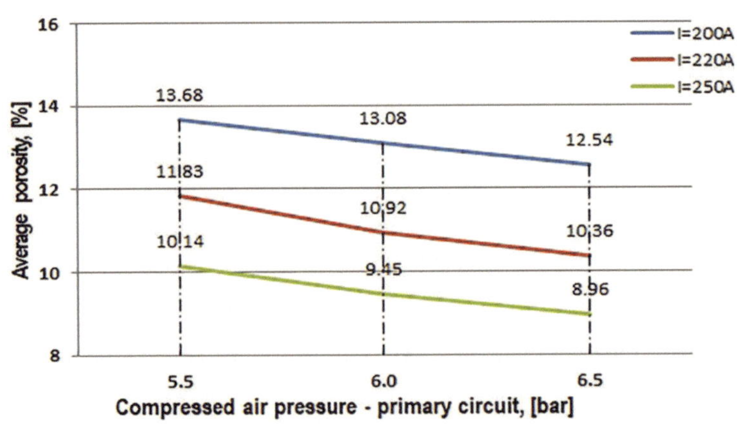

Figure 19.
Average porosity variation of 97MXC coatings with the compressed air pressure.

for the same value of the intensity of the electric current, when the compressed air pressure increases, the average porosity of the deposits decreases by up to 12%. The effect of increasing the temperature of the arc achieved by increasing the electric current determines the reduction of the average porosity by a maximum of 28%. It can be stated that between the two process parameters: compressed air pressure and electric current intensity, the latter has a "dramatic" influence on the deposition density of 97MXC.

3. Conclusions

The studies, carried out in this paper, demonstrate that qualitative deposits of hard alloys, containing WC and TiC, can be obtained by arc spray process, using a classic spray device equipped with a conical nozzle system.

The increase of the electric arc temperature, due to the increase of the current intensity favors the decomposition of some chemical compounds (TiC, FeB, WC), respectively the appearance of hard chemical compounds of type W_2C and of complex carbides of type eta: FeW_3C, Fe_3W_3C and Fe_6W_6C; the accentuated melting of the additional material and the formation of small particles with low inertia, capable to fix in the surface roughness determine the increase of the adhesion of the deposits of hard alloys and the reduction of the porosity in relatively large limits (up to 28%).

The pressure of the compressed air passing through the primary circuit determines the increase of the velocity of the sprayed particles [33], respectively their better fixation in the asperities of the substrate surface. This aspect justifies the adhesion increasing, the average porosity reducing of the coatings (by up to 12%), and the increasing of the microhardness of the coatings - at the increasing of the compressed air pressure.

The presence, inside the coatings of hard chemical compounds, type WC and W_2C - incorporated in the carbide matrix $(FeW)_xC$, as well as compounds TiC, FeCr, and FeB, permits improving the wear behavior of deposits, underlined through the stability of the friction torque, the increasing of the coefficient of friction and obtaining a low wear rate.

Conflict of interest

The authors declare no conflict of interest.

Author details

Stefan Lucian Toma[1*], Radu Armand Haraga[1], Daniela Lucia Chicet[2], Viorel Paleu[3] and Costica Bejinariu[1]

1 Department of Materials Engineering and Industrial Safety, Faculty of Materials Science and Engineering, Gheorghe Asachi Technical University of Iasi, 700050 Iasi, Romania

2 Materials Science Department, Materials Science and Engineering Faculty, Gheorghe Asachi Technical University of Iasi, 700050 Iasi, Romania

3 Mechanical Engineering, Mechatronics and Robotics Department of Mechanical Engineering Faculty, Gheorghe Asachi Technical University of Iasi, 700050 Iasi, Romania

*Address all correspondence to: stefan-lucian.toma@academic.tuiasi.ro

References

[1] Sacriste D., Goubot N., Dhers J., Ducos M., and Vardelle A., An evaluation of the electric arc spray and (HPPS) processes for the manufacturing of high power plasma spraying MCrAlY coatings, J. Therm. Spray Technol., 2001, **352**, p 35-358.

[2] Pawlowski L., in: 2nd ed., The Science and Engineering of Thermal Spray Coatings, Ed. Wiley, Chichester, England, 2008.

[3] Toma B F, Baciu R E, Bejinariu C, Cimpoieșu N, Ciuntu B M, Toma S L, Burduhos-Nergis D P and Timofte D, Researches on the improvement of the bioactivity of tio2 deposits, obtained by magnetron sputtering – DC, IOP Conf. Series: Mat. Science and Engineering, 2018, 374: 012017.

[4] Newbery, A.P., Grant, P.S., Neiser, R.A., The velocity and temperature of steel droplets during electric arc spraying, Surf. Coat. Technol. 195 (2005) 91-101.

[5] Bémer, D., Subra, I., Morele, Y., Charvet, A., Thomas, D., Experimental study of granular bed filtration of ultra fine particles emitted by a thermal spraying process, Journal of AerosolScience 2013, 63:25-37.

[6] Toma SL, Bejinariu, C, Gheorghiu, DA, Baciu, C, The improvement of the physical and mechanical properties of steel deposits obtained by thermal spraying in electric arc, Advanced Materials Research, 2013, 814, 173-179, DOI: 10.4028/www.scientific.net/AMR.814.173.

[7] Katranidis V., Kamnis S., Allcock B., Gu S., Effects and interplays of spray angle and stand-off distance on the sliding wear behavior of HVOF WC-17Co coatings, J Therm Spray Tech, 2019, 28, pp. 514-534.

[8] Audisio, S., Caillet, M., Galerie, A. and Mazille, M., Préparation d'une Surface (Surface Preparation), Traitements de Surface et Protection contre la Corrosion (Surface Treatments and Protection against Corrosion), Les Éditions de Physiques, Paris, France, 1987, p 169-174

[9] Toma S.L., Bejinariu C., Eva L., Sandu I.G., Toma B.F., Influence of process parameters on the properties of TiO_2 films deposited by a D.C. magnetron sputtering system on glass support, Key Engineering Materials, 2015, 660: 86-92.

[10] Tillmann W., Walther F., Luo W.F., Haack M., Nellesen J., Knyazeva M., In Situ Acoustic Monitoring of Thermal Spray Process Using High-Frequency Impulse Measurements, J. of Therm. Spray Technol., 2018, 27: 50-58.

[11] St. Toma, Contribution Regarding the Steels Metallization Through Activated Thermal Spraying, PhD Thesis, Gheorghe Asachi Technical University Iasi, Romania, 2009.

[12] Zimmermann, S., Gries, B., Fischer, J., Lugscheider, E. (Eds.), Thermal Spray 2008: Crossing Borders: Proceedings of the International Thermal Spray Conference 2008, DVS Deutscher Verband für Schweißen, Düsseldorf, 2008.

[13] Arif, Z.U., Shah, M., and Rehman, E., Tariq, A., Effect of spraying parameters on surface roughness, deposition efficiency, and microstructure of electric arc sprayed brass coating, International Journal of Advanced and Applied Sciences, 2020, 7(7): 25-39.

[14] Berger L.M., Saaro S., Naumann T., Wiener M., Weihnacht V., Thiele S., Suchánek J., Microstructure and properties of HVOF-sprayed chromium

alloyed WC-Co and WC-Ni coatings, Surf. Coat. Technol. 2008, Vol. 202, pp 4417.

[15] Nurisna Z.; Triyono; Muhayat N, Wijayanta, AT., Effect of layer thickness on the properties of nickel thermal sprayed steel, Book Series AIP Conference Proceedings, 2016, Vol. 1717, Art. Nr. 040012.

[16] Tillmann, W. and Abdulgader, Wire Composition: Its Effect on Metal Disintegration and Particle Formation in Twin-Wire Arc-Spraying Process, Journal of Thermal Spray Technology, 2013, 22: 352-362.

[17] http://ctwhardfacing.co.uk/celcoat-services/wire-arc-spray-coatings/(Accesed: 05.10.2020)

[18] Toma S.L, The influence of jet gas temperature on the characteristics of steel coating obtained by wire arc spraying, Surf & Coat. Technol., 2013, 220:261-265.

[19] Z. Lu, J. Cao, H.F. Lu, L.Y. Zhang, K.Y. Luo, Wear properties and microstructural analyses of Fe-based coatings with various WC contents on H13 die steel by laser cladding, Surface & Coatings Technology, 2019, 369: 228-237.

[20] Tillmann, W., Luo, W., Selvadurai, U., Wear analysis of thermal spray coatings on 3D surfaces, J. Therm. Spray Technol. 2014, 23: 245-251.

[21] Toma S.L., Badescu M., Ionita I., Ciocoiu M. and L. Eva, Influence of the Spraying Distance and Jet Temperature on the Porosity and Adhesion of the Ti Depositions, Obtained by Thermal Spraying in Electric Arc - Thermal Activated, Applied Mechanics and Materials, 2014, 657: 296-300.

[22] Planche MP, Liao H, Coddet C, Relationships between in-flight particle characteristics and coating microstructure with a twin wire arc spray process and different working conditions, Surf & Coat. Technol., 2004, 220: 215-226.

[23] Gedzevicius I, Valiulis AV Analysis of wire arc spraying process variables on coatings properties, Analysis of wire arc spraying process variables on coatings properties, 2006, 175:206-211.

[24] Cazac, A.M., Bejinariu, C., Ionita, I., Toma, S.L., Rodu, C., Design and Implementation of a Device for Nanostructuring of Metallic Materials by Multiaxial Forging Method, Applied Mechanics and Materials, 2014, 657: 193-197, DOI: 10.4028/www.scientific.net/AMM.657.193.

[25] Sharifahmadian O, Salimijazi HR, Fathi MH, Mostaghimi J, Pershin L, Relationship between surface properties and antibacterial behavior of wire arc spray copper coatings Surf & Coat. Technol., 2013, 233: 79-79.

[26] Toma SL, Gheorghiu DA, Radu S, Bejinariu C, The influence of the diffusion on adherence of the 60T deposits obtained through thermal spraying in electric arc, Applied Mechanics and Materials, 2013, 371: 193-197, DOI: 10.4028/www.scientific.net/AMM.657.193.

[27] Zhao XY, Xiang, M, Zhou HC, Zhang WH, Numerical Investigation and Analysis of the Unsteady Supercavity Flows with a Strong Gas Jet, Journal Of Applied Fluid Mechanics, 2020, 13, 1323-1337.

[28] Masoumeh, G., Shahrooz, S., Mahmood, G., Ahmad, S. E., Investigation of stand-off distance effect on structure, adhesion and hardness of copper coatings obtained by the APS technique, J. Theor Appl Phys, 2018, 12, 85-91.

[29] Grigorenko GM, Adeeva LI, Tunik AY, Korzhik VN, Doroshenko LK,

Titkov YP, Chaika AA, Structurization of coatings in the plasma arc spraying process using B4C + (Cr, Fe)(7)C-3-cored wires, Powder Metallurgy and Metal Ceramics, 2019, 58:312-322.

[30] Giovanni Bolelli, Alberto Colella, Luca Lusvarghi, Stefania Morelli, Pietro Puddu, Enrico Righetti, Paolo Sassatelli, Veronica Testa, 2020, TiC–NiCr thermal spray coatings as an alternative to WC-CoCr and Cr3C2–NiCr, Wear, 450-451, 203273.

[31] Haraga, R.A., Bejinariu, C., Cazac, A., Toma B.F, Baciu, C., Toma, St.L. 2019, Influence of surface roughness and current intensity on the adhesion of high alloyed steel deposits - obtained by thermal spraying in electric arc, IOP Conference Series: Materials Science and Engineering, 572, DOI:10.1088/1757-899X/572/1/012056.

[32] Toma SL, Savin G, Toma BF, Bejinariu C, Ionita I, Vizureanu P, Badaru G, Sandu AV, Cazac a, Burduhos ND, System of concentric nozzles for metal spraying guns for drawable metallic materials, has conical stiffening shoulder that is provided on conical nozzle, inclined under angle relative to longitudinal axis of conical nozzle, Gheorghe Asachi Technical University Iasi, Patent Number RO134208-A2, 2020.

[33] Toma SL, Bejinariu C, Baciu R, Radu S. The effect of frontal nozzle geometry and gas pressure on the steel coating properties obtained by wire arc spraying, Surf & Coat. Technol., 2013, 220. DOI.org/10.1016/j.surfcoat.2012.11.011.

[34] Bejinariu C, Munteanu C, Florea CD, Istrate B, Cimpoesu N, Alexandru A, Sandu AV, Electro-chemical Corrosion of a Cast Iron Protected with a Al2O3 Ceramic Layer, Revista de chimie, 2018, 69(12): 3586-3589.

[35] Coddet, C., Montavon, G., Ayrault-Costil, S., Freneaux, O., Rigolet, F., Barbezat, G., Folio, F., Diard, A. and Wazen, P., Surface Preparation and Thermal Spray in a Single Step: The PROTAL Process—Example of Application for an Aluminum-Base Substrate Journal of Thermal Spray Technology,1999, 8(2): 235.

[36] Wigren, J., Grit Blasting as Surface Preparation before Plasma Spraying, Surf. Coat. Technol., Vol 34, 1988, p 101-108.

[37] Chicet D., Tufescu A., Paulin C., Panturu M., Munteanu C., The Simulation of Point Contact Stress State for APS Coatings, IOP Conference Series: Materials Science and Engineering, 2017, 209, 17578981.

[38] Calin, MA; Curteza, A; Toma, S, Agop, M, Morphological properties of polyamide 6-cnt nanofibers obtained by electrospinning method, Metalurgia International, 2013, 18: 19-22.

[39] Gray, H., Wagner, L. and Lütjering, G., Influence of Surface Treatment on the Fatigue Behavior of Ti-Alloys at Room and Elevated Temperatures, Sixth World Conference on Titanium, P. Lacombe, R. Tricot, and G. Béranger, Ed., Les Éditions de Physique, Paris, France, 1998, p 1895-1900.

[40] Paulin C., Chicet D.L., Istrate B., Panturu M., Munteanu C., Corrosion behavior aspects of Ni-base self-fluxing coatings, IOP Conference Series: Materials Science and Engineering, 2016, 147, 17578981.

[41] Nanu, C, Poeata, I, Popescu, C; Eva, L; Toma, BF; Toma, SL, The Influence of the Characteristics of Plastic Materials Used in the Performance of the Thoraco-Lumbar Orthoses, Materiale Plastice, 2018, 55: 85-90.

[42] Panturu M., Chicet D., Paulin C., Lupescu S., Munteanu C., 2017,

Microstructural aspects of TBC's deposited on internal combustion engine valve materials, Materials Science Forum, 907, 02555476.

[43] He, D.J., Fu, B.J., Jiang, J.M., Li, X.J., Microstructure and wear performance of arc sprayed Fe-FeB-WC coatings, J. Therm. Spray Technol. 17 (2008) 757-761.

[44] Tillmann, W., Hagen, L., Kokalj, D., Embedment of eutectic tungsten carbides in arc sprayed steel coatings, Surface & Coatings Technology, 2017, 331:153-162.

[45] Paleu, V., Georgescu, S., Baciu, C., Istrate, B., & Baciu, ER. Preliminary experimental research on friction characteristics of a thick gravitational casted babbit layer on steel substrate. IOP Conference Series: Materials Science and Engineering, 2016, 147, 012028. doi:10.1088/1757-899x/147/1/012028.

[46] Paulin, C., Chicet, D., Paleu, V., Benchea, M., Lupescu, Ş., & Munteanu, C. Dry friction aspects of Ni-based self-fluxing flame sprayed coatings. IOP Conference Series: Materials Science and Engineering, 2017, 227, 012091. doi:10.1088/1757-899x/227/1/012091.

[47] Haraga, R.A., Chicet Dl, Cimpoiesu N, Toma SL, Bejinariu C. Influence of the Stand-off Distance and of the Layers Thickness on the Adhesion and Porosity of the 97MXC Deposits Obtained by Arc Spraying Process, IOP Conference Series: Materials Science and Engineering, 2020, 877(1)/ DOI:10.1088 / 1757-899X / 877/1/012020.

[48] Bejinariu, C, Burduhos-Nergis, DP, Cimpoesu, N, Bernevig-Sava, A, Toma, SL, Study on the anticorrosive phosphated steel carabiners used at personal protective equipment Quality-access to success, 2019, 20: 71-76 Supplement 1.

Section 3

Metal Thermo-Mechanical Processing

Chapter 8

High-Chromium (9-12Cr) Steels: Creep Enhancement by Conventional Thermomechanical Treatments

Javier Vivas, David San-Martin, Francisca G. Caballero and Carlos Capdevila

Abstract

There is a worldwide need to develop materials for advanced power plants with steam temperatures of 700°C and above which have the capacity to achieve high efficiency and low CO_2 emissions. This request involves the development of new grades of 9-12Cr heat-resistant steels, with a nanostructured martensite, mainly focusing on the long-term creep rupture strength of base metal and welded joints, creep-fatigue properties, and microstructure evolution during exposure at such elevated temperatures. The main shortcomings of actual 9-12Cr high-chromium steels are that the creep resistance is not enough to fulfill the engineering requirements at temperatures higher than 600°C and the material undergoes a cyclic softening. Creep strength at high temperature could be improved by a microstructural optimization through nano-precipitation, guided by computational thermodynamics, and thermomechanical control process optimization.

Keywords: creep-resistant steels, thermomechanical treatment, creep fracture behavior, microstructural degradation, small punch creep tests, ausforming

1. Introduction

There is a worldwide need for the sustainability of current energy sources in order to ensure the viability of future generations, as long as these sources are environmentally friendly. In this sense, power plant designs for the future should ensure a cost-efficient reduction of CO_2 emissions and improvements in efficiency of fuel consumption.

The essential function of a power station is to convert energy from fuel (fossil or nuclear) into electrical energy. In the steam power plant, this conversion involves consuming the fuel to produce heat which is then used to produce steam to drive a turbine. The mechanical energy of the turbine is then converted to electrical energy by an alternator. The steam temperature on the entrance of the turbine is essential to increase the efficiency of the conventional steam cycle.

The maximum steam temperature and pressure are limited by the performance of certain components. The main components which are critical are steam headers, superheater and reheater tubing in boilers, turbine valve chest, rotors and casings, main steam and reheat pipework, generator rotors, and bolts used for high-temperature applications. The boiler components are limited by corrosion and creep. Pipework also suffers creep as well as weld cracking and thermal fatigue. Turbine components are subjected to creep and fatigue (both thermal and mechanical).

Therefore, the development of improved structural materials to increase in thermal efficiency has been the driving force to develop new generations of 9-12Cr ferritic/martensitic (FM) steels [1–3]. The most relevant in-use properties that heat-resistant steels employed to manufacture components in power plants should fulfill are good mechanical properties, fabricability, corrosion resistance, and creep strength. As indicated above, creep strength has been the most studied and has led to innumerable research activities, aiming at improving the creep strength in 9-12Cr FM steel developments [4–6]. The disadvantage of these steels is their loss of strength beyond 600°C, so they need to be optimized to guarantee their use in the future power plants. In this chapter one of the most promising ideas described is applying a thermomechanical treatment (TMT) instead of a conventional treatment. The main contribution of the TMT is the ausforming, which, as other authors have reported, allows increasing considerably the number density of the thermally stable precipitates, i.e., MX nanoprecipitates. Consequently, the creep strength has improved greatly.

2. Brief remarks on the evolution of 9Cr FM steel for the power-generation industry

As it has been reviewed by Klueh in his seminal work on high-chromium FM steels [7], the design and production of 9-12Cr FM steels began in 1912 when Krupp and Mannesmann produced a 12 wt. % Cr steel containing 2–5 wt. % Mo. This type of steel was used for steam turbine blades, and it is still in use under the designation of X22CrMoV12. The 2¼Cr-1Mo bainitic steel grade normally known as ASTM Grade 22[1] (with nominal composition of Fe-2.25Cr-1.0 Mo-0.3Si-0.45Mn-0.12C) was firstly introduced in fossil fuel power plants in the 1940s and is nowadays widely used. The 9Cr-1Mo FM steel grade (known as Grade 9) is a natural evolution from Grade 22, seeking a better corrosion resistance and, hence, increasing the chromium addition. These two steel grades are the reference steels for heat-resistant application in power plants. Since then, the steady need of pushing up the operating conditions in conventional fossil-fired power-generation systems led to the development of several "generations" of steels with improved elevated-temperature strengths. The evolution of steel compositions (**Figure 1**), which began with G22 and G9 (zeroth generation) with 100,000 h creep rupture strengths at 600°C of about 40 MPa, has allowed for increased operating steam temperatures and pressures [1–3, 8–12]. Three generations of steels have been introduced since the introduction of G22 and G9, and a fourth generation is in development.

The strategy adopted for improved corrosion and oxidation resistance for elevated-temperature operating conditions was the addition of carbide formers such as vanadium and niobium to add precipitate strengthening. Hence, the zeroth generation containing mainly 9-12Cr evolved to the 12Cr-MoV steels introduced in the power plants in the mid-1960s for thin- and thick-walled power station

[1] Grade 22 and the other commercial steels are given designations by ASTM (e.g., Grade 9 is 9Cr-1Mo, and Grade 91 is modified 9Cr-1Mo). The steels are further distinguished as T22 or T91 for tubing, P22 and P91 for piping, F22 and F91 for forgings, etc. The "G" designation will mainly be used here.

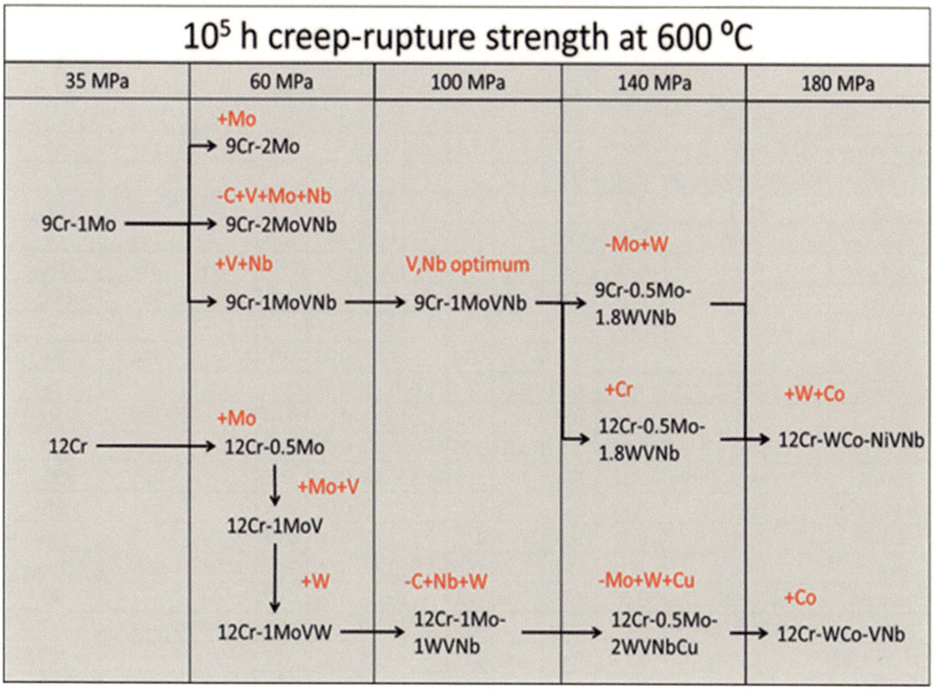

Figure 1.
Flowchart showing the evolution of 9-12Cr FM steels [7].

components. Their creep strength is based on solution hardening and on the precipitation of $M_{23}C_6$ carbides. These steels have been applied successfully in power stations over several decades [10]. These steels had increased 10^5 h rupture strengths at 600°C of up to 60 MPa (**Table 1**).

The second generation, developed in the late 1970s, is based on the modified 9Cr-1Mo, designated as G91 and HCM12 (see **Table 1**), which were developed for manufacturing of pipes and vessels for fast breeder reactors [10]. In this steel class, C, Nb, and V contents were optimized, N (0.03–0.05 wt. %) was added, and the maximum operating temperature increased to 593°C. The new steels have a duplex structure (tempered martensite and δ-ferrite). These steels have 10^5 h rupture strengths at 600°C of about 100 MPa. Of these latter steels, G91 has been used most extensively in the power-generation industry in all new power plants with operational temperatures up to 600°C [7]. The responsible mechanism for this substantial increment of creep strength as compared with 12Cr-MoV steels is the formation of thermally stable V and Nb carbonitrides. Besides lowering the Cr content down to 9 wt. %, tempered martensite microstructure also contributes to the higher creep strength [13, 14].

The Japanese steel development program led by Nippon Steel achieved the development of the P92 steel (NF616). This steel grade, designated as Grade 92, presents a further increases in stress rupture by the addition of 0.003 wt. % B and 1.8 wt. % W and reducing the Mo content from 1 to 0.5 wt. % [15, 16]. The addition of B ensures thermally stable $M_{23}(C,B)_6$ precipitates, whereas the higher W content leads to a higher amount of precipitated Laves phase [17, 18]. Grade 92, firstly introduced in the 1990s along with equivalent steel such as E911, fulfills the niche of steam operational temperature of 620°C for 10^4 h creep rupture strengths at 140 MPa.

Finally, the goal for the next steel generation being developed at present is pushing the limit of operation temperature above 650°C. This so-called fourth generation differs from the previous ones mainly by the addition of 3.0 wt. % Co

Country	Steel	Chemical composition (wt.%)									Rupture strength at 600°C (MPa)	
		C	Cr	Mo	Ni	W	V	Nb	N	B	10^4 h	10^5 h
	Basics											
Germany	X22CrMoV12-1	0.22	12.0	1.0	0.5	-	0.3	-	-	-	103	59
UK	9Cr-1Mo (Grade 9)	0.10	9.5	1.0	0.2	-	-	-	-	-	-	-
UK	H46	0.16	11.5	0.65	0.7	-	0.3	0.3	0.05	-	118	62
France	54T5	0.19	11.0	0.8	0.4	-	0.2	0.45	0.05	-	144	64
Japan	TAF	0.18	10.5	1.5	0.1	-	0.2	0.15	0.01	0.035	216	(150)
USA	11%CrMoVNbN	0.18	10.5	1.0	0.7	-	0.2	0.08	0.06	-	165	(85)
	Advanced											
USA	G91	0.1	9.0	1.0	<0.4	-	0.22	0.08	0.05	-	124	94
Japan	HCM12	0.1	12.0	1.0	-	1.0	0.25	0.05	0.03	-	-	75
Japan	TMK2	0.14	10.5	0.5	0.5	1.8	0.17	0.05	0.04	-	185	90
Europe	X18CrMoVNB 91	0.18	9.5	1.5	0.05	-	0.25	0.05	0.01	0.01	170	122
Europe	X12CrMoWVNbN	0.12	10.3	1.0	0.8	0.8	0.18	0.05	0.06	-	165	90
EU Cost501	E911	0.11	9.0	0.95	0.2	1.0	0.20	0.08	0.06	-	139	98
Japan	P92	0.07	9.0	0.50	0.06	1.8	0.20	0.05	0.06	0.003	153	113
Japan	P122	0.1	11.0	0.4	<0.40	2.0	0.22	0.06	0.06	0.003	156	101
Japan	HCM2S	0.06	2.25	0.2	0.2	-	0.25	0.05	0.02	0.003	-	80
Germany	7CrMoTiB	0.07	2.40	1.0	1.0	-	0.25	-	0.01	0.004	-	60
EU Cost522	FB2	0.13	9.0	1.5	0.15	-	0.20	0.05	0.02	0.0085	-	125
EU Cost522	CB2	0.12	9.0	1.5	0.15	-	0.20	0.06	0.02	0.011	-	125

Table 1.
Chemical composition and creep rupture strength at 600°C of the 9-12Cr heat-resistant steels from 1950 to 2005 [19].

as an austenite stabilizer because of the adverse effect of nickel on creep. They have projected 10^5 h creep rupture strengths at 600°C of 180 MPa [7]. In these steels with about 0.1 wt. % carbon, molybdenum has been further reduced or eliminated, and tungsten (2.6–3.0 wt. %) has been increased compared to third-generation compositions. In **Table 1**, an overview of the historical development of the 9-12Cr heat-resistant steels from 1950 to 2005 is shown.

2.1 Creep deformation

Creep deformation is a thermally activated process, and the rate of deformation (creep rate) is extremely temperature sensitive. In metals, creep deformation becomes important at temperatures greater than about $0.4T_M$, where T_M is the absolute melting temperature [20]. In the case of 9-12Cr FM steels, this temperature is approximately 450°C. Clearly, power plant materials operate in the temperature regime where creep process is significant. The creep properties of the material used limit the operating temperature of many power plant components, such as the turbines. Development of materials with an increased creep resistance is central to the use of power plants with higher steam temperatures.

Creep deformation can occur by a variety of different mechanisms. The mechanism that dominates depends on the stress and temperature conditions as well as the microstructure of the material.

In the case of power plant steels, the stress levels are relatively high, and the temperatures (compared with melting point) are relatively low. In the case of creep deformation, it is controlled primarily by dislocation movement and the thermal energy available for dislocations to overcome obstacles. A deformation mechanism map gives information about which mechanism will dominate for a particular set of conditions. Such a diagram for a G91 steel is shown in **Figure 2**. For the exposure conditions, for this material, a power law creep (dislocation creep) is expected to dominate.

Power law creep involves the movement of dislocations, and the creep rate is a result of the balance between work hardening and recovery. Work hardening results in an increase in the dislocation density, while recovery leads to a reduction in the dislocation density. If the dislocation density remains constant, then the creep rate is given by Norton's law [21]:

$$\dot{\varepsilon} = A\,(\sigma_a + \sigma_l + \sigma_u + \sigma_g)^4 \tag{1}$$

In this equation σ_a is the applied stress, σ_l, σ_u, and σ_g are the internal stresses due to solution, precipitation, and grain boundary hardening, respectively. The most effective method for reducing the creep rate is therefore to form a suitable distribution of particles which are also able to act as barriers to dislocation motion.

A study of the possible creep mechanisms suggests microstructures would be expected to have good creep resistance under conditions used in power plant. In general, creep-resistant alloys are based on a matrix which is a solid solution. The presence of misfitting solute atoms in solid solution makes the passage of dislocations through the matrix more difficult. However, the majority of the creep resistance, at least in the early stages of service, is derived from precipitate particles. Ideally these particles should be small, and they should be widely and homogeneously distributed in large numbers through the matrix. The particles need to be stable at operating temperatures for which the alloy is designated, and they should be resistant to coarsening, as this will reduce their effectiveness as strengtheners. In general, excessive work hardening and very fine grain sizes, which provide strengthening at ambient temperatures, are considered detrimental in high-temperature alloys. This is because both of them provide easy diffusion paths and therefore lead to an increase in the creep rate.

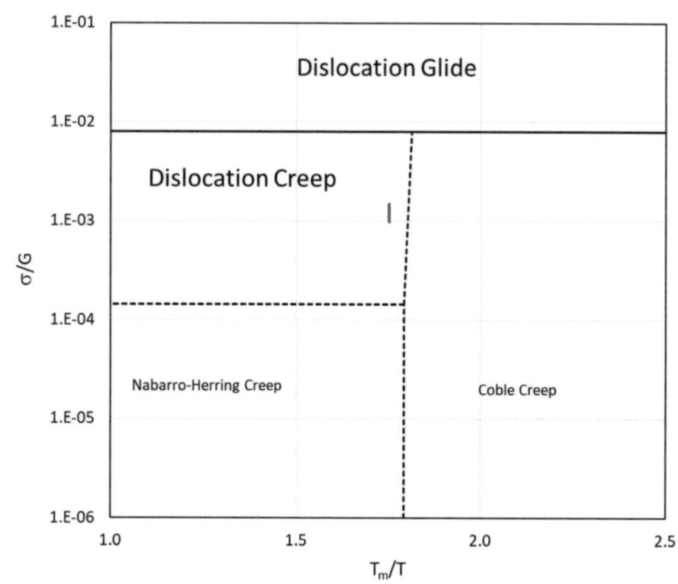

Figure 2.
Deformation mechanism map (D-MAP) of T91 steel calculated from experimental reported values [22, 23]. The red line indicates the experimental conditions considered in this work.

2.2 Creep and microstructural evolution

The new environmental regulations and commercial needs of the industry are the driving force for the development of new heat-resistant steels that push forward the operational limits of current steels. In this framework, the high-Cr FM steels applied as structural materials in fossil-fired and in nuclear power plants need to implement the operating temperatures above 650°C [1–3, 8–12]. The mechanism responsible for creep strengthening in these steels is the solid-solution and dispersion strengthening.

In the particular case of the so-called 9Cr FM steels, the creep degradation is a consequence of the thermal evolution of their hierarchal martensitic microstructure constituted by prior austenite grains, martensitic packets, blocks, and laths [24]. The microstructural degradation during creep consists of the coarsening of the lath structure [12, 24]. Such coarsening is governed by the subgrain boundary formation and evolution inside the laths, which can be prevented at high temperatures, and virtually frozen, by the dispersion of proper precipitates. The precipitates pin boundary migration and dislocation motion, slowing down the degradation of the martensitic microstructure and hence reducing creep rates [25, 26].

There are two main actors for the microstructural stability driven by precipitation in 9Cr FM steels: The first one is the coarse $M_{23}C_6$ carbides located mainly at the grain boundaries either from the prior austenitic grains or from the blocks or martensite lath boundaries. The second one is the V- and Nb-rich MX carbonitrides. Contrary to M23C6, those MX precipitates are homogeneously distributed within martensite laths. Therefore, the ideal situation would consist of reducing the presence of the $M_{23}C_6$ carbides to the minimum since their fast coarsening induces crack formation at the particle-matrix interface and promoting the formation of MX carbonitrides (nanometric in size), since they will delay the lath coarsening as mentioned above; it has been studied extensively [27–29].

Thermomechanical processing of 9Cr FM steels has been revealed as a promising tool to promote a high number density of MX carbonitrides [30–38]. TMT involves different steps that need to be optimized to produce the most favorable precipitate microstructure for elevated-temperature strength.

2.3 Creep tests

The creep behavior of a material may be characterized by a number of different parameters which can be measured by performing the appropriate creep test. For metallic materials most creep tests are conducted in uniaxial tension with a dumb-bell-shaped specimen similar to that used for tensile testing. The tests are carried out at a constant temperature and under either a constant load or stress. Applying a constant stress is more useful if the test is being employed to provide information about a creep mechanism.

The conventional treatments (AR) and TMT considered in this work were carried out on 10 mm in length and 5 mm in diameter cylindrical samples using a DIL 805A/D plastodilatometer (TA instruments) as described elsewhere [39, 40]. Due to the limited amount of material available after the TMT is carried out in the plastodilatometer, the creep properties were investigated by means of the small punch creep test (SPCT) performed at 700°C as it has been previously reported [41, 42]. The SPCT samples were cut transversally, from cylindrical specimens, with a thickness of 600 μm and a diameter of 8 mm. Then, the disks were ground on both sides down to a final thickness of 500 μm. In the setup of the SPCT, the lower and upper dice are connected via a thread to ensure the clamping of the sample. The load is applied by a ceramic punch ball which is in contact with the sample. A plunger rod is used to transmit the dead weight load to the punch ball. All these components are made of Al_2O_3 ceramics. The clamping device is surrounded by an electrical heater and a thermal insulation. The upper plate carrying the additional dead weight is guided by two pillars with ball bearings. The temperature is measured in the lower die directly under the sample. The displacement is measured by a capacitive sensor between the upper plate and the thermal insulation with an accuracy of ±1 μm. A load cell is placed between the upper plate and the plunger rod.

The disk deflection vs. time resulting from the SPCTs might be divided into three different regions similarly to conventional strain vs. time creep curves obtained from uniaxial testing. However, the failure in SPCTs occurs away from the load line with cracks propagating in a circumferential direction due to membrane stretching. Therefore, the first part of the disk deflection vs. time curve corresponds to the loading region where the spherical indenter loads on a very small contact area of the sample are. Since the stresses will be higher than the yield stress of the material, local plasticity and an initial large deformation are produced. This large deformation is accumulated in a short period of time. The second stage corresponds to the steady-state region, which coincides with most of the sample life, where the disk deflection rate reached almost a minimum. Finally, the third stage consists of an acceleration of disk deflection and fracture region. The interpretation of this behavior is that once a crack propagates to a critical length, the sample is no longer in balance, leading to an increase in deflection rate and to a reduction in the structure stiffness in the tertiary region. Another explanation might be due to the localized necking without crack presence. The deformation mechanism in the tertiary region is a mixture among accumulation of creep damage, geometric softening, and crack growth effect.

3. Effect of thermomechanical control processing on microstructure

As it has been introduced in previous sections, the pioneer commercial 9-12Cr steels present an upper service temperature of 540°C, which was successfully increased in the late 1970s up to 595°C with the introduction of vanadium and niobium microalloying in the composition of the steel. This steel was used as a benchmark for the development of steels with upper-use temperatures of 600–620°C.

However, it is difficult keep pushing the higher operating temperature too much. Therefore, to continue to exploit the advantages of ferritic steels, oxide dispersion-strengthened (ODS) steels [43–46] were introduced. The first successful alloy was presented in the 1960s, and, since then, it has been an active research field. ODS steels are strengthened by small oxide particles, but the complicated and expensive manufacturing route avoided the full implantation as structural material in the current power plants.

Despite being around for about 40 years, the ODS steels are still in the development stage because of having mechanical property anisotropy [43, 45, 47–49]. Therefore, an alternative strategy to achieve a high number density of precipitates is needed. In this section, we present preliminary results that allow us to conclude that conventional thermomechanical control processing strategy is adequate to achieve dispersion-strengthened steels.

3.1 Microstructure after conventional heat treatment

Lath martensite is a particular microstructure that ensures microstructural stability. Furuhara and Miyamoto [50] described the variety of crystalline size in lath martensite structures. A hierarchy of lath martensite structure is clearly identified particularly in low-carbon steels. A prior austenite (γ) grain is divided into "packets," each of which consists of a group of martensite laths with the same parallel close-packed plane relationship in the Kurdjumov-Sachs (K-S) orientation relationship, denoted as "CP group" recently. In general, a packet is partitioned into several blocks, each of which contains laths of a single variant of the K-S relationship. Blocks and packets are mostly surrounded by high-angle boundaries, whereas lath boundaries inside a block are of low-angle type.

The microstructure resulting from conventional industrial heat treatment consists of tempered martensite, which presents elongated subgrains with an average size of 0.25–0.5 μ m (**Figure 3**). Two types of precipitates, $M_{23}C_6$ carbides rich in chromium and MX carbonitrides rich in V or Nb, are present in the microstructure. The size of $M_{23}C_6$ carbides is around 100–200 nm, and they are precipitated on subgrain boundaries and prior austenitic grain boundaries. The size of MX carbonitrides is much smaller than $M_{23}C_6$ carbides, 20–50 nm, and they are in the matrix [51]. The purpose of this work is to produce a dispersion of nanosized precipitates by a controlled TMT, bearing in mind that a high number density of fine MX precipitates (Nb-MX and V-MX) should display superior high-temperature performance.

3.2 Effect of austenitization temperature

The effect of austenitization temperature on the temper microstructure of G91 steel is analyzed in this section. **Figure 4** schematically illustrates the two alternative processing routes considered:

- High austenitization temperature (HAT): In order to achieve an almost complete solid solution in austenite of most of the potential MX precipitate formers, the austenitization condition set will imply an elevated temperature.

- Thermomechanical treatment: The combined effect of the elevated austenitization temperature and a subsequent deformation will be studied with the aim of optimizing the MX-nanoprecipitate distribution during tempering of the martensitic microstructure.

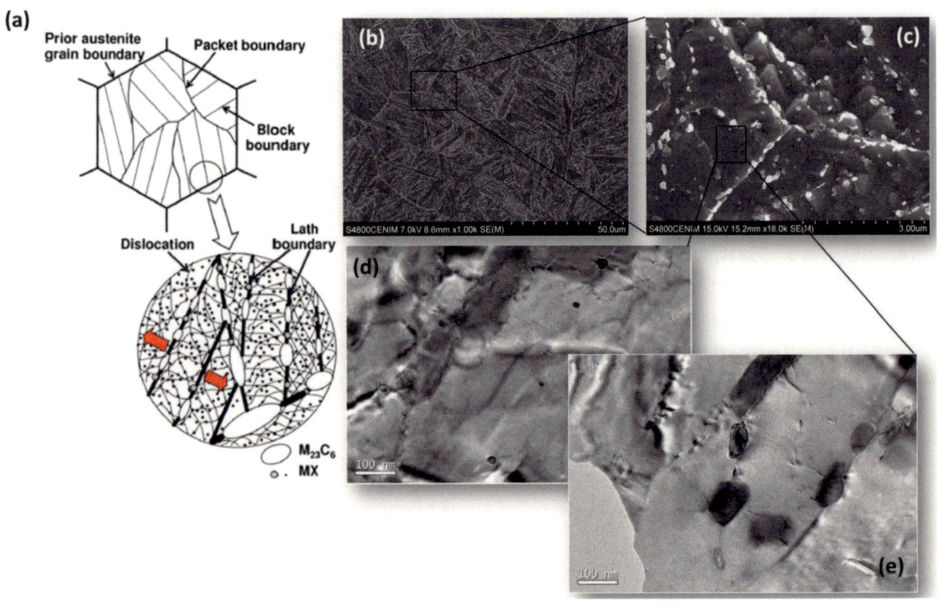

Figure 3.
(a) Resulting hierarchy microstructure achieved by conventional heat treatment; (b) and (c) SEM micrographs of the as-received state; (d) and (e) TEM micrographs. Arrow heads point out the location of the $M_{23}C_6$ carbides on lath boundaries and MX carbonitrides within the laths [37].

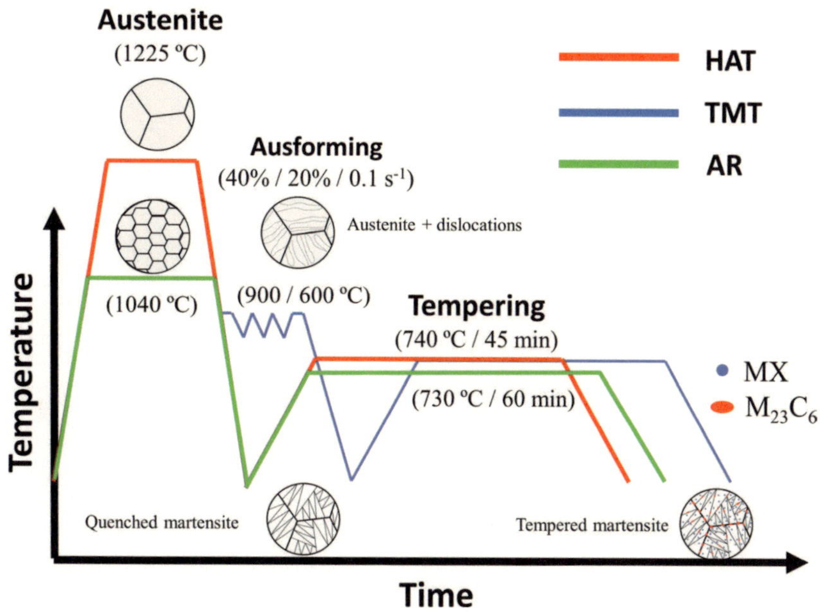

Figure 4.
Thermomechanical treatments investigated in this study [40].

For the sake of comparison, **Figure 4** also includes the industrial manufacturing conditions for G91 steel named as-received (AR) condition. The goal of exploring the effect of austenitization temperature on the microstructure is to enhance the precipitation of nanoparticles during the subsequent tempering stage indicated in **Figure 4**. As it was mentioned above, the main cause for creep softening in conventional G91 is due to the recovery of the martensitic lath microstructure because of mechanisms, such as the dislocation movement, controlled by diffusion [12, 27]. The dislocation pinning by nanosized MX precipitates can delay this phenomenon, since they present an enhanced ripening resistance [8, 52–54]. The goal of undergoing such elevated temperatures in the HAT treatment as compared to conventional austenitization heat treatment (AR treatment) is to dissolve all the primary carbides in the microstructure and drive to solid solution all the potential carbide former elements. Therefore, the martensite formed after quenching from such elevated austenitization temperature keeps in solid solution most of the precursor elements of MX carbides (M = Nb, V; X = C, N) that might precipitate during the subsequent tempering.

It is important to consider that the austenitization temperature has to be high enough to eliminate as much as possible the primary carbides formed during the casting process, but lower than the delta ferrite formation temperature, in order to avoid the detrimental effect of this phase from a long-term creep property point of view. Computational thermodynamic calculations by means of Thermocalc® determine the optimum austenitization temperature in 1225°C (**Figure 5**).

The interest of TMT relies on the role that austenite deformation has on refining the martensitic microstructure [55, 56]. Depending on the deformation temperature, several are the mechanisms that affect the austenite microstructure, and hence, that could be transferred to the martensite upon quenching. If deformation temperature is above the non-recrystallization temperature, the freshly formed austenite microstructure will present a significantly reduced grain size that would induce the concomitant martensitic microstructural refinement. Similarly, by applying plastic deformation to the austenite at temperatures below the non-recrystallization temperature, which is the so-called ausforming processing [57],

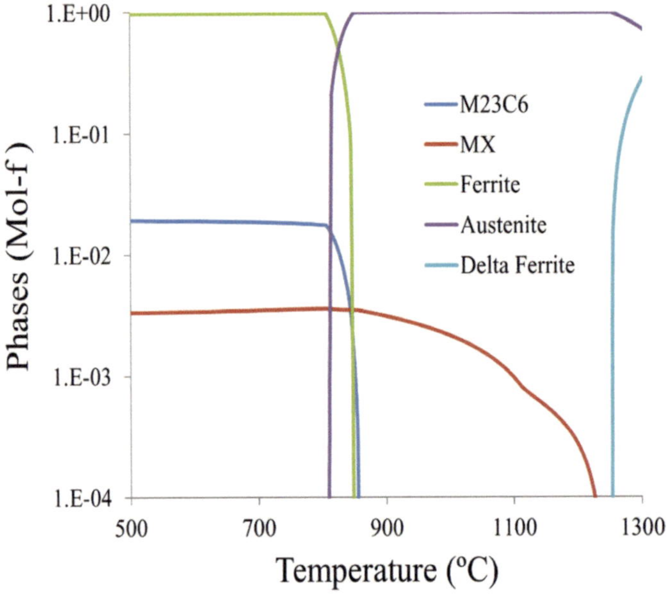

Figure 5.
Temperature evolution of phase mole fraction in G91 calculated by Thermocalc® [38].

an austenitic microstructure with a high population of deformation bands will be formed. This would directly induce the preferential formation of some specific martensitic variants upon austenite transformation (martensite variant selection), leading to the development of strong transformation texture.

Figure 6 illustrates the IPF maps, SEM and TEM micrographs after HAT and TMT processing routes, and the reference (AR) condition. The first conclusion obtained is the coarsening of the block size (white arrows in **Figure 6**) in HAT and TMT conditions as compared with AR condition, because of the high austenitization temperature. Block widths of 2.7 ± 0.2 μm for AR condition were obtained; meanwhile, values of 4.12 ± 0.37 μm for HAT and 3.21 ± 0.27 μm for TMT were measured. The coarser the prior austenite grain, the coarser the block size. However, it is worth noting that finer block size is observed after TMT than with HAT, which is consistent with the fact that thermomechanical processing increases the low-angle substructure and decreases the block size of as-quenched martensite.

The dislocation density after HAT and TMT was measured by XRD [40]. The results show a dislocation density of $(14 \pm 0.1) \times 10^{14}\ m^{-2}$ and $(28 \pm 0.1) \times 10^{14}\ m^{-2}$ after austenitization and ausforming, respectively. One might conclude from these results that the dislocation density in the as-quenched martensite after the TMT is substantially increased as compared with conventional treatment. A similar effect of the ausforming on the dislocation density was reported by other authors [58, 59]. Finally, TEM examination of the microstructure allowed us to determine the lath width of the martensitic microstructure. Values of 360 ± 35 nm for AR condition, 350 ± 20 nm for HAT condition, and 318 ± 32 nm for TMT condition were obtained, which are significantly finer than those reported after conventional treatments, i.e., lath size ranging from 300 to 500 nm [60].

The distribution of $M_{23}C_6$ precipitates in the tempered martensitic microstructure is also worth analyzing. **Figure 6** illustrates the distribution of $M_{23}C_6$ carbides after AR, HAT, and TMT processing routes. Coarse and closely spaced $M_{23}C_6$ carbides, about 70 to 500 nm, were observed. The number density and average particle size of these carbides were determined by studying several SEM micrographs to determine values of $6.19 \times 10^{19}\ m^{-3}$ and 141 ± 3 nm for AR condition, $8.24 \times 10^{19}\ m^{-3}$

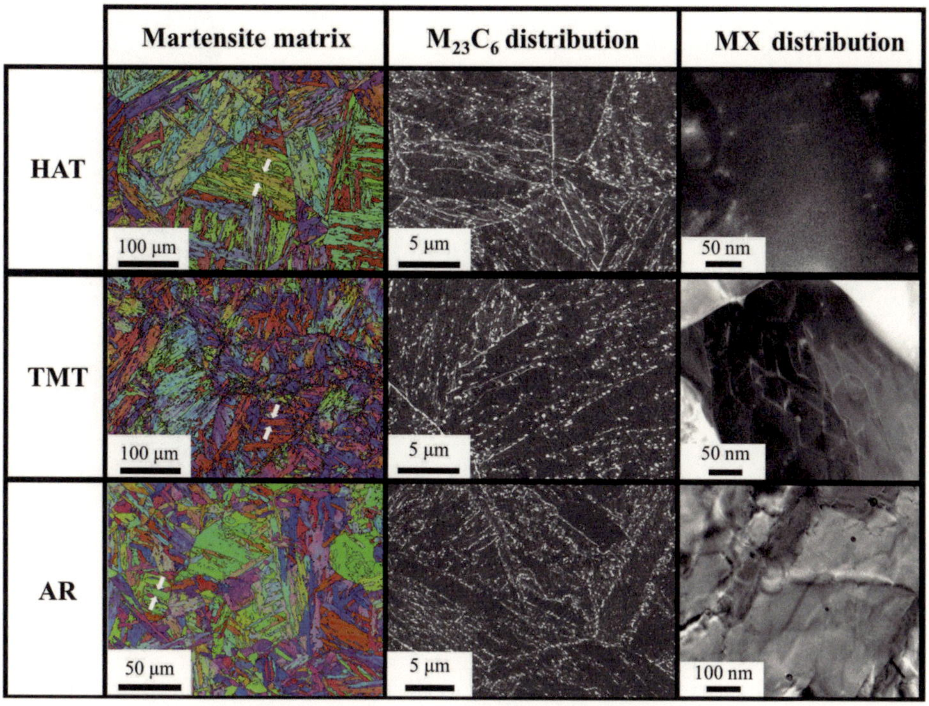

Figure 6.
Martensite matrix, $M_{23}C_6$ precipitate, and MX-nanoprecipitate distributions after the different thermomechanical and heat [40].

and 124 ± 3 nm for HAT condition, and 4.11 × 10^{19} m^{-3} and 143 ± 5 nm for TMT steel. These values are very similar to those reported by Klueh et al. for the steel after conventional heat treatment [35].

On the contrary, the finely dispersed MX nanoprecipitates present inside the martensitic laths and associated with dislocations are also observed in **Figure 6**. Therefore, this result suggests the role of dislocations as potential nucleation sites for MX nanoprecipitates. Hence, the importance of ausforming in generating a homogeneous distribution of nanosized MX particles in the microstructure might be also foreseen. These spherical MX nanoprecipitates had a mean particle size of 12 ± 1 nm with a number density of 7.20 × 10^{21} m^{-3} for HAT steel and 9 ± 1 nm with a number density of 1.86 × 10^{22} m^{-3} for TMT steel. The MX precipitates are, in both cases, significantly smaller than those measured after AR condition, i.e., particle size of 25 ± 5 nm with a number density of 8.14 × 10^{19} m^{-3}. The size values obtained after HAT and TMT are smaller, and the number density higher, than measurements reported in the literature after conventional heat treatments [61].

Figure 7 shows the disk deflection versus time curves obtained for the three conditions studied (AR, HAT, and TMT) at 700°C with a load of 200 N. The curves exhibit the three stages of creep that were described in previous sections. The first stage corresponds to the loading region where the spherical indenter loads the sample, and the mode of deformation is by bending. The second stage is characterized by a decrease in deflection rate and corresponds to the steady-state region with a minimum disk deflection rate. Finally, the third stage consists of an acceleration of disk deflection and fracture region. In the secondary and tertiary stages, stretching is the prominent deformation mode. Once a crack propagates to a critical length, the sample is no longer in balance, leading to an increase in deflection rate and to a reduction in the structure stiffness in the tertiary region until the final fracture.

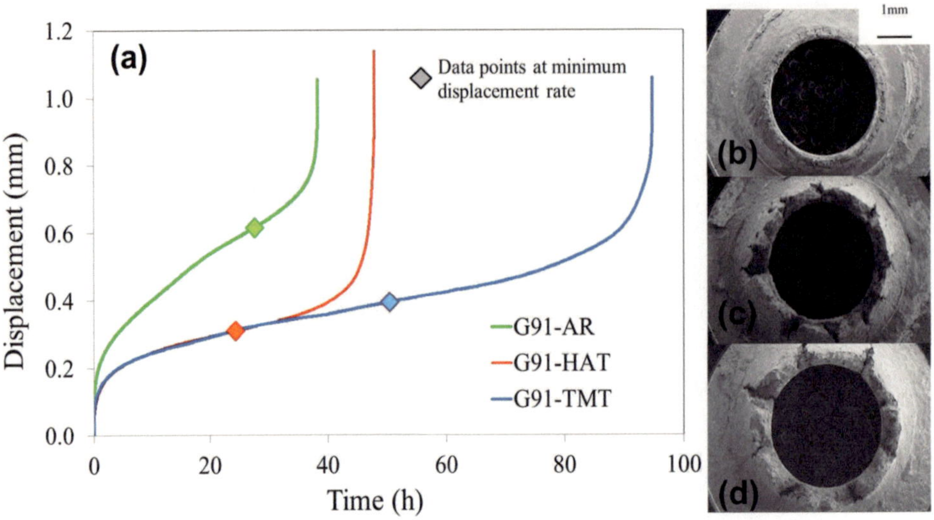

Figure 7.
(a) SPCT curves measured for the samples after the different thermomechanical and heat treatments and the creep fracture micrographs for the (a) AR, (b) HAT, and (c) TMT [40].

As indicated above, the minimum disk deflection rate (δ_d) is an important parameter that can be evaluated by SPCT. The evolution of disk deflection rate with the applied load might be described by an equivalent expression to the conventional Norton's power law for creep, which is similar to the expression used in Eq. (1):

$$\delta_d = A \cdot F^n \tag{2}$$

where A is a temperature-dependent constant, F is the force applied on the specimen, and *n* is the force exponent. One might conclude, therefore, from **Figure** 7 that the creep strength has significantly improved after the TMT condition. The time to rupture was 2.5 and 1.24 times greater than AR condition, from 38 to 95 h and 48 h for the TMT and HAT, respectively. The δ_d was 2.9 $\mu m \cdot h^{-1}$ for the TMT sample, while for the HAT sample, it was 3.7 $\mu m \cdot h^{-1}$. These minimum disk deflection rates were significantly slower than the minimum disk deflection rate measured for the G91 in the AR condition, which was 9.5 $\mu m \cdot h^{-1}$.

The results obtained suggest that the increase in the number density of MX precipitates enhances the strengthening capability at high temperature, since they are able to pin more effectively the dislocations. Hence minimum creep rate is reduced and the onset of tertiary creep is retarded. The differences in minimum disk deflection rate and time to rupture between the sample after TMT and HAT support the importance of ausforming on improving creep resistance.

3.3 Effect of ausforming

The next stage in the TMT after austenitization is the ausforming as shown in **Figure 4**. The effect of ausforming on low-carbon lath martensitic microstructure has been already described by Miyamoto et al. [62]. The authors reported that martensite variants with habit planes that are nearly parallel to the close-packed primary and secondary slip planes in austenite transform preferentially, i.e., martensite habit planes such as (575)γ that are nearly parallel to (111)γ and (−111) γ in asutenite [63]. Since strain is accumulated preferentially in (111)γ and (−111) γ slip planes during ausforming, this results in an increasing number of dislocation that might be transferred to martensite $(011)_M$ planes. Therefore, ausforming might increase the dislocation density in the resulting martensitic microstructure.

On the other hand, Takahashi et al. [64] reported recently the formation of Nb-cottrell atmospheres in low-carbon Nb-microalloyed steels. The authors explained that this mechanism is based on the fact that segregation energy of Nb to edge dislocation core was almost the same as the energy for grain boundary segregation. Besides, the large attractive interaction between Nb and dislocation core was due to its large atomic size. Therefore, such interaction between Nb atoms and dislocations retards the recovery of dislocation at high temperatures and, hence, stabilizes the microstructure at high temperatures. It might be expected that Nb presents the same behavior in the studied steel, preventing recovery after ausforming and promoting the fine and homogeneous MX carbonitride precipitation during tempering accordingly.

In this work the role of ausforming temperature by selecting 600 and 900°C, at a constant deformation of 20% (**Figure 4**), is explored. As mentioned above, the dislocation densities were estimated by XRD in fresh martensite after each ausforming condition studied [41]. Values of $(2.8 \pm 0.1) \times 10^{15}\ m^{-2}$ and $(1.9 \pm 0.1) \times 10^{15}\ m^{-2}$ were obtained for the ausforming at 600 and 900°C, respectively. These results show that the lower the ausforming temperature, the higher the dislocation density introduced in austenite is, which might be due to the fact that some of the dislocations in fresh martensite are inherited from deformed austenite as it was mentioned above.

On the other hand, Bhadeshia and Takahashi reported [65] an expression that allows to estimate the dislocation density (ρ_d):

$$\mathrm{Log}\ \rho_d = 9.28480 + \frac{6880}{T} - \frac{1780360}{T^2} \tag{3}$$

This expression is valid only when the martensite start temperature (T) is between the range 297 and 647°C.

Extracting the data of the martensite start temperature from a previous work [38], the estimation of the dislocation density obtained after the different ausforming conditions can be estimated. In this sense, ausformed samples at 600°C present a martensite start temperature of 338°C; introducing this value in Eq. (3), a dislocation density of $5.97 \times 10^{15}\ m^{-2}$ is calculated. Similarly, for the material ausformed at 900°C with a martensite start temperature of 374°C, the dislocation density calculated is $4.62 \times 10^{15}\ m^{-2}$. These results are in the same order of magnitude than those measured by X-ray diffraction, which demonstrate that the ausforming increases the dislocation density in the martensite.

During the final stage (tempering), MX carbonitrides and $M_{23}C_6$ carbides precipitate, and the recovery of dislocations takes place. Because of the higher dislocation density of ausformed samples, the number density of finer MX increases, and these precipitates are found homogeneously distributed within laths, as it can be seen in **Figure 8(a)** pointed out by white arrows.

The number density of MX precipitates (N) was determined through the direct measurements of spacing (λ) between MX carbonitrides from several TEM micrographs as indicated by Eq. (4):

$$N = 1/\lambda^3 \tag{4}$$

Figure 8(b) shows the size distribution of the precipitates in ausformed material. In the material ausformed at 600 and 900°C, the average size of MX carbonitrides was 5.6 nm and 7.4 nm, respectively. The number density of MX carbonitrides was $9.39 \times 10^{22}\ m^{-3}$ for the material ausformed at 600°C and $6.4 \times 10^{22}\ m^{-3}$ for the material ausformed at 900°C. On the other hand, the reported values of the size and number density of MX carbonitrides after the conventional processing were 30 nm

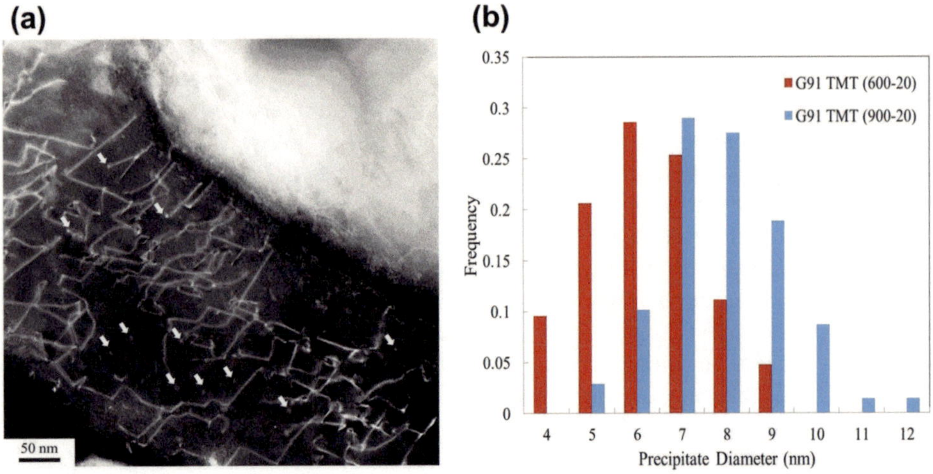

Figure 8.
(a) MX carbonitrides (white arrows) within laths after thermomechanical treatment ausformed (20%) at 900°C; (b) size distribution of MX precipitates in the TMT samples for the two ausforming temperatures: 600 and 900°C [41].

and 10^{20} m^{-3}, respectively [61]. It might be concluded that ausforming promotes a refining of precipitates, up to five times as compared with conventional processing, as well as an increase in number density up to two orders of magnitude. In fact, these number densities and precipitate sizes are very similar compared to those corresponding to oxides present in oxide dispersion-strengthened (ODS) steels [66, 67].

The elevated number density of nanosized MX precipitates has a direct impact on creep response of this material as it can be clearly observed in **Figure 9**. This figure shows characteristic SPCT curves at 200 N, exhibiting the variation of specimen deflection with time. It might be concluded from this figure that introducing an ausforming step improves the δ_d significantly, and most precisely, the lower the ausforming temperature, the lower the δ_d is, and, hence, the better the creep resistance is.

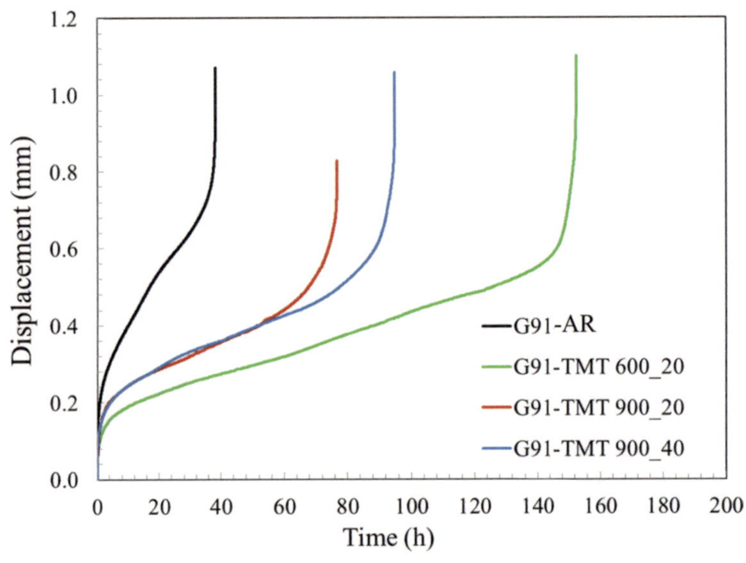

Figure 9.
SPCT curves for all samples tested at 700°C with a load of 200 N [39].

4. Failure mechanism: post-creep characterization of SPCT samples

Scanning electron microscopy (SEM) images of fractured SPCT specimens for different conditions are shown in **Figure 10**. Radial cracks can be observed in all the TMT samples (**Figure 10a–c**). This is an evidence of the loss of ductility and indicates a brittle fracture, which is a change in rupture ductility in comparison to the conventionally treated sample. Those samples do not show radial cracks (**Figure 10d**). Besides, a higher reduction in thickness is evident in the conventionally treated sample in comparison to the TMT ones, suggesting a ductile fracture behavior.

To clarify the failure mechanisms, the fractured samples were cut and prepared adequately. **Figure 11(a)** and **(b)** shows the SEM images for the TMT samples ausformed at 600°C with a deformation of 20% and ausformed at 900°C with a deformation of 40%. It is worth noting in those images the existence of cavities nearby coarse particles, which are located at the vicinity of PAGBs. The EDS spectrum shown in **Figure 11(c)** allows us to conclude that these particles are $M_{23}C_6$ carbides with M = (Fe, Cr, Mo).

The greater size of the $M_{23}C_6$ carbides at the vicinity of PAGB contributes to the inhomogeneous and localized deformation experienced by the TMT samples at these locations during creep. The local creep concentration close to PAGB would be promoting the nucleation of cavities that lead to the intergranular fracture with the brittle behavior.

Figure 12 shows different inverse pole figure (IPF) maps for all the samples under study before and after SPCT. It should be pointed out that, contrary to the lath boundaries that are not correctly indexed due to the step size used for the EBSD mapping, the block boundaries before and after SPCT are clearly disclosed. It is observed that the microstructures of the samples exhibit the characteristic lath-like morphology of the martensitic microstructure. However, such morphology is

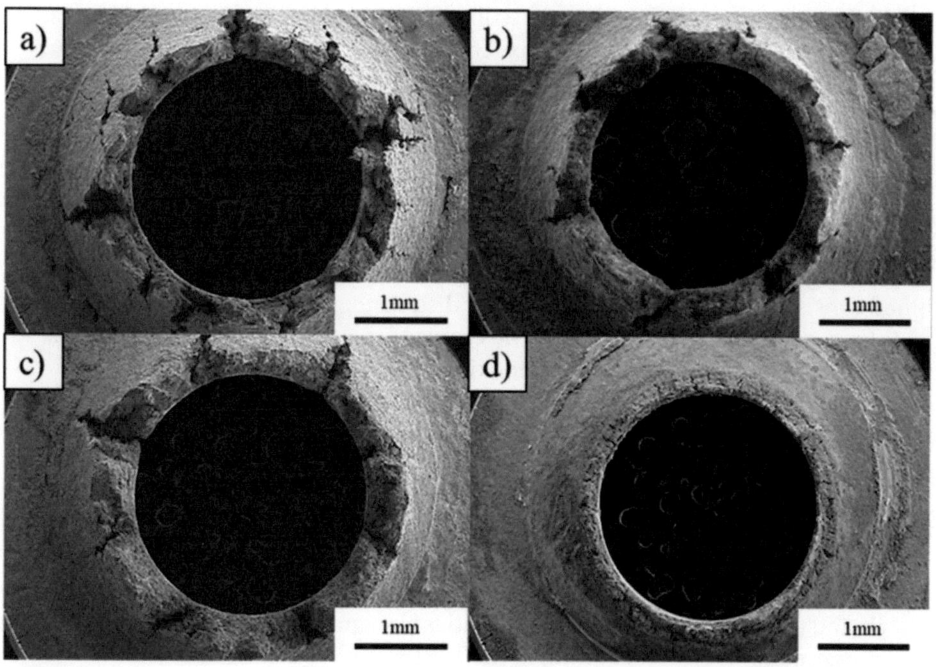

Figure 10.
Scanning electron microscopy images of the SPCT fracture surfaces for samples tested at 700°C with a load of 200 N: (a) G91-TMT 900_20; (b) G91-TMT 600_20; (c) G91-TMT 900_40; and (d) G91-AR [39].

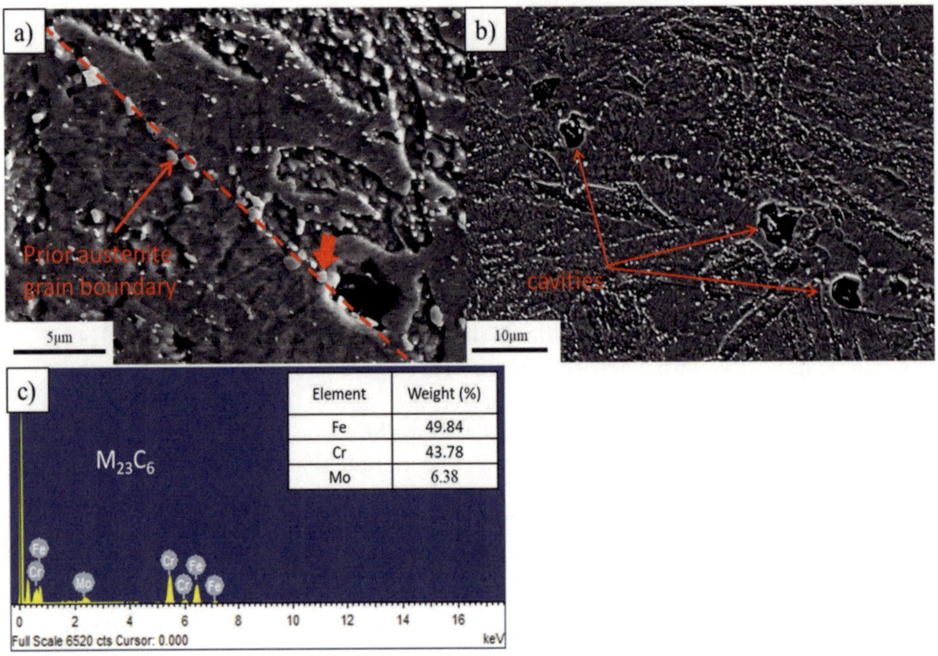

Figure 11.
Scanning electron microscopy images: (a) $M_{23}C_6$ precipitates located at a prior austenite grain boundary in sample G91-TMT 600_20. The prior austenite grain boundary in this image has been highlighted with a dash line as a guide to the eye; (b) cavities associated with coarse $M_{23}C_6$ precipitates have nucleated at a prior austenite grain boundary in sample G91-TMT 900_40. Cavities have been pinpointed with arrows and (c) EDS analysis of the particle marked with a red arrow in image (a), close to a cavity [39].

blurred in samples ausformed at 600°C because of the high deformation accumulated in the austenite during ausforming [38]. After the SPCT, it is observed that the original lath-like morphology has partially disappeared, and it has evolved towards a fine-grained equiaxed ferritic matrix. One might conclude from the microstructural observations made after SPCT that newly formed equiaxed grains are distributed homogeneously in the conventionally treated sample (AR), while these grains are located mainly nearby the prior austenite grain boundaries in the TMT samples, which is consistent with the fact that it is in these samples where the deformation accumulated is larger during creep.

Therefore, taking into account the results shown previously in the SEM micrographs (**Figure 11(a)** and **(b)**), the microstructural degradation would be a combined consequence of the accumulation of dislocations at the low-angle boundaries and the stress concentration close to the coarse $M_{23}C_6$ carbides, which lead to the progressive loss of the lath-like martensitic microstructure, which evolves to an equiaxed ferritic matrix. As it has been discussed above in the case of the TMT samples, the nucleation of cavities takes place close to $M_{23}C_6$ precipitates located at the prior austenite grain boundaries. The coalescence of the cavities formed surrounding the $M_{23}C_6$ carbides would initiate the cracks, and they will propagate along the prior austenite grain boundaries.

The more homogeneous distribution of the $M_{23}C_6$ precipitates in the conventionally treated sample favors the apparition of equiaxed grains in the whole martensitic matrix and develops the nucleation of cavities intragranularly, which provokes the transgranular fracture. Besides, in the TMT samples, the high austenitization temperature produces an enormous prior austenite grain sizes with concomitant large grain boundary surfaces, facilitating an earlier formation of the critical crack length that causes the brittle fracture [68, 69].

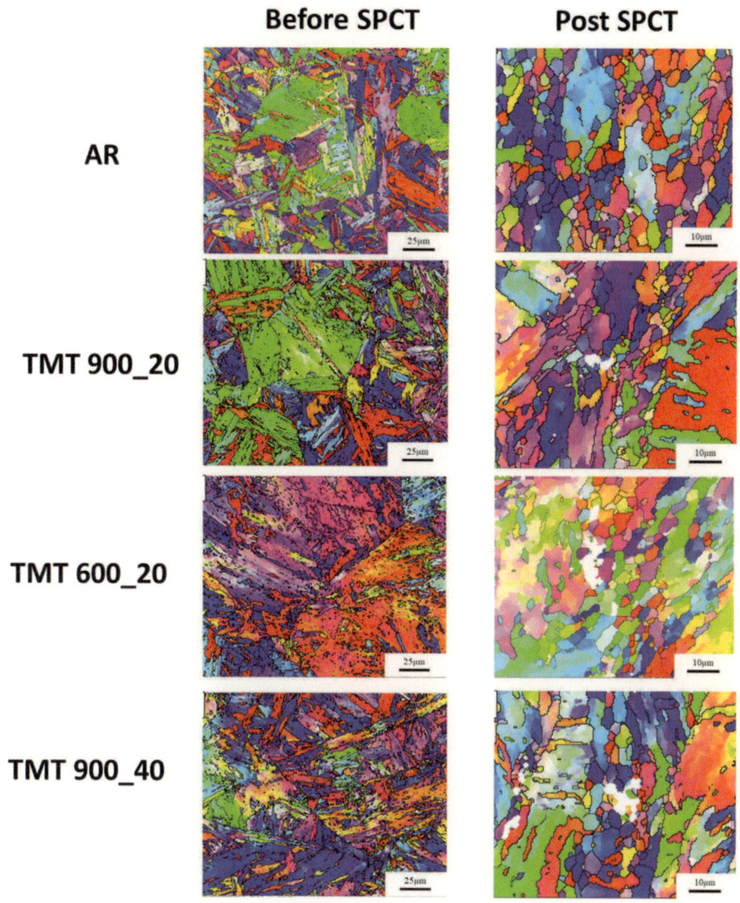

Figure 12.
Representative inverse pole figure (IPF) maps of the initial and after SPCT microstructures: (a) G91-AR; (b) G91-TMT 900_20; (c) G91-TMT 600_20; and (d) G91-TMT 900_40. Cavities are in white in post-SPCT microstructures [39].

5. Conclusions

Effect of austenitization temperature: compared to the conventional heat treatments, the use of a higher austenitization temperature (1225°C rather than 1040°C), combined with an ausforming processing step at 900°C, allows the increase of the number density of MX precipitates up to three orders of magnitude after the tempering step, which raises the strengthening capability of the MX at 700°C up to 6.5 times. These microstructures have reduced considerably the minimum disk deflection rate and showed greater time to rupture during the SPCT carried out at 700°C. By contrast, such elevated austenitization temperature induces an important drop in ductility.

Effect of ausforming: the SPCT was applied to evaluate the creep behavior of G91 steel after different TMT and heat treatments. The minimum disk deflection rate was lower, and the time to rupture was longer for G91 after the TMT than with the conventional G91 heat treatment (AR). The improvement in creep rupture strength is attributed to the fine and homogeneous distribution of MX carbonitrides. The number density and average precipitate size of MX carbonitrides after the TMT are similar to the oxide particles in ODS steels. These latter steels possess high creep strength due to the high number density of oxides distributed in the matrix. Considering the MX carbonitrides as a substitute for oxides, 9Cr FM steels after the TMT are a potential replacement of ODS steels, which are fabricated by expensive powder metallurgy and mechanical alloying processing routes.

Creep failure: based on the results presented above and taking into account the different stages of the TMT, the loss of creep ductility that enhances the change in fracture mechanism would be promoted by the coarsening of $M_{23}C_6$ carbides at the vicinity of the prior austenite grains. The coarse $M_{23}C_6$ carbides located on prior austenite grain boundaries favor the nucleation of the cavities at the vicinity of the prior austenite grains. Besides, in the TMT samples, the high austenitization temperature produces an enormous prior austenite grain sizes with concomitant large grain boundary surfaces, facilitating an earlier formation of the critical crack length that causes the brittle fracture.

Acknowledgements

We acknowledge support for the publication fee by the CSIC Open Access Publication Support Initiative through its Unit of Information Resources for Research (URICI).

Author details

Javier Vivas, David San-Martin, Francisca G. Caballero and Carlos Capdevila*
National Center for Research in Metals (CENIM), Spanish Council for Scientific Research (CSIC), Madrid, Spain

*Address all correspondence to: ccm@cenim.csic.es

References

[1] Klueh RL, Gelles DS, Jitsukawa S, Kimura A, Odette GR, van der Schaaf B, et al. Ferritic/martensitic steels—Overview of recent results. Journal of Nuclear Materials. Part 1(0). 2002;**307-311**:455-465

[2] Zinkle SJ, Was GS. Materials challenges in nuclear energy. Acta Materialia. 2013;**61**(3):735-758

[3] Klueh RL, Nelson AT. Ferritic/martensitic steels for next-generation reactors. Journal of Nuclear Materials. 2007;**371**(1-3):37-52

[4] Hald J. Microstructure and long-term creep properties of 9-12% Cr steels. International Journal of Pressure Vessels and Piping. 2008;**85**(1-2):30-37

[5] Pasternak J, Dobrzański J. Properties of welded joints on superheater coils made from new generation high alloy martensitic steels connected to austenitic creep-resisting steels and supper alloy grades, for supercritical parameters. Advanced Materials Research. 2011:466-471

[6] Skelton RP, Gandy D. Creep-fatigue damage accumulation and interaction diagram based on metallographic interpretation of mechanisms. Materials at High Temperatures. 2008;**25**(1):27-54

[7] Klueh RL, Harries DR. High-Chromium Ferritic and Martensitic Steels for Nuclear Applications. West Conshohocken: American Society for Testing and Materials; 2001. p. 228

[8] Abe F. Bainitic and martensitic creep-resistant steels. Current Opinion in Solid State and Materials Science. 2004;**8**(3-4):305-311

[9] Klueh RL, Ehrlich K, Abe F. Ferritic/martensitic steels: promises and problems. Journal of Nuclear Materials. Part A(0). 1992;**191-194**:116-124

[10] Masuyama F. History of power plants and Progress in heat resistant steels. ISIJ International. 2001;**41**(6):612-625

[11] Mayer KH, Masuyama F. The Development of Creep-Resistant Steels. Woodhead Publishing; 2008. pp. 15-77

[12] Abe F. Coarsening behavior of lath and its effect on creep rates in tempered martensitic 9Cr–W steels. Materials Science and Engineering A. 2004;**387**:565-569

[13] Fujita T. Heat-resistant steels for advanced power plants. Advanced Materials & Processes. 1992;**141**(4):42-45

[14] Vaillant JC, Vandenberghe B, Hahn B, Heuser H, Jochum C. T/P23, 24, 911 and 92: New grades for advanced coal-fired power plants-properties and experience. International Journal of Pressure Vessels and Piping. 2008;**85**(1-2):38-46

[15] Ennis PJ, Czyrska-Filemonowicz A. Recent advances in creep-resistant steels for power plant applications. Sadhana. 2003;**28**(3-4):709-730

[16] Wasilkowska A, Bartsch M, Messerschmidt U, Herzog R, Czyrska-Filemonowicz A. Creep mechanisms of ferritic oxide dispersion strengthened alloys. Journal of Materials Processing Technology. 2003;**133**(1-2):218-224

[17] Abe F. Effect of boron on creep deformation behavior and microstructure evolution in 9% Cr steel at 650°C. International Journal of Materials Research. 2008;**99**(4):387-394

[18] Eggeler G, Dlouhy A. Boron segregation and creep in ultra-fine grained tempered martensite ferritic steels. Zeitschrift für Metallkunde. 2005;**96**(7):743-748

[19] Abe F, Kern TU, Viswanathan R, editors. Creep-Resistant Steels. Materials. Cambridge: Woodhead Publishing Limited and CRC Press; 2008

[20] Reed-hill R, Abbachian R. Physical Metallurgy Principles. Boston: PWS Kent; 1992

[21] Norton FH. Creep of Steel at High-Temperatures. New York: Mc-Graw Hill; 1929

[22] Langdon TG, Mohamed FA. A simple method of constructing an Ashby-type deformation mechanism map. Journal of Materials Science. 1978;**13**(6):1282-1290

[23] Zhang XZ, Wu XJ, Liu R, Liu J, Yao MX. Deformation-mechanism-based modeling of creep behavior of modified 9Cr-1Mo steel. Materials Science and Engineering A. 2017;**689**:345-352

[24] Abe F. Precipitate design for creep strengthening of 9% Cr tempered martensitic steel for ultra-supercritical power plants. Science and Technology of Advanced Materials. 2008;**9**(1):15

[25] Abe F, Horiuchi T, Taneike M, Sawada K. Stabilization of martensitic microstructure in advanced 9Cr steel during creep at high temperature. Materials Science and Engineering A. 2004;**378**(1-2):299-303

[26] Prat O, García J, Rojas D, Sanhueza JP, Camurri C. Study of nucleation, growth and coarsening of precipitates in a novel 9%Cr heat resistant steel: Experimental and modeling. Materials Chemistry and Physics. 2014;**143**(2):754-764

[27] Abe F, Nakazawa S, Araki H, Noda T. The role of microstructural instability on creep behavior of a martensitic 9Cr-2W steel. Metallurgical Transactions A. 1992;**23**(2):469-477

[28] Maddi L, Barbadikar D, Sahare M, Ballal AR, Peshwe DR, Paretkar RK, et al. Microstructure evolution during short term creep of 9Cr–0.5Mo–1.8W steel. Transactions of the Indian Institute of Metals. 2015;**68**(2):259-266

[29] Ghassemi-Armaki H, Chen RP, Maruyama K, Igarashi M. Contribution of recovery mechanisms of microstructure during long-term creep of Gr.91 steels. Journal of Nuclear Materials. 2013;**433**(1-3):23-29

[30] Klueh RL, Hashimoto N, Maziasz PJ. Development of new nano-particle-strengthened martensitic steels. Scripta Materialia. 2005;**53**(3):275-280

[31] Hollner S, Fournier B, Le Pendu J, Cozzika T, Tournié I, Brachet JC, et al. High-temperature mechanical properties improvement on modified 9Cr–1Mo martensitic steel through thermomechanical treatments. Journal of Nuclear Materials. 2010;**405**(2):101-108

[32] Tan L, Busby JT, Maziasz PJ, Yamamoto Y. Effect of thermomechanical treatment on 9Cr ferritic–martensitic steels. Journal of Nuclear Materials. 2013;**441**(1-3):713-717

[33] Li S, Eliniyaz Z, Sun F, Shen Y, Zhang L, Shan A. Effect of thermo-mechanical treatment on microstructure and mechanical properties of P92 heat resistant steel. Materials Science and Engineering A. 2013;**559**(0):882-88

[34] Tan L, Yang Y, Busby JT. Effects of alloying elements and thermomechanical treatment on 9Cr reduced activation ferritic–martensitic (RAFM) steels. Journal of Nuclear Materials. 2013;**442** (1-3, Supplement 1):S13-SS7

[35] Klueh RL, Hashimoto N, Maziasz PJ. Newnano-particle-strengthenedferritic/ martensitic steels by conventional

thermo-mechanical treatment. Journal of Nuclear Materials. Part A(0). 2007;**367-370**:48-53

[36] Song M, Sun C, Fan Z, Chen Y, Zhu R, Yu KY, et al. A roadmap for tailoring the strength and ductility of ferritic/martensitic T91 steel via thermo-mechanical treatment. Acta Materialia. 2016;**112**:361-377

[37] Vivas J, Celada-Casero C, San Martín D, Serrano M, Urones-Garrote E, Adeva P, et al. Nano-precipitation strengthened G91 by thermo-mechanical treatment optimization. Metallurgical and Materials Transactions A. 2016:1-8

[38] Vivas J, Capdevila C, Jimenez J, Benito-Alfonso M, San-Martin D. Effect of ausforming temperature on the microstructure of G91 steel. Metals. 2017;**7**(7):236

[39] Vivas J, Capdevila C, Altstadt E, Houska M, Sabirov I, San-Martín D. Microstructural degradation and creep fracture behavior of conventionally and thermomechanically treated 9% chromium heat resistant steel. Metals and Materials International. 2019;**25**(2):343-352

[40] Vivas J, Capdevila C, Altstadt E, Houska M, San-Martín D. Importance of austenitization temperature and ausforming on creep strength in 9Cr ferritic/martensitic steel. Scripta Materialia. 2018;**153**:14-18

[41] Vivas J, Capdevila C, Altstadt E, Houska M, Serrano M, De-Castro D, et al. Effect of ausforming temperature on creep strength of G91 investigated by means of small punch creep tests. Materials Science and Engineering A. 2018;**728**:259-265

[42] Vivas J, Rementeria R, Serrano M, Altstadt E, San Martín D, Capdevila C. Effect of ausforming on creep strength of G91 heat-resistant steel. In: Jeandin M, Ionescu M, Richard C, Shabadi R, Chandra T, editors. 10th International Conference on Processing and Manufacturing of Advanced Materials, 2018. Zurich: Trans Tech Publications Ltd; 2018. pp. 400-406

[43] Capdevila C, Aranda MM, Rementeria R, Chao J, Urones-Garrote E, Aldazabal J, et al. Strengthening by intermetallic nanoprecipitation in Fe-Cr-Al-Ti alloy. Acta Materialia. 2016;**107**:27-37

[44] Capdevila C, Chen YL, Jones AR, Bhadeshia H. Grain boundary mobility in Fe-base oxide dispersion strengthened PM2000 alloy. ISIJ International. 2003;**43**(5):777-783

[45] Capdevila C, Miller MK, Chao J. Phase separation kinetics in a Fe–Cr–Al alloy. Acta Materialia. 2012;**60**(12):4673-4684

[46] Capdevila C, Miller U, Jelenak H, Bhadeshia H. Strain heterogeneity and the production of coarse grains in mechanically alloyed iron-based PM2000 alloy. Materials Science & Engineering, A: Structural Materials: Properties, Microstructure and Processing. 2001;**316**(1-2):161-165

[47] Capdevila C, Bhadeshia H. Manufacturing and microstructural evolution of mechanically alloyed oxide dispersion strengthened superalloys. Advanced Engineering Materials. 2001;**3**(9):647-656

[48] Capdevila C, Chao J, Jimenez JA, Miller MK. Effect of nanoscale precipitation on strengthening of ferritic ODS Fe-Cr-Al alloy. Materials Science and Technology (United Kingdom). 2013;**29**(10):1179-1184

[49] Capdevila C, Miller M, Russell K. Aluminum partitioning during phase separation in Fe–20%Cr–6%Al ODS alloy. Journal of Materials Science. 2008;**43**(11):3889-3893

[50] Miyamoto G, Iwata N, Takayama N, Furuhara T. Variant selection of lath martensite and bainite transformation in low carbon steel by ausforming. Journal of Alloys and Compounds. 2013;**577**(SUPPL. 1):S528-SS32

[51] Klueh RL. Elevated-Temperature Ferritic and Martensitic Steels and their Application to Future Nuclear Reactor. Knoxville: Oak Ridge National Laboratory; 2004

[52] Abe F, Taneike M, Sawada K. Alloy design of creep resistant 9Cr steel using a dispersion of nano-sized carbonitrides. International Journal of Pressure Vessels and Piping. 2007;**84**(1-2):3-12

[53] Abe F. Creep rates and strengthening mechanisms in tungsten-strengthened 9Cr steels. Materials Science and Engineering A. 2001;**319-321**:770-773

[54] Taneike M, Abe F, Sawada K. Creep-strengthening of steel at high temperatures using nano-sized carbonitride dispersions. Nature. 2003;**424**(6946):294-296

[55] Chiba T, Miyamoto G, Furuhara T. Variant selection of lenticular martensite by ausforming. Scripta Materialia. 2012;**67**(4):324-327

[56] Miyamoto G, Iwata N, Takayama N, Furuhara T. Mapping the parent austenite orientation reconstructed from the orientation of martensite by EBSD and its application to ausformed martensite. Acta Materialia. 2010;**58**(19):6393-6403

[57] Tamura I, Tsuzaki K, Maki T. Morphology of lath martensite formed from deformed austenite in 18% Ni maraging steel. Journal de Physique (Paris), Colloque. 1982;**321**:288-293

[58] Brookfield DJ, Li W, Rodgers B, Mottershead JE, Hellen TK, Jarvis J, et al. Material properties from small specimens using the punch and bulge test. The Journal of Strain Analysis for Engineering Design. 1999;**34**(6):423-436

[59] Kennett SC, editor. Strengthening and Toughening Mechanisms in Low-C Microalloyed Martensitic Steel as Influenced by Austenite Conditioning. Golden, CO, USA: Colorado School of Mines; 2014

[60] Mitsuhara M, Yamasaki S, Miake M, Nakashima H, Nishida M, Kusumoto J, et al. Creep strengthening by lath boundaries in 9Cr ferritic heat-resistant steel. Philosophical Magazine Letters. 2016;**96**(2):76-83

[61] Tan L, Snead LL, Katoh Y. Development of new generation reduced activation ferritic-martensitic steels for advanced fusion reactors. Journal of Nuclear Materials. 2016;**478**:42-49

[62] Miyamoto G, Iwata N, Takayama N, Furuhara T. Quantitative analysis of variant selection in ausformed lath martensite. Acta Materialia. 2012;**60**(3):1139-1148

[63] Wakasa K, Wayman CM. The morphology and crystallography of ferrous lath martensite. Studies of Fe-20%Ni-5%Mn—I. Optical microscopy. Acta Metallurgica. 1981;**29**(6):973-990

[64] Takahashi J, Kawakami K, J-i H, Kimura K. Direct observation of niobium segregation to dislocations in steel. Acta Materialia. 2016;**107**(Supplement C):415-422

[65] Bhadeshia HKDH. Bainite in Steels: Theory and Practice. London: Maney Publishing; 2015

[66] Toualbi L, Cayron C, Olier P, Malaplate J, Praud M, Mathon MH, et al. Assessment of a new fabrication route for Fe–9Cr–1W ODS cladding

tubes. Journal of Nuclear Materials. 2012;**428**(1):47-53

[67] Heintze C, Bergner F, Ulbricht A, Hernández-Mayoral M, Keiderling U, Lindau R, et al. Microstructure of oxide dispersion strengthened Eurofer and iron–chromium alloys investigated by means of small-angle neutron scattering and transmission electron microscopy. Journal of Nuclear Materials. 2011;**416**(1):35-39

[68] Anderson TL, editor. Fracture Mechanics : Fundamentals and Applications. 3rd ed. Boca Raton, FL: Taylor & Francis/CRC Press; 2005

[69] Plesiutschnig E, Beal C, Paul S, Zeiler G, Sommitsch C. Optimised microstructure for increased creep rupture strength of MarBN steels. Materials at High Temperatures. 2015;**32**(3):318-322

Chapter 9

Strengthening of High-Alloy Steel through Innovative Heat Treatment Routes

Nicky Kisku

Abstract

Heat treatment route is an important route for the development of high-strength alloy steel. Many heat treatment processes are applied depending on alloy compositions and desired mechanical properties. There are various high-strength alloy steels, namely, austenitic stainless steel (16–26 wt%Cr, 0.07–0.15 wt%C, 8–10 wt%Ni, rest Fe), where the heat treatment adopted is the low-temperature plasma nitriding so as to achieve a strength in a range of 800–1000 MPa. In twinning-induced plasticity (TWIP) steel (>20 wt%Mn, <1 wt%C, <3 wt%Si, <3 wt%Al, rest Fe), high-temperature thermomechanical heat treatment provides a strength greater than 1000 MPa. High-speed steel (18 wt%W, 4 wt%Cr, 1 wt%V, 0.7 wt%C, 5–8 wt%Co, rest Fe) suits best for high-speed machining purpose, owing to secondary hardening. Besides, high-temperature annealing is performed with majorly ferritic structure to achieve a maximum bending strength of 4700 MPa. Furthermore, in Hadfield steel (11–14 wt%Mn, 1–1.4 wt%C), a fully austenitic phase is obtained with a strength level of 1000 MPa. High-alloy tool steel (5 wt%Mo, 6 wt%W, 4 wt%Cr, 0.3 wt%Si, 1 wt%V, rest Fe) is provided with austenitizing, quenching, and tempering treatment to achieve a maximum hardness of 1200–1400 HV.

Keywords: high alloy steel, heat treatment, strengthening mechanism, mechanical properties

1. Introduction

Steel, because of its numerous applications, is the most important material among any engineering materials. It is mostly used in tools, automobiles, buildings, infrastructure, machines, ships, trains, appliances, etc., due to its low cost and high tensile strength. Primarily, steel is an alloy of iron and carbon, along with some other elements. The prime material of steel is iron. Iron is commonly found in the Earth's crust in the form of ore, generally an iron oxide, i.e., magnetite or hematite. The extraction of iron from iron ore is done by removing oxygen and then reacting it with carbon to form carbon dioxide. This process is called smelting. Iron has the ability to have two crystalline forms, i.e., face-centered cubic (FCC) and body-centered cubic (BCC), depending on the operation temperature. Fe-C mixture is also added with other elements to produce steel with enhanced properties. Manganese and nickel (Ni) in steel are added to increase its tensile strength and promote stable

IntechOpen

austenite phase in Fe-C solution, chromium (Cr) increases hardness and melting temperature, and titanium (Ti), vanadium (V), and niobium (Nb) also increase the hardness. There are two types of steel depending on the alloying elements. If the alloying elements are above 10%, it is referred to as high-alloy steel, and in case of alloying element with 5–10%, it is referred to as medium-alloy steel. If the alloying element in the steel is below 5%, it is called low-alloy steel. The density of steel varies from 7.1 to 8.05 g/cm^3 according to the alloying constituents.

When 0.8% of carbon-contained steels (identified as a eutectoid steel) are cooled, austenitic phase (FCC) of the combination tries to revert to the ferrite phase (BCC). The carbon is no longer contained in the FCC austenite structure, which causes excess of carbon. The alternative method to remove carbon from austenite is the precipitation of the solution like cementite and parting behind a neighboring phase of ferrite BCC iron with small quantity of carbon. A layered structure called pearlite is produced when the two, ferrite and cementite, precipitate at the same time. In case of hypereutectoid composition (>0.8% carbon), the carbon will predominantly precipitate out in the form of large inclusions of cementite on the austenite grain boundaries until the amount of carbon in the grains has reduced to the eutectoid composition (0.8% carbon), at which stage the pearlite formation takes place. For steels that have less than 0.8% carbon (called hypoeutectoid), it results in ferrite formation initially in the grains unless the residual content reaches 0.8%, at which stage pearlite formation takes place. No bulky cementite inclusion occurs in the boundaries in hypoeutectoid steel. The cooling process is assumed to be very slow due to the above reasons, hence letting adequate time for the transmission of carbon. Increased rate of cooling does not allow the carbon to migrate for the formation of carbide in the grain boundaries. Rather it will form large amount of finer structure pearlite; hence the carbide is further extensively dispersed and performs to prevent slip of defects inside those grains, ensuing in hardening of the steel. At very high rate of cooling, the carbon has no time to transfer; as a result it is confined inside the austenite and transforms to martensite. The martensite phase is the supersaturated type of carbon, the most strained as well as stressed phase which is exceptionally hard although brittle. Considering the carbon content, the martensite phase obtains various forms. Carbon below 0.2% obtains a ferrite (BCC) form, whereas at higher level of carbon, it acquires a body-centered tetragonal (BCT) structure. Thermal activation energy is not acquired for the conversion from austenite into martensite.

Martensite has a lesser density (as it expands at the time of cooling) than austenite does. As a result the conversion among them consequences a variation in amount. During the above process, growth occurs. Internal stresses as of this growth usually acquire the compressed crystal form of martensite and elongated form on the left over ferrite, along with a significant quantity of shear on the constituents. When quenching is not appropriately done, it can cause crack on cooling due to the internal stresses in a part. They cause interior work hardening and other microscopic imperfections. It is ordinary for quench cracks to appear when steel is water quenched, even though they may not always be visible.

2. Role of major alloying element in steel

2.1 Carbon

The carbon steels are composed of carbon and iron by means of carbon up to 2.1 wt%. At the same time, when the carbon content increases, steel has the capability to become harder as well as stronger by heat treating, though it undergoes less ductility. In spite of heat treatment, a higher carbon content also decreases weldability. In carbon steels, the higher carbon content lowers the melting point.

The classifications of carbon steel are on the basis of carbon content:

Low-carbon steel: carbon wt% is in the range of 0.05–0.30 (called plain carbon steel) [1].

Medium-carbon steel: 0.3–0.6% is the approximate carbon content [1]. It helps in balancing ductility and strength and also has superior wear resistance; it is used in automobiles [2, 3].

High-carbon steel: carbon content lies from 0.60 to 1.00% [1]. It has very high strength and is used for tools, edged tools, springs, and wires [4].

Ultrahigh-carbon steel: it has carbon% between 1.25 and 2.0 [1]. It can be tempered to immense hardness. It is used in various purposes like axles, punches, or knives.

2.2 Detailed study of low- and high-carbon steel

2.2.1 Mild- or low-carbon steel

Mild steel, well known as plain carbon, is at present the common variety of steel as it is cost-effective and offers material properties for a lot of applications. It contains carbon wt% in the range of 0.05–0.30, building it more malleable and ductile. It has comparatively low tensile strength, other than being contemptible and simple to produce; surface hardness can be improved by carburizing. Due to its ductile nature, the failure from yielding is less risky, so it is best applicable (e.g., structural steel). The density of mild or low steel is ~7.85 g/cm^3 [5] and Young's modulus is ~200 GPa [6]. Low-carbon steels include a smaller amount of carbon than other steels and are easy to handle as it is more deformable.

2.2.2 Higher-carbon steels

Carbon steels that successfully experience heat treatment contain carbon in between 0.30 and 1.70 wt%. The impurities of different elements also have a considerable consequence on the superiority of the ensuing steel. Small amount of sulfur content makes steel brittle and crumble on operational temperatures. Manganese is added to enhance the hardenability of the steels. The name "carbon steel" can be employed in terms of the steel that is not stainless steel; in addition to it, carbon steel can be involved in alloy steels. Current modern steels are prepared with various mixtures of alloying elements to execute in various applications. The steel is alloyed along with additional elements, typically manganese, molybdenum (Mo), nickel, or chromium up to 10 wt%, in order to develop the hardenability. High-strength low-alloy steel has small additions (<2 wt%) of added elements, usually 1.5 wt% manganese, to offer extra strength.

3. Alloying elements and their effects on steels

Alloy steel reflects a category of steel facilitated with the addition of different elements. In general, all steels are referred to as alloy steel, while the plain steel is composed of iron added up to 2.06 wt% carbon. However, the term "alloy steel" commonly refers to steels that are alloyed with elements other than carbon. The total wt% of the alloying elements can be up to 20% to provide the material enhanced properties like better wear resistance, strength, or ductility. Low-alloyed steels are distinguished by their lower content of alloys with total content below 5%, whereas in the case of high-alloyed steel, the total sum of elements can be in the range of 5–20%, with improved properties. Apart from the above alloyed steels, there are even unalloyed steels that carry very small quantity of alloys. High-alloyed steel contributes to high strength, toughness, hardness, and creep resistance at specific heat treatment temperature. It also advances machinability and corrosion resistance. In addition, it even strengthens the properties of other alloying elements.

3.1 Austenite-stabilizing alloying element

The accumulation of certain alloying elements, such as manganese and nickel, can stabilize the austenitic structure, facilitating heat treatment of low-alloy steels. In the extreme case of austenitic stainless steel, much higher alloy content makes this structure stable even at room temperature. On the other hand, such elements as silicon, molybdenum, and chromium tend to destabilize austenite, raising the eutectoid temperature.

Austenite is only stable above 910°C (1670°F) in bulk metal form. However, FCC transition metals can be grown on a face-centered cubic or diamond cubic [7]. The epitaxial growth of austenite on the diamond (100) face is feasible because of the close lattice match, and the symmetry of the diamond (100) face is FCC. More than a monolayer of γ-iron can be grown because the critical thickness for the strained multilayer is greater than a monolayer [7]. The determined critical thickness is in close agreement with theoretical prediction.

As the names suggest, austenite stabilizers are elements, which make austenite (of iron) stable at lower temperature, that would occur in pure iron. With enough amount of austenite stabilizer, you can have austenite stable at room temperature. Effectively, they decrease the austenitizing temperature of iron, in the Fe-C diagram.

Examples: Mn, Ni, C etc.

Manganese: in alloy steel, manganese is typically used in combination with sulfur and phosphorus. Manganese helps reduce brittleness and improves forgeability, tensile strength, and resistance to wear. Manganese reacts with sulfur, resulting in manganese sulfides which prevent the formation of iron sulfides. Manganese is also added for better hardenability as it leads to slower quenching rates in hardening techniques. Excess oxygen can be removed in molten steel by using manganese.

Nickel: austenitic stainless steels are most known for their high content in nickel and chromium. It is used to increase strength, hardness, impact toughness, and corrosion resistance. Nickel-alloyed steels are often found in combination with chromium, resulting in an even higher hardness.

3.2 Ferrite-stabilizing alloying element

By decreasing eutectoid composition and increasing eutectoid temperature, ferrite stabilizers are the elements which stabilize ferrite phase. Cr and Si are examples for ferrite stabilizers. Ferrite stabilizers are also called carbide former element. **Stabilizing ferrite** decreases the temperature range, in which austenite exists.

The elements, with the same crystal structure as that of ferrite (body-centered cubic—BCC), increase the A_3 temperature and lower the A_4 point. An increase in the amount of carbides in the steel is caused by decreasing the solubility of carbon in austenite by these elements. The following elements have ferrite-stabilizing effect: chromium, tungsten (W), aluminum (Al), molybdenum, silicon, and vanadium. Examples of ferritic steels are transformer sheet steel (3% Si) and F-Cr alloys.

Chromium: chromium is one of the most common alloying metals for steel because of its high hardness and corrosion resistance. Pure chromium is a gray, brittle, and hard metal with a melting point of 1907°C (3465°F) and a high-temperature resistance. In steel, hardenability is increased by the alloying chromium. Higher chromium contents up to 18% result in enhanced corrosion resistance. For example, stainless steel, which is one of the most popular steel alloys, uses at least 10.5% chromium, enhancing its resistance against water, heat, or corrosion damage. Chromium oxide does not spread and fall away from the material in contrast to iron oxide in unprotected carbon steel. It creates a film of dense chromium oxide on the surface that blocks out any further corrosion attacks.

Molybdenum: it is a silvery-white metal that is ductile and highly resistant to corrosion. It has one of the highest melting points of all pure elements—together with the elements tantalum (Ta) and tungsten. **Molybdenum** is also a micronutrient essential for life.

3.3 Carbide-forming alloying elements

Carbide-forming elements form hard carbides in steels. Steel hardness and strength are increased by hard (often complex) carbides formed by the elements like tungsten, niobium, molybdenum, chromium, vanadium, titanium, zirconium (Zr), and tantalum. Examples of steels containing relatively high concentration of carbides are high-speed steel and shot work tool steels. During reaction with nitrogen in steel, carbide-forming elements also form nitrides.

Tungsten is a rare metal found naturally on the Earth almost exclusively combined with other elements in chemical compounds rather than alone. It was identified as a new element in 1781 and first isolated as a metal in 1783. Its important ores include wolframite and scheelite.

The free element is remarkable for its robustness, especially the fact that it has the highest melting point of all the elements discovered, at 3422°C (6192°F, 3695 K). It also has the highest boiling point, at 5930°C (10,706°F, 6203 K). Its density is 19.25 times that of water, comparable to that of uranium and gold, and much higher (about 1.7 times) than that of lead. Polycrystalline tungsten is an intrinsically brittle and hard material (under standard conditions, when uncombined), making it difficult to work. However, pure single-crystalline tungsten is more ductile and can be cut with a hard steel.

4. Evolution of high-alloy steel

Alloy steel is added with a choice of elements in total amounts between 10 and 50 wt% to expand its mechanical properties. Alloyed steels are categorized into two groups: low- and high-alloy steels. The simplest form of steel is iron with carbon alloy (~0.1–1%). Common alloying elements comprise manganese (the most frequent one), chromium, nickel, molybdenum, silicon, aluminum, vanadium, titanium, niobium, and boron (B). Alloyed steels have improved properties such as strength, hardenability, toughness, hardness, wear resistance, corrosion resistance, and hot hardness [8]. To achieve these better-quality properties, the metal may require various heat treatment processes. Several of these are utilized in highly requiring applications, like in the turbine blades used in jet engines, in nuclear reactor, in spacecraft, etc. Iron, owing to its ferromagnetic nature, discovers major applications wherever the response to magnetism is important, like in transformers and electric motors.

4.1 Categorization of alloy steel and their heat treatments

Alloy steels are categorized into low- and high-alloy steels. High-alloy steels would be more than 10 wt% of alloying elements in steel groups [1, 5, 8, 9]. The majority of alloy steels lie under the group of low alloy. The most common alloy elements include chromium, manganese, nickel, molybdenum, vanadium, tungsten, cobalt, boron, and copper.

4.1.1 Low-alloy steel

Low-alloy steels are a group of ferrous materials that show improved mechanical properties compared to plain carbon steels, because of the alloying elements such as nickel, molybdenum and chromium. Through the development of specific alloys, low-alloy steel provides desired mechanical properties. Microstructure consists of ferrite and pearlite. Its properties are relatively soft and weak, although they have high ductility and toughness. Its various applications are auto-body components, structural shapes, sheets, etc. [2, 3, 5, 6, 10–12].

Some of the compositions of low-alloy steels are the following:

Cr 0.50% or 0.80% or 0.95%, Mo 0.12% or 0.20% or 0.25% or 0.30%, rest Fe
Mo 0.20% or 0.25% or 0.25% Mo or 0.042% S, rest Fe
Mo 0.40% or 0.52% C, rest Fe Ni 1.82%, Cr 0.50% to 0.80%, Mo 0.25% Cu, rest Fe Several low-alloy steels underwent normalizing and tempering in the manufacturing industries; however there is an increase affinity to a quenching and tempering action. Low-alloy steels are weldable, but pre-welding or post-welding heat treatment is essential to evade weld zone cracking issues.

4.1.2 High-alloy steels

In high-alloy steel, the entire alloying element content is above 10 wt%. In stainless steels, the principally alloying element is Cr (≥11 wt%). It is greatly resistant to corrosion. Nickel and molybdenum addition adds to corrosion resistance. An important property of the highly alloyed steel is the capability of alloying elements to promote the creation of a certain multiple phases and stabilize it. These elements are grouped into four major classes as discussed in the previous section: (1) austenite-forming, (2) ferrite-forming, and (3) carbide-forming.

Some varieties of the high-alloy steels are the following:

a. Stainless steels: Fe-18Cr-8Ni-1Mn-0.1C characteristically is γ-alloy. It stabilizes austenite for its rising temperature range, where austenite subsists. It elevates the austenite-forming temperature (A1) and reduces the A3 temperature. Mostly, this type of steels underwent solution annealing type of heat treatment primarily specified for austenitic stainless steels. The main requirement for this treatment is to dissolve all the precipitated phases, mainly chromium-rich carbides, where the precipitate of $M_{23}C_6$ occurs in the range of 673–1173 K. For other stainless steels, it is recommended to maintain the solution annealing temperature in the range of 1273–1393 K.

b. Tool steel: it provides necessary hardness with simpler heat treatment and retains hardness at high temperature. The primary alloying elements are Mo,

W, and Cr. These elements have wear resistance, high strength, and toughness but have low ductility. One of the primary heat treatments provided for tool steel is tempering that requires cautious preparation. Various complex tool steels like the high-speed steel need twice over tempering to convert austenite to martensite completely. High-speed steel (18 wt%W, 4 wt%Cr, 1 wt%V, 0.7 wt%C, 5–8 wt%Co, rest Fe) suits best for high-speed machining purpose, owing to secondary hardening. Besides, high-temperature annealing is performed with majorly ferritic structure to achieve a maximum bending strength of 4700 MPa. These types of steels achieve utmost hardness after first tempering, which is followed by second tempering that lowers the hardness to the desired working level. In some cases, the third temper is needed for secondary hardening of steels to make sure that some new martensite produced as a consequence of austenite conversion in tempering is efficiently tempered. This is a subject of individual selection and includes minimum extra cost.

c. High-entropy alloy steel: the essential elements of the high-entropy steels are Fe, Co, Ni, Cr, Cu, and Al. The cast microstructure expands from FCC to BCC phase along with the increase in Al content. The hardness in BCC phase is greater than FCC phase; in addition to it, the corrosion resistance is also superior in BCC phase. Some of the high-entropy alloy steels like Al-Fe-Cr-Co-Ni-Ti alloy coating was equipped by laser cladding, and the effects of annealing temperature (873, 1073, and 1473 K) on structure and its properties were studied. The consequences illustrate that the intermetallic precipitation compounds in the coating are efficiently repressed through laser cladding by means of fast solidification, and the microstructure of the coating forms dendrite structure of BCC, having superior hardness (~698 HV). As a result, the grain size of the coating rises somewhat, and the microhardness reduces slightly, following various annealing temperatures at a range of 1073–1373 K. This specifies that the elevated temperature stability of the structure and microhardness of the coating are superior. Al and Fe are improved in dendritic boundary, while Co, Ni, Ti, and Cr are enhanced in interdendritic boundary. In addition, the degree of segregation rises with the enhancement of annealing temperature.

d. Twinning-induced plasticity (TWIP) steel: in TWIP steel (>20 wt%Mn, <1 wt%C, <3 wt%Si, <3 wt%Al, rest Fe) high-temperature thermomechanical heat treatment provides a strength greater than1000 MPa. The examination of the solution heat treatment of hot-rolled TWIP steel of the three various compositions (Fe-30Mn-3Si3Al, Fe-25Mn-4Si-2Al, and Fe-30Mn-4Si-2Al) reflected that prolonging the time of holding temperature can enhance the elongation through no change observed in strength. Prolonging the holding time facilitates both the production of additional annealing twins to amplify their areas of boundary and the boost in the number of twin boundaries that are favorable for the corrosion resistance creep and fracture.

e. Hadfield steel: in Hadfield steel (11–14 wt%Mn, 1–1.4 wt%C), a fully austenitic phase is obtained with a strength level of 1000 MPa. High-alloy tool steel (5 wt%Mo, 6 wt%W, 4 wt%Cr, 0.3 wt%Si, 1 wt%V, rest Fe) is provided with austenitizing, quenching, and tempering treatment to achieve a maximum hardness of 1200–1400 HV. The heat treatment processing of Hadfield manganese steel means dissolving the carbide precipitates at higher temperature, followed by fast cooling to attain austenitic carbide-free grains which is desired to be the preferred microstructure for the commercial applications.

5. Innovative heat treatment (processing)-structure–property correlation in high-alloy steel

High-temperature homogenization, complete annealing, normalizing, tempering, etc. are the usual methods in heat treatment process of steel. But there are certain modified ways of processing routes in order to enhance the mechanical properties [13–31]. The main objective of heat treatment in steel is to upgrade the mechanical properties like strength, toughness, impact resistance, etc. It is to be noted that thermal and electrical conductivities are changed to some extent, whereas Young's modulus remains unchanged. Iron has a better solubility for carbon in the austenitic phase, so the steel is heated at which the austenite phase persists.

Some of the newly introduced high-alloyed steels like TWIP steel show excellent mechanical properties, depending on the adoption of advanced heat treatment processes. In some processes the fabricated steels are first homogenized to ~1373 K for 1 hour, followed by hot rolling at 1273 K. The steels are then cooled in the furnace and then rolled at room temperature (as shown in **Figure 1**). Due to the above heat treatment, the presence of duplex phases of austenite and ferrite is observed. The rolling effect contributes in grain size reduction and hence helps in enhancing the strength of the steel. Additionally, due to the high-temperature rolling, there is also an occurrence of twins on the austenitic grains that also increases the strength of the metal. The above modification in the microstructure resulted in the improved tensile properties with 1000 MPa ultimate tensile strength and up to 60% elongation [13].

Recently, Mazaheri et al. suggested a cold rolling, followed by various intercritical annealing techniques for the production of ultimate ultrarefined-grained steel [22]. The microstructure contains ferrite-martensite duplex steel with excellent mechanical properties. In this processing route, the fabricated steel was first heated to austenitizing temperature, i.e., 880°C for 1 hour. Then it was annealed intercritically at ~770°C for 100 minutes trailed by water quenching (as shown in **Figure 2**). The steel was water cooled to acquire the desired microstructure of ferrite and

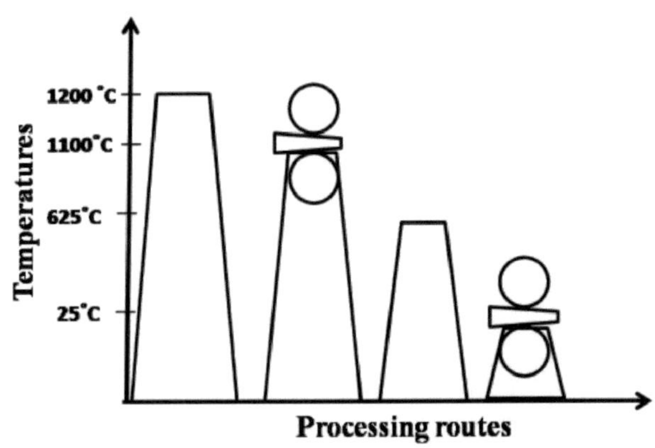

Figure 1.
Illustrating the processing routes of TWIP steel.

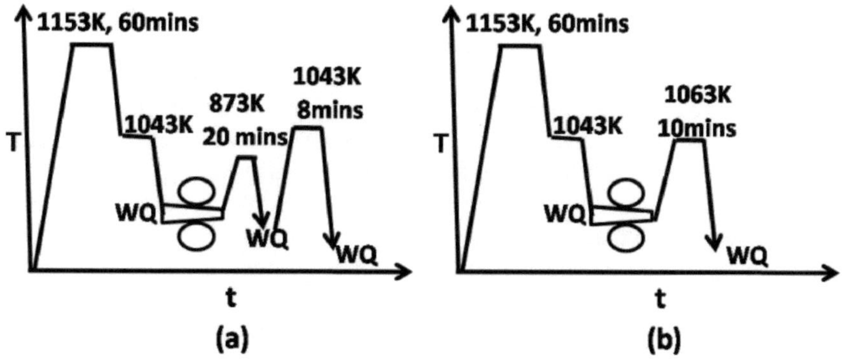

Figure 2.
Thermomechanical processing routes of dual-phase steel.

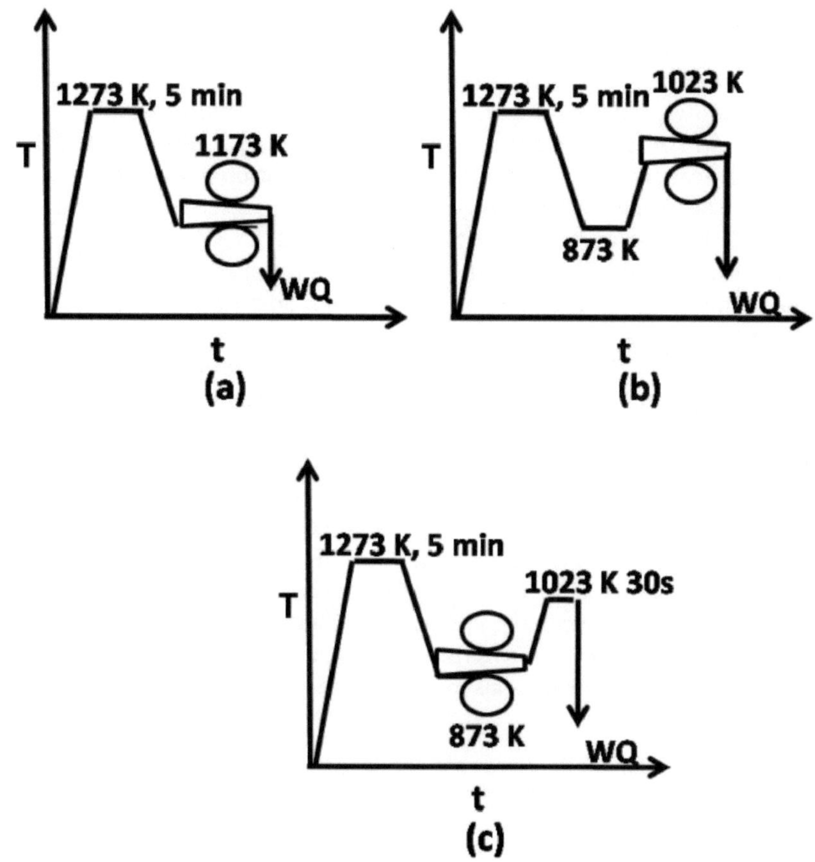

Figure 3.
Various heat treatment processes owing to different ways of thermomechanical treatments in steel.

martensite structures, and on further annealing the aimed ultrafined-grained microstructure was achieved. The achieved strength (UTS) is ~1600 MPa with 30% elongation [13].

The temperature of deformation also plays a vital role in influencing the refinement of the microstructure through hot deformation. In **Figure 3a** the martensitic phase is dominated, resulting in ultrafined grains due to dynamic recrystallization

(DRX) of ferrite grains. In the processing of steel as shown in **Figure 3b**, the martensitic content is above 30% which contributes to the strength of the steel by the varying the degree of deformation. As compared to the routes of **Figure 3a** and **b** with **Figure 3c**, the DRX is not necessary for the formation of ultrafined grains; the warm temperature deformation followed by intercritical annealing can also result in the formation of similar structure. Therefore, the warm rolling and high rate of intercritical annealing and high rate of cooling significantly affect the microstructural properties of the steel.

There are various strengthening mechanisms affecting the strength of the steel. By following specific thermomechanical treatment, the occurrence of twins enhances the strength of the steel. Twinning-induced plasticity steels are FCC crystal-structured steels. The appearance of the crystallographic twins greatly depends on the stacking fault energy (SFE), and the SFE of the steel is controlled by the rate of heating treatment. Temperature is directly proportional to SFE. Low SFE (below 20 mJ/m^2) results in the conversion of austenite to martensite (i.e., TRIP effect), whereas high SFE (above 20 mJ/m^2) gives TWIP effect (formation of twins). The dislocation generated during the deformation is obstructed by the twins and, therefore, increases the strength of the steel [32, 33].

Thus by adopting this technique, the microstructural modification takes place by the combined effect of mechanical and thermal energy. There are also iterative thermomechanical processes where percent of deformation is applied prior to heat treatment (**Table 1**). This process also contributes to the resistance of corrosion with respect to the orientation of the grain [3, 14, 21, 23].

The above heat treatments are aimed to enhance the specific properties of the high-alloyed steel to get rid of unwanted properties. Some of the microstructures evolved during processing are given in **Figure 4**.

The behavior of steel in exterior load describes its mechanical properties. Plastic deformations are supported by the movement of dislocation and the presence of twins, and precipitates hinder the motion of dislocations and thereby increase the strength of the steel. Mechanical properties are associated with the yield stress, separating the elastic and plastic regions, where the activity of dislocation extends [15–17, 30–32]. Pinning of dislocations by random obstruction is controlled by the misfit and size of the particles. In general, larger SFE promotes dislocation gliding, which enables the dislocation to move freely. On the other hand, the smaller SFE increases the area between the two partials, thereby making the motion of dislocation difficult and resulting in the piling up of dislocation. For the duration of the dislocation union, the partials must reconnect to prevail over the obstruction

Steel type	Maximum forging temperature (°C)	Burning temperature (°C)
Carbon steel	1200	1349
Nickel steel	1249	1380
Chromium steel	1200	1370
Nickel-chromium steel	1249	1370
Stainless steel	1280	1380
TWIP steel	1200	1350
High-speed steel	1280	1400

Table 1.
Various steels corresponding to different ranges of deformation temperature.

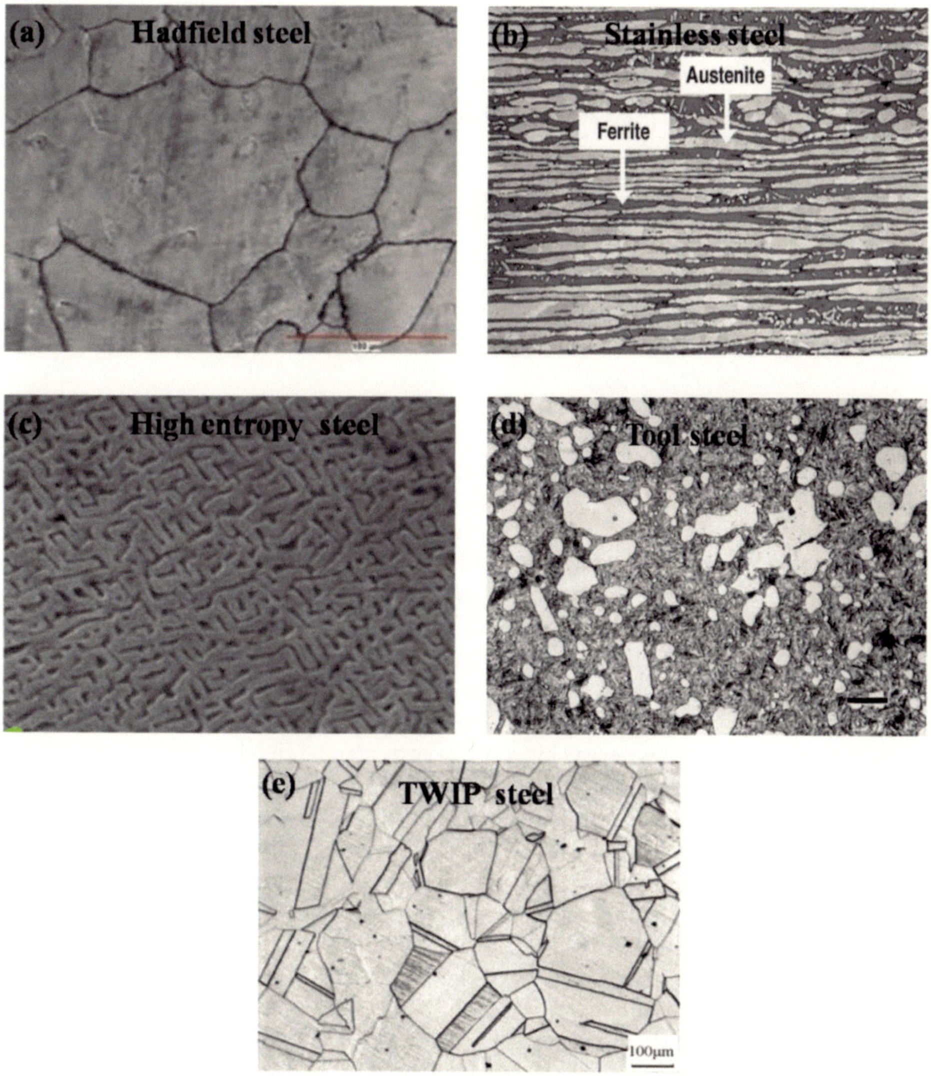

Figure 4.
Various microstructures of high-alloy steels.

[30–34]. The opposition of steel to plastic deformation reduces with rising SFE, and for this reason the SFE should be lowered to reinforce the strength. Based on the observation, SFE is regulated by alloyed elements in the steel for preferred enhanced properties like strength, hardness, or rate of work hardening.

6. Application of high-alloy steel

High-alloy steels have vast applications such as:

- Stainless steel: it has excellent corrosion properties and is used in structural applications, refrigerator, freezers, food packaging, etc.

- Tool steel: used in dies, shear blade, rollers, cutting tools, etc.

- TWIP steel: used in automobiles, ship building, infrastructure, railways, aircrafts, etc.
- High-entropy steel: used as structural material in low-temperature applications due to its high toughness.
- Hadfield steel: used in railways, structural applications, shafts, gears, housing, cables, etc.
- High-speed steel: used as cutting tool materials due to its high hardness like drilling machine, blades, etc.

7. Conclusions

High-alloyed steels are complex alloys, along with desired chemical composition and multiple phased microstructures through various heat treatment processes. Various strengthening mechanisms through controlled heat treatment techniques are adopted to achieve excellent mechanical properties. The chapter examines the advanced methods used in the field of heat treatment routes for high-alloyed steel and focuses on their structure-property relation. The high-alloy steels acquire its enhanced mechanical properties from the modified microstructures of austenite, ferrite, martensite, and some carbides. Ferrite and austenite provide the formability, whereas martensite provides strength to the steel in addition to the low-temperature transforming phases like bainite and retained austenite to achieve better combinations of mechanical properties. The advanced thermomechanical treatments used for high-alloy steels aim to explore the possible phases that contribute to the mechanical properties. In thermomechanical routes aims on heat treatment as the microstructural qualities required for the steels are mainly achieved by post-deformation controlled heat treatment processes. From the above discussions, it can be concluded that the microstructure and its properties are based on variation in chemical composition and processing conditions. Determined by latest demands for the performance of the high-alloy steel in various applications, the progress of thermomechanical processing is introduced.

8. Futuristic development of high-alloy steel

High-alloy steel has undergone significant evolution through time. Around 70% is used in various applications. These steels are highly demanding as they display various environmental, chemical, physical, and mechanical properties. Here the different proportions of alloying element in steel provide various mechanical properties. As can be seen from the foregoing, high-alloy steel plays an important role in the building and construction industries as well as in automotive industries. High-alloy steel offers economy, high performance, corrosion resistance, high strength, durability, lightweight and high performance under extreme conditions, and its wide variety of products for desirable applications.

Author details

Nicky Kisku
Indian Institute of Technology, Kharagpur, India

*Address all correspondence to: kiskunicky@gmail.com

References

[1] Degarmo EP, Black JT, Kohser RA. Materials and Processes in Manufacturing. 10th ed. United States of America: Wiley; 2007. pp. 150-155

[2] Subramanyam DK, Grube GW, Chapin JH. Austenitic Manganese Steel Castings. 9th ed. United States of America: ASM Metals Handbook; 1985. pp. 251-256

[3] Sun B, Aydin H, Fazeli F, Yue S. Microstructure evolution of a medium Mn steel during thermomechanical processing. Metallurgical and Materials Transactions A. 2016;**47**:1782-1791

[4] Grassel O, Kruger L, Frommeyer G, Meyer LW. High strength Fe-Mn- (Al, Si) steel. International Journal of Plasticity. 2000;**16**:1391-1409

[5] Jack KH. Heat treatment Process. In: Proceeding of heat treatment (Heat treatment'73); 12-13 December 2018; London, UK; 2018. pp. 39-50

[6] Sun G, Hu S, Gao Y. Influence of direct annealing heat treatment on the mechanical properties of as cast TWIP steels. Journal of Materials Engineering and Performance. 2017;**26**:1981-1985

[7] Aleshina EA, Sizova O. V: Formation of structural-phase states of the surface of Hadfield steel. Steel in Translation. 2007;**12**:989-990

[8] Smith WF, Hashemi J. Foundations of Material Science and Engineering. 4th Ed. McGraw-Hill; 2001. p. 394

[9] Groover MP. Fundamentals of modern manufacturing: Materials, processes and systems. 3rd ed. United States of America: John Wiley & Sons, Inc.; 2008. pp. 105-106

[10] Stagno E. Behaviour of sintered 410 low carbon steels towards ion nitriding. Journal of Alloys and Compounds. 2015;**24**:2122-2134

[11] Lee D, Kim J, Lee S, Lee K, De Cooman B. C: Microstructures and mechanical properties of Ti and Mo micro-alloyed medium Mn steel. Journal of Materials Engineering and Performance. 2017;**706**:1-14

[12] Marandi A, Haghdadi N, Eskandari M. The prediction of hot deformation behavior in Fe-21Mn-2.5Si-1.5Al steel. Materials Science and Engineering A. 2012;**554**:72-78

[13] Zhao J, Jiang Z. Thermomechanical processing of advanced high strength steels. Progress in Materials Science. 2018;**94**:174-242

[14] Yeh JW, Chen SK. Nanostructured high-entropy alloys with multiple principal elements: Novel alloy design concepts and outcomes. Advanced Engineering Materials. 2019;**5**:299-303

[15] Mazaheri Y, Kermanpur A, Najafizadeh A, Saeidi N. Effects of initial microstructure and thermomechanical processing parameters on microstructures and mechanical properties of ultrafine grained dual phase steels. Materials Science and Engineering A. 2017;**612**:54-62

[16] Speich GR, Demarest VA, Miller RL. Formation of austenite during intercritical annealing of dual-phase steels. Metallurgical Transactions A. 1981;**12**:1419-1428

[17] Judd RR, Paxton HW. Kinetics of austenite formation from a spheroidized ferrite-carbide aggregate. Transactions of the Metallurgical Society of the American Institute of Mechanical Engineers. 1968;**242**:206-215

[18] Azizi-Alizamini H, Militzer M, Poole WJ. Formation of ultrafine grained dual phase steels through rapid heating. ISIJ International. 2011;**51**:958-964

[19] Movahed P, Kolahgar S, Marashi SPH, Pouranvari M, Parvin N. The effect of intercritical heat treatment temperature on the tensile properties and work hardening behavior of ferrite–martensite dual phase steel sheets. Materials Science and Engineering A. 2009;**518**:1-6

[20] Sun S, Pugh M. Properties of thermomechanically processed dual-phase steels containing fibrous martensite. Materials Science and Engineering A. 2016;**335**:298-308

[21] Sodjit S, Uthaisangsuk V. Micro-structure based prediction of strain hardening behavior of dual phase steels. Materials and Design. 2012;**41**:370-379

[22] Mazaheri Y, Kermanpur A, Najafizadeh A. A novel route for development of ultrahigh strength dual phase steels. Materials Science and Engineering A. 2014;**619**:1-11

[23] Asadi M, De Cooman BC, Palkowsk H. Influence of martensite volume fraction and cooling rate on the properties of thermomechanically processed dual phase steel. Materials Science and Engineering A. 2012;**538**:42-52

[24] Mukherjee K, Hazra SS, Militzer M. Grain refinement in dual-phase steels. Metallurgical and Materials Transactions A: Physical Metallurgy and Materials Science. 2009;**40**:2145-2159

[25] Beladi H, Kelly GL, Shokouhi A, Hodgson PD. The evolution of ultrafine ferrite formation through dynamic stain-induced transformation. Materials Science and Engineering: A. 2004;**371**:343-352

[26] Hurley PJ, Hodgson PD. Formation of ultra-fine ferrite in hot rolled strip: Potential mechanisms for grain refinement. Materials Science and Engineering A. 2017;**302**:206-214

[27] Choi JK, Seo DH, Lee JS, Um KK, Choo WY. Formation of ultrafine ferrite by strain-induced dynamic transformation in plain low carbon steel. ISIJ International. 2018;**43**:746-754

[28] Hong SC, Lim SH, Lee KJ, Lee KS. Determination of dynamic ferrite transformation during deformation in austenite. In: Zhu YT, Langdon TG, Mishra RS, Semiatin SL, Saran MJ, Lowe TC, editors. Ultrafine-Grained Materials II. Warrendale, PA: TMS; 2002. pp. 267-274

[29] Beladi H, Kelly GL, Shokouhi A, Hodgson PD. Effect of thermome-chanical parameters on the critical strain for ultrafine ferrite formation through hot torsion testing. Materials Science and Engineering A. 2017;**367**: 152-161

[30] Yang Z, Wang R. Formation of ultra-fine grain structure of plain low carbon steel through deformation induced ferrite transformation. ISIJ International. 2003;**43**:761-766

[31] Adamczyk J, Grajcar A. Structure and mechanical properties of DP-type and TRIP-type sheets obtained after the thermomechanical processing. Journal of Materials Processing Technology. 2005;**162-163**:267-274

[32] Calcagnotto M, Ponge D, Raabe D. Microstructure control during fabrication of ultrafine grained dual-phase steel: Characterization and effect of intercritical annealing parameters. ISIJ International. 2012;**52**:874-883

[33] Rao MP, Sarma VS, Sankaran S. Development of high strength and ductile ultra fine grained dual

phase steel with nano sized carbide precipitates in a V-Nb microalloyed steel. Materials Science and Engineering A. 2018;**568**:171-175

[34] Karmakar A, Karani A, Patra S, Chakrabarti D. Development of bimodal ferrite-grain structures in low-carbon steel using rapid intercritical annealing. Metallurgical and Materials Transactions A: Physical Metallurgy and Materials Science. 2013;**44**:2041-2052

Chapter 10

Heat Treatment of Metastable Beta Titanium Alloys

Sudhagara Rajan Soundararajan, Jithin Vishnu, Geetha Manivasagam and Nageswara Rao Muktinutalapati

Abstract

Heat treatment of metastable beta titanium alloys involves essentially two steps—solution treatment in beta or alpha+beta phase field and aging at appropriate lower temperatures. High strength in beta titanium alloys can be developed via solution treatment followed by aging by precipitating fine alpha (α) particles in a beta (β) matrix. Volume fraction and morphology of α determine the strength whereas ductility is dependent on the β grain size. Solution treatment in (α + β) range can give rise to a better combination of mechanical properties, compared to solution treatment in the β range. However, aging at some temperatures may lead to a low/nil-ductility situation and this has to be taken into account while designing the aging step. Heating rate to aging temperature also has a significant effect on the microstructure and mechanical properties obtained after aging. In addition to α, formation of intermediate phases such as omega, beta prime during decomposition of beta phase has been a subject of detailed studies. In addition to covering these issues, the review pays special attention to heat treatment of beta titanium alloys for biomedical applications, in view of the growing interest this class of alloys have been receiving.

Keywords: beta titanium alloys, heat treatment, duplex aging, precipitation hardening, intermediate phases, fatigue behavior

1. Introduction

High specific strength and excellent corrosion resistance of titanium-based materials make them an attractive choice for application in various industries such as aerospace, biomaterials, and automotive [1, 2]. Alloying of pure titanium opens a new horizon to develop a variety of products with exceptional properties. Based on the alloying elements and phases present at room temperature, Ti alloys are broadly classified into α, α + β, and β alloys. Compared to α and α + β alloys, β alloys have advantages such as excellent higher specific strength, sufficient toughness, excellent corrosion resistance, better biocompatibility, good fatigue resistance, and good formability.

Mo_{eq} is a well-accepted measure to characterize the β-phase stability for a given composition and equation for the deriving the Mo_{eq} is shown in Eq. (1) [3, 4].

$$\begin{aligned} Mo_{eq.} = {} & 1.0(\text{wt.\% Mo}) + 0.67\,(\text{wt.\% V}) + 0.44\,(\text{wt.\% W}) + 0.28\,(\text{wt\% Nb}) \\ & + 0.22\,(\text{wt.\% Ta}) + 2.9\,(\text{wt.\% Fe}) + 1.6(\text{wt.\% Cr}) - 1.0(\text{wt.\% Al}) \end{aligned} \quad (1)$$

Beta titanium alloys with a Mo_{eq} between 10 and 30 are metastable and hence heat treatable and deeply hardenable [4]. Mo_{eq} of the various beta titanium alloys along with their commercial name is shown in **Figure 1**.

Unlike α + β alloys, in β alloys, higher beta stabilizer content results in complete retention of beta phase upon air cooling or water quenching from solution treatment temperature (above β transus temperature). This difference in transformation can be related to the difference in electron density; the α + β alloys have <4 el/atom, while β alloys have higher electron density, for example, it is 4.148 el/atom for Ti-15-3 alloy [5]. Beta alloys are more workable because of the higher stacking fault energy of the BCC phase, which supports the formation of multiple and cross slips upon deformation, thereby preventing crack formation [6].

In most of the commercial beta alloys, metastable or thermodynamically unstable β phase with BCC crystal structure is formed upon quenching after solution treatment. Hence, subsequent aging leads to the precipitation of the α phase from the beta matrix. To understand the transformation in detail, modeling of precipitation of α by decomposition of β is performed under the framework of Johnson-Mehl-Avrami for Ti-15-3 a metastable beta alloy [7] and Ti-5Mo-2.6Nb-3Al-0.2S/β21s [8], both being β alloys. The transformation of β to α + β upon aging is a slow diffusion-controlled growth of alpha plates in the beta matrix. Hence, the aging time decides the α precipitation fraction [7]. Various microstructures and correspondingly mechanical properties are feasible through heat treatment of β titanium alloys, thereby making them a wide spectrum of candidate materials for a wide range of applications. Beta Ti alloys with less beta stabilizing element were found to have faster precipitation reactions. For example, VT22 alloy exhibited higher precipitation kinetics compared to the Ti-15-3 and Timetal LCB [9]. Devaraj et al. reported that superior strength is achieved through the micro and nanoscale precipitation of α phase in a beta matrix of Ti-1Al-8V-5Fe [10]. As already mentioned, heat treatment of β titanium alloys is comprised of two steps, that is, solution treatment and aging. The solution treatment can be subdivided into α + β and β solution treatment based on the temperature (i.e., α + β solution treatment T < β transus temperature and β solution treatment temperature T > β transus

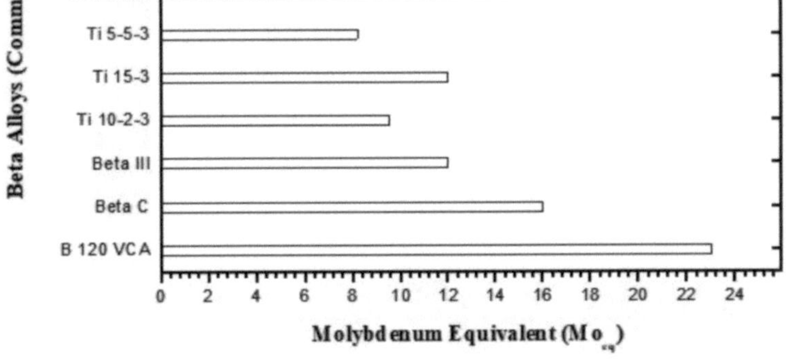

Figure 1.
Mo_{eq} of various beta Ti alloys.

temperature). In beta alloys, Ti-6Cr-5Mo-5V-4Al and Ti-5Al-5Mo-5V-3Cr, solution treatment in α + β range followed by aging yielded a better strength–ductility combination compared to solution treatment in the β range followed by aging [11, 12]. Further, if the α + β solution treatment is preceded by rolling in the α + β range even better strength–ductility combination was obtained; this was attributed to the formation of finer β grains in Ti-3.5Al-5Mo-6V-3Cr-2Sn-0.5Fe [13]. Similarly, α + β rolling followed by α + β solution treatment of Ti-10V-2Fe-3Al resulted in improvement of fracture toughness [14]. Deformation or cold working in-between solution treatment and aging could lead a path for obtaining homogeneous precipitation because the dislocations serve as a precursor for precipitation [9]. Zhan et al. also reported that dislocations formed during the high strain rate deformation of the metastable β alloy Ti-6Cr-5Mo-5V-4Al have acted as nucleation sites for the α laths/precipitation at elevated temperature [15]. Similarly, in Ti-15-3 alloy cold working before duplex aging is found to be advantageous in forming finer precipitates [16]. Ti-15-3 alloy exhibits lower precipitation kinetics compared to Timetal LCB and VT22 alloy [17]. Cold working before aging or two-step/duplex aging can be used to increase the precipitation kinetics in Ti-15-3. However, intervening cold work also leads to a significant loss in ductility. Aging of Ti-15-3 alloy with deformed microstructure for prolonged times leads to dislocation rearrangement and formation of subgrains [18]. Grain boundary α and precipitation free zones may occur in aged condition; they play an important role in degrading the tensile and fatigue properties. The authors have reviewed various processing techniques of the beta titanium alloys elsewhere [19]. The heat treatment of beta titanium alloy for biomedical application [20, 21] and heat treatment of additively manufactured beta Ti alloy [22] have also been reported in the literature.

2. Solution treatment

Solution treatment comprises of heating the sample from 20 to 30°C above the beta transus temperature (super-transus) or ~50°C below the beta transus temperature (sub-transus) for a specified time and rapid cooling of the sample to room temperature. Hence, beta transus temperature (β_{trans}) plays a vital role in selecting heat treatment temperatures. This beta transus temperature is strongly influenced by the alloying element (i.e., alpha stabilizers will rise the $\beta_{trans,}$ beta stabilizers will lower the $\beta_{trans,}$ and neutral elements will hardly do the changes in the β_{trans}). The equation (Eq. (2)) to find the beta transus temperature is given below [23].

$$T_\beta = 882 + 21.1\,[Al] - 9.5\,[Mo] + 4.2\,[Sn] - 6.9\,[Zr] - 11.8\,[V] - 12.1\,[Cr] - 15.4\,[Fe] + 23.3\,[Si] + 123\,[O] \tag{2}$$

Beta transus temperature for some of the important beta titanium alloys is listed in **Table 1**.

Solution treatment temperature and the cooling rate strongly influence the mechanical properties realized after subsequent aging treatment. Depending on the requirement, metastable beta alloys such as Ti-13V-l1Cr-3Al and Ti-15Mo-3Al-3Nb-0.2Si are supplied in the solution-treated condition to ease the down-stream cold working operations [4]. Schematic representation of super- and sub-transus solution treatment and the aging process is shown in **Figure 2**. Super-transus solution treatment is done above the β_{trans} temperature and sub-transus solution treatment below the β_{trans} temperature. In alloys, such as Ti-5Al-5Mo-5V-3Cr, both super-transus and sub-transus solution treatments were found to be useful in

S.No	Alloy name	Commercial name	β_{trans} temperature (°C)
1	Ti-13V-11Cr-3Al	B 120 VCA	650
2	Ti-3Al-8V-6Cr-4Mo-4Zr	Beta C	795
3	Ti-15V-3Cr-3Sn-3Al	Ti 15-3	760
4	Ti-11.5Mo-6Zr-4.5Sn	Beta III	745
5	Ti-10V-2Fe-3Al	Ti 10-2-3	800
6	Ti-1Al-8V-5Fe	Ti 1-8-5	825
7	Ti-12Mo-6Zr-2Fe	TMZF	743
8	Ti-4.5Fe-6.8Mo-1.5Al	TIMETAL LCB	800
9	Ti-5V-5Mo-1Cr-1Fe-5Al	VT22	850
10	Ti-8V-8Mo-2Fe-3Al	Ti 8-8-2-3	775
11	Ti-6V-6Mo-5.7Fe-2.7Al	TIMETAL 125	704

Table 1.
β_{trans} temperature of the beta titanium alloy [16].

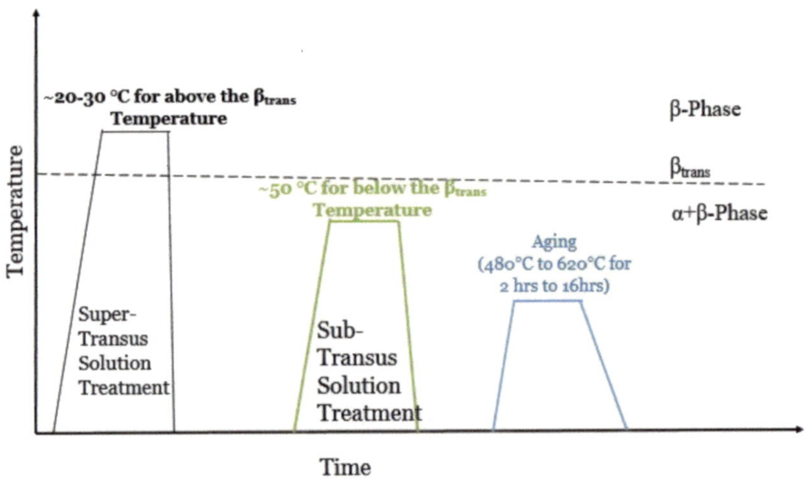

Figure 2.
Schematic representation of solution treatment and the aging process.

practice [24]. However, prolonged solution treatment above β transus may lead to a remarkable loss of mechanical properties owing to the coarsening of β grains [25]. Selection of the solution treatment temperature will also have a strong influence on the morphology and distribution of the alpha precipitation. For example, in Ti-1Al-8V-5Fe (Ti185), sub-transus solution treatment results in higher yield strength and tensile strength and this enhancement is ascribed to the nanoscale α precipitation in the β matrix [10].

3. Aging

During age hardening, solution-treated alloy will be heat treated in the temperature range of 480–620°C for 2–16 h. This heat treatment leads to precipitation of fine alpha phase in the beta matrix, and these precipitations hinder the movement of dislocations, making deformation difficult. This phenomenon is referred to as precipitation hardening. Coherency strains between α precipitates and the β matrix induce a strengthening effect [5]. Comparative examination of the microstructure of solution-treated (800°C/0.5 h) (ST) and microstructure of solution-treated and aged (ST + A) (500°C/8 h) Ti-15-3 samples has clearly revealed the presence of α precipitates in the latter. Precipitation leads to a significant increase in mechanical properties like tensile strength (79%) and hardness (44%) [26]. Age hardening is more effective for beta alloys compared to the α + β alloy owing to the capability of the former to form finer and homogenous α precipitates [27]. The sequence of precipitation of α is dependent on the Mo_{eq} of the alloy. The sequence of precipitation is given below:

$$\text{For lower } Mo_{eq} \text{ (solute} - \text{lean alloy), for example Ti} - 10V - 2Fe - 3Al,$$
$$\beta -> \beta + \omega_{iso} -> \beta + \omega_{iso} + \alpha -> \beta + \alpha.$$
$$\text{For higher } Mo_{eq} \text{ (solute} - \text{rich alloy), for example Ti} - 13V - 11Cr - 3Al,$$
$$\beta -> \beta + \beta' -> \beta + \beta' + \alpha -> \beta + \alpha.$$

3.1 Single aging

To produce optimum strength–ductility combination, post solution treatment, aging or soaking the material in the temperature range of 200–650°C (well below the β_{trans} temperature) for a specified time followed by air or furnace cooling is performed [3]. Single aging comprises of heating to the desired temperature, holding for a specified time, and cooling in air or furnace. In Ti-3.5Al-5Mo-6V-3Cr-2Sn-0.5Fe beta alloy, single-step aging at 440°C for 8 h resulted in high tensile strength of 1637 MPa [28]. In Ti-15-3 alloy, aging at 520°C for 10 h yields a good combination of fatigue life and fracture toughness [29]. Similarly, in the same Ti-15-3 alloy, single-step aging led to a significant increase in the microhardness and fatigue life compared to the solution-treated condition [30].

3.2 Duplex aging

Dual-step aging or duplex aging unlocks the room for further betterment in the mechanical properties through finer and homogeneous α precipitation compared to single-step aging. Many researchers have reported the advantage of duplex aging over single-step aging of beta Ti alloys; most studied is the low-high combination, that is, a low temperature for first step aging and a somewhat higher temperature for second step aging [31–35]. Enhancement of the material behavior during unidirectional and cyclic/fatigue loading could be achieved through duplex aging. In Ti-3Al-8V-6Cr-4Mo-4Zr alloy (also known as Ti 38-644, Beta C), duplex aging resulted in more homogenous alpha precipitation [3]. Precipitates were found to be finer and microstructure was also free of precipitate-free zones (PFZs) and grain boundary α (GB_{α}); this led to a significant improvement in fatigue life of Ti 38-644 [33]. In Ti-15-3 alloy, finer and more homogenous distribution of α precipitates was achieved through duplex aging compared to the single-step aging [34, 36]. In addition to an increase in the mechanical strength (i.e., YS and UTS), increase in ductility was also achieved by duplex aging of Ti-15-3 alloy [17]. Santhosh et al. [37] reported that duplex aging of Ti-15-3 sample leads to (i) a refined and homogenous distribution of the α precipitates in β matrix; (ii) a higher α phase fraction in β matrix; (iii) freedom from PFZs; and (iv) much less GB_{α}, compared to single-aged Ti-15-3 sample. Similarly, duplex aging of the β-C alloy leads to a substantial increase of fatigue behaviour by producing the completely recrystallized β microstructure with homogenously distributed α precipitates and reducing the grain boundary α [32]. Finer and homogenous α precipitation resulting from duplex aging enhanced the fatigue limit of beta alloy Ti-5Al-5Mo-5V-3Cr [38]. In a pre-strained Ti-10Mo-8V-1Fe-3.5Al, two-step aging was found more effective and yielded higher strength than conventional aging [39]. In contrast to the proceeding instances, Kazanjian et al. reported that multi-step aging made little difference to fatigue crack growth compared to the single-step aging [40]. In addition to the single and duplex aging, triplex aging or aging performed in three steps was attempted by some researchers on Ti-15-3 beta alloy; they found no significant benefit in either tensile strength or ductility of the material [41]. Duplex aging was also found to result in an enhancement of thermal stability during the elevated temperature application [24].

4. Influence of the rate of heating to the aging temperature

During the heat treatment of metastable beta titanium alloys, heating rate adopted to attain the desired aging temperature has an influential role in the α precipitation [42–45]. In Timetal LCB, lower heating rate (0.25 ks^{-1}) yielded an optimum combination of strength and ductility with a finer and homogenous α precipitation compared to the faster heating rate (20 ks^{-1}) [42]. However, this heating rate will vary from alloy to alloy. For example, a similar heating rate of 0.25 ks^{-1} produced coarser and non-uniform alpha precipitation in the Ti-15-3 alloy and the same authors reported 0.01 ks^{-1} as the optimum heating rate for this alloy [42]. Wu et al. [45] reported a significant increase in microhardness of the Ti-15-3 alloy when a lower heating rate was used. They attributed it to the homogenous alpha precipitation. In addition, the lower heating rate yielded a microstructure free of grain boundary α.

5. Grain boundary α

Grain boundary alpha (GB_α) is found detrimental by serving as a nucleus for crack initiation along α/β interfaces during the monotonic as well as cyclic loading. When the thickness of these GB_α exceeds several microns, ductility and fatigue crack inititation and propagation are detrimentally affected. Crack is found to propagate with little resistance along the GB_α in Ti-8Mo-8V-2Fe-3A1 [46]. In addition to tensile ductility, GB_α also has a strong negative influence over the fatigue behaviour of the β titanium alloys [32, 47]. In fatigue loading, the preferred site for the crack initiation will be the grain boundary decorated with α and inclined at 45° to the axis of loading [48]. This inclined GB_α provides potential sites for slip localization as well as fracture inititation. Similarly, subsurface crack initiation induced by the well-developed GB_α is commonly observed in the highly β stabilized Ti alloy β-C [32]. One of the strategies to improve the endurance limit is by properly designing a duplex aging heat treatment step compared to single aging, in order to facilitate more uniform α precipitation. Duplex aging of Ti-15V-3Al-3Cr-3Sn alloy at 250°C/24 h + 500°C/8 h resulted in a microstructure almost free of GB_α, and this was also reported as one of the important reasons for the notable increase in fatigue life in high cycle regime after duplex aging [34]. Presence of GB_α supports the intergranular fracture and reduces the ductility of the material [25, 28, 49, 50]. In aged Ti-10V-2Fe-3Al, soft zones were observed along the grain boundaries due to the GB_α; these zones preferentially undergo plastic deformation upon loading [51].

Moreover the fractographic studies have revealed the presence of a band of intense deformation originating from the grain boundary triple point and spreading into the grain interior. This was ascribed to the accommodation deformation required for continous GB_{α}, which has been stopped at triple point leading to high localized stress concentrations.

6. Precipitation-free zones (PFZs)

Non-uniform distribution of precipitates can occur during certain heat treatment conditions forming regions in microstructure free of precipitates usually near proximity of grain boundary. Uneven precipitation of α upon certain aging conditions may result in such zones where precipitation will not occur, and such zones are termed as precipitation-free zones (PFZs). The preferential α phase nucleation along beta grain boundaries can result in depletion of solute atoms near grain beta boundary region eventually resulting in the formation of PFZs. The hardness of this PFZ is less than the precipitation-hardened surrounding matrix. Hence, PFZs act as sites for strain localization during loading and reduce the tensile strength and ductility as the strength difference between PFZs and aged matrix is higher [34, 50]. In the case of fatigue loading, the presence of PFZs can act as crack nucleation sites imposing a deleterious effect in Ti-3Al-8V-6Cr-4Mo-4Zr [33] and Ti-15-3 [34] by slip localization leading to early crack initiation. To avoid the formation of PFZs and to improve the monotonic and fatigue loading behaviour, duplex aging is developed; results are promising [32–34].

7. Intermediate phases

The intermediate phases, such as isothermal ω phase and β′ phase, are formed during low-temperature aging, with the aging temperature generally in the range of 200–450°C [3]. Moreover, the omega phase can also form athermally. The ω phase provides nuclei for the α precipitation in the subsequent high-temperature aging (second step of duplex aging), thereby promoting the finer and homogenous distribution of the α phase [3]. The above statement is proven in Ti-7333 near beta alloy, isothermal ω phase formed during aging has assisted the precipitation of the α phase in the beta matrix [52]. During the first step of the dual-step aging of Ti-5Al-5Mo-5V-3Cr-0.3Fe, ~10% volume fraction of ω phase was reported by Coakley et al. [53] and this ω phase contributed to a ~15% hike in microhardness compared to the solution-treated or quenched sample. However, the ω phase leads to the embrittlement/loss of ductility in Ti-Mo alloys due to the inhomogeneous slip distribution caused by the interaction of dislocation and ω phase/particles upon deformation [54]. Researchers also reported ω precipitation during low-temperature aging of Ti-15-3 alloy [7, 34]. However, the embrittlement effect of the ω phase could be efficiently compensated by processing to realize fine β grains [51]. Researchers also reported dynamic precipitation of ω phase under cyclic loading condition [55].

Similarly, stress-induced ω phase is observed in a metastable beta alloy during the dynamic compression deformation [56]. The ω phase is hexagonal in leaner beta alloys and trigonal in heavily stabilized beta alloys [56]. Other than the ω phase, the metastable phase β′ forms as an intermediate phase during the aging of some beta Ti alloy. β′ phase with a BCC crystal structure forms if the distortion is less due to the higher concentration of alloy. Similarly, ω phase with hexagonal crystal structure forms when the distortion in BCC lattice is higher, which is the case with less concentrated alloys [27]. In line with the preceding discussions, isothermal ω is proven to be assisting the alpha nucleation in the Ti-20V [57]. In addition to ω phase, α″ (martensite) phase was also observed in the initial microstructure of the solution-treated Ti-10V-2Fe-3Al and the author reported that stable α could be formed from this α″. This is in addition to the ω phase serving as the nucleus for α formation upon aging [58]. The intermediate martensitic phases α′ and α″ are suppressed when alloying is done with sufficient beta stabilizer content. This leads to enhancement of the hardenability of the beta alloy [4].

8. Annealing

In general, annealing is performed to eliminate the deleterious residual stress (stress-relieving annealing) and to ease the fabrication process (recrystallization annealing). Schematic representation of the beta alloy and alpha-beta alloy microstructure in annealed condition is shown in **Figure 3**.

Deleterious tensile residual stresses are induced during various thermo-mechanical processing steps and fabrication techniques like welding. Sources of residual stresses are given in **Table 2**.

The residual stress gets superimposed on to the service stress, leading to a significant reduction of the life of the component. For example, Ti-5Al-5Mo-5V-3Cr metastable beta alloy was subjected to the stress-relief annealing at 650–750°C for 4 h followed by air cooling [24]. Stress-relief annealing is an intermediate step of thermo-mechanical processing. This annealing is not meant for altering the microstructure. Hence, extreme care should be taken to select the temperature and time combination (i.e., higher temperature annealing should be performed for a shorter time and lower temperature annealing should be performed for a longer time). Post

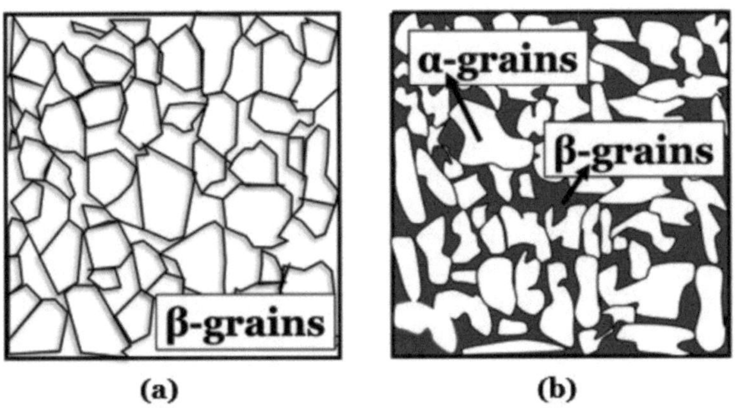

Figure 3.
Schematic representation of typical (a) beta (β) alloy and (b) alpha-beta (α-β) alloy microstructures in annealed condition.

Treatment/processes	Remarks
Aging	Solid-state reactions such as precipitation, phase transformation
Quenching	Non-uniform thermal expansion or contraction
Fabrication	Grinding, polishing, milling, and welding

Table 2.
Sources of residual stresses.

heat treatment, oil/water quenching will not be carried out to avoid insidious residual stress formation. In metastable beta alloys, mostly the heat treatment procedure generally starts with solution treatment and this is followed by aging. In solution treatment, quenching is performed and unwanted residual stresses will get induced due to non-uniformed thermal expansion. Often, the stress-relieving annealing is combined with aging, as the temperature involved is about the same. For example, in Ti-10V-2Fe-3Al, isothermal aging at 495 and 525°C for 8 h leads to precipitation (age hardening) as well as relief of the stresses induced during the solution treatment [24]. In addition to the stress-relieving annealing, recrystallization annealing is also performed to enhance the fabricability of beta titanium alloys, more specifically, if a significant reduction in cross-section is involved, for example sheet formation [4]. Cold workability of Ti-15V-3Cr-3Al-3Sn and Ti-7Mo-5Fe-2A alloys is notably increased by the annealing treatment [59]. In variance to the foregoing discussion, annealing does not increase the cold workability of the Ti-5Mo-5V-5Al-1Fe-1Cr (VT-22) and Ti-7V-4Mo-3Al (TC6) [59]. Cold working is directly related to the formation of sub-grain and cell structure. Little or no influence of the annealing on the cold workability of VT22 and TC6 is attributed to the poorly defined sub-grain and cell structure [59].

9. Mechanical properties influenced by heat treatment

9.1 Tensile, microhardness, and impact properties

The volume fraction of the beta phase in solution-treated alloy plays an important role in determining the tensile strength achieved through heat treatment process [60]. The optimum combination of tensile strength and ductility could be achieved through adequate knowledge of the aging temperature and holding time. For example, in Ti-3.5Al-5Mo-6V-3Cr-2Sn-0.5Fe beta alloy, aging at 440°C for 8 h leads to the peak strength of 1697 MPa with 5.6% of ductility. On the other hand, with the same holding time (8 h), 18% ductility along with a considerable decrease in the tensile strength is obtained by increasing the aging temperature to 560°C; the difference is attributed to the variation in the size of the acicular α precipitates [28]. The influence of aging on Young's modulus and ductility of Ti-15-3 alloy was clearly brought out by Naresh Kumar et al. [23]. Hardenability of the beta Ti alloy is proportional to the content of the beta stabilizer. For example, the beta alloy Ti-5Al-2Sn-2Zr-4Mo-4Cr possesses an excellent hardenability; it can be hardened uniformly up to 150 mm of thickness [60]. Single-step aging has increased microhardness of Ti-15-3 alloy by 40% compared to the as-received/solution-treated condition [30]. In a similar way, finer precipitation kinetics associated with duplex aging process yields a higher hardness value in Ti −15-3 alloy [36]. In Ti-5Al-5Mo-5V-3Cr-0.3Fe, duplex aging (300°C/8 h + 500°C/2 h) was adopted; a ~15% increase at first stage and 90% increase at second stage in the microhardness was observed. The remarkable increase in the microhardness in the second stage is ascribed to the precipitation of α phase [53]. Aging after α + β solution treatment resulted in a considerable increase in the hardness of β CEZ alloy, but the impact property deteriorated [61].

9.2 Fatigue behavior

In beta alloys, precipitate-free zones and grain boundary α also have control over the fatigue behavior [32]. Precipitation-free zones can be a fatigue crack nucleation site and reduce fatigue life. Similarly, the presence of soft zones associated with grain boundary α also reduces the resistance to fatigue crack propagation. Hence, duplex aging treatment yielding homogeneous alpha precipitation in beta grains and essential freedom from precipitation-free zone and grain boundary alpha is promising to improve the fatigue life of β alloys. In Ti-15-3 alloy, aging at 500°C at 8 h leads to a ~24% of surge in the fatigue strength compared to the solution-treated alloy and the α platelets precipitated during the aging strongly influence the fatigue behavior [26]. Dual-step aging (300°C/2 h + 608°C/8 h) was found to improve the fatigue limit of Ti-5Al-5Mo-5V-3Cr by yielding a microstructure with finer and homogenous alpha precipitation [38]. In Ti-3Al-8V-6Cr-4Mo-4Zr beta alloy, duplex aging led to a loftier hike in fatigue strength and a marginal increase in the fatigue crack growth behavior [32]. Tsay et al. described the prominent influence of the aging temperature upon the fatigue crack growth rate (FCGR); they concluded that the coarser α platelets resulting from longer aging time resist the fatigue crack growth effectively [62]. On the other hand, with a coarser lamellar microstructure, fatigue life in high cycle regime will not be attractive [41].

10. Heat treatment of biomedical beta titanium alloys

Beta titanium alloys are appropriate materials for a wide range of biomedical applications encompassing orthopedic and dental implants, vascular stents, intracranial aneurysms and maxillofacial prostheses. In particular, metastable biocompatible beta titanium alloys have gained substantial interest in this regard and it is highly imperative to tailor the microstructure and properties of these components or devices by suitable thermo-mechanical processing route. The vast majority of the processing routes include a homogenization treatment (for an uniform microstructure without cast dendritic structures), a forming operation (hot/cold rolling or forging), solution treatment, and aging. Since the present context is focusing on heat treatment, the following section will discuss about solution treatment and aging of some relevant metastable beta titanium alloys for cardiovascular stent and orthopedic applications.

10.1 Heat treatment of beta titanium alloys for cardiovascular stent applications

Nitinol is one of the widely used materials for vascular stent applications due to its unequivocal superelasticity properties associated with a reversible stress-induced transformation. However, recently there is a growing distress related with the nitinol implant materials over nickel ion release, which can elicit nickel hypersensitivity, toxicity and carcinogenicity. These mounting concerns have stimulated intensive research for the development of Ni-free biocompatible and corrosion-resistant titanium-niobium (TiNb) based alloy systems for these applications. Titanium-niobium (TiNb) based alloy systems are capable of exhibiting superelasticity functionalities based on the allotropic transformation between parent β (disordered bcc) phase and an orthorhombic α″ (martensite) phase.

Compared to nitinol alloys, Ni-free TiNb alloys possess inferior superelastic properties at room temperature, particularly in terms of inadequate recovery strain (less than 4%) due to a low critical stress for slip deformation. As depicted in schematic **Figure 4a**, a material with superelastic property exhibits a two-stage yielding. The initial yield stress corresponds to the critical stress for inducing martensitic transformation leading to superelasticity, whereas the second yield relates to the critical stress for slip-induced plastic deformation. In the case of TiNb alloys, the apparent martensitic yield stress increases with an increase in temperature;

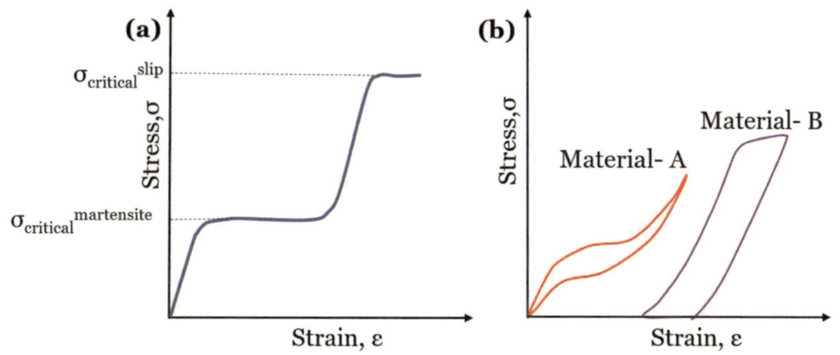

Figure 4.
Schematic representation of (a) two-stage yielding phenomenon exhibited by superelastic materials during monotonic loading and (b) cyclic loading unloading test in which material-A exhibits superelasticity and material-B shows plastic deformation occurring prior to martensitic transformation.

hence higher stresses are required to induce martensite transformation, which can be above the critical slip-inducing stress, leading to plastic deformation with no superelasticity as shown schematically in **Figure 4b**.

Heat treatment is an efficient strategy to improve the critical stress for slip deformation in TiNb alloys. Stable superelasticity and higher recovery strain (4.2%) were obtained by aging a Ti-26Nb alloy by a low-temperature annealing treatment (600°C) followed by aging (300°C). This was attributed to the precipitation of dense and finer ω during aging heat treatment consequently leading to a higher critical stress for slip deformation [63]. A high-temperature, low-duration annealing (900°C/5 min) treatment on a Ti-Zr-Nb-Sn-Mo alloy exhibited nearly perfect superelasticity with a relatively high recovery strain of 6-6.2% [64, 65]. A well-developed {001}β<110>β type recrystallization texture due to the presence of Sn resulted in these desirable large recovery strains and solid solution strengthening by Mo addition developed higher tensile strength values. It is also noteworthy to mention here that one of the drawbacks associated with thermal treatment-assisted microstructural evolution is the chemical stabilization of β phase (due to β stabilizer enrichment) adversely affecting superelastic properties. To counteract this, short-duration aging treatments have been developed, which can yield ultra-fine grain β grains (1–2 μm) with concurrent improvement in superelastic properties [66].

10.2 Heat treatment of beta titanium alloys for orthopedic implant applications

The usage of beta titanium alloys for orthopedic implants can be attributed to their inherent biocompatible compositions and lower elastic modulus values compared to conventional orthopedic materials. Compared to conventional CP titanium and Ti-6Al-4V, beta Ti alloys exhibit lower modulus values reducing clinical complications associated with stress shielding. Solution treatment in beta phase often results in a retained beta phase along with non-equilibrium omega (ω) or martensitic (α″) phase precipitation. As a lower elastic modulus is essential for reducing the clinical complications associated with bone tissue resorption, these metastable phases play a predominant role in determining the implant efficacy. Among these, omega phase precipitation is associated with an increase in strength, reduction in ductility, and in most instances an undesirable increment in modulus values. Moreover, in the case of solution-treated and aged condition, volume fraction, size, and morphology of α precipitates are dependent on ω precipitation. In contrast, orthorhombic α″ martensite or hexagonal α′ in a beta matrix can significantly reduce modulus values, improve the ductility, even though with a corresponding reduction in strength. Compared to the low-strength solution-treated conditions, cold working/oxygen content increase/subsequent aging can result in strengthening associated with ω and/or α precipitation. For example, aging of low-modulus biomedical ternary alloys (Ti-35Nb-7Zr-5Ta and Ti-29Nb-13Ta-4.6Zr) in the temperature range of 300–400°C induced ω, 400–475°C ω-α mixture, and high temperature aging above 475°C revealed α precipitation without any ω [67, 68]. It should also be taken in to account that an increased oxygen content in these alloys suppressed ω formation while promoting α precipitation.

Heat treatment of newly designed Sn-based β titanium alloys (Ti-32Nb-2Sn and Ti-32Nb-4Sn) exhibited a single β phase microstructure after solution treatment at 950°C for 0.5 h followed by quenching; subsequent aging resulted in alpha phase precipitation [69]. Higher aspect ratio of precipitated alpha led to age hardening after aging at 500°C for 6 h; aging at 600°C, on the other hand, delitioriosly affected mechanical properties due to matrix softening and relatively coarser alpha precipitates. The presence of Sn even in smaller amounts can suppress the ω/α″ precipitation. The abrasion resistance of Ti-10V-1Fe-3Al ($\beta_{transus}$ = 830°C) and Ti-10V-2Cr-3Al ($\beta_{transus}$ = 830°C) was investigated under different microstructures established by various heat treatments [70]. α + β solution treatment resulted in near spherical or rod-like α, β annealing led to metastable β grains and acicular martensite phase, β + (α + β) produced flake α phase or Widmanstatten α phase and aging at a low and medium temperatures generated high density of nano ω phase precipitates. This study concluded that a dual phase mixture of β and flake-shaped alpha is an appropriate microstructure for improving the abrasion resistance.

11. Conclusions

Metastable beta titanium alloys have exclusive properties like the ease of fabrication, excellent biocompatibility, and good corrosion resistance. Hence, a steady progress has been there in the application of these alloys in aerospace industries and other high-technology industrial segments. Metastable beta titanium alloys are evolving as a potential candidate even for biomedical and automotive industries. As the β_{trans} temperature of the metastable beta alloys is significantly lower when compared to α and α + β alloys, the cost of processing is considerably lower. Possibility of tailoring the properties through heat treatments based on the requirement is an important and outstanding property of the metastable beta titanium alloys. However, sound knowledge in the process-structure–property correlation is required. Heat treatments should be designed appropriately to avoid embrittlement due to intermediate phases such as ω and premature failure due to the grain boundary alpha (GB_{α}). In this chapter, we have attempted to provide insights into the heat treatment of metastable beta titanium alloys and optimization of the heat treatment parameters to achieve maximized material performance under monotonic and cyclic loading conditions.

Acknowledgements

The authors would like to express their gratitude to the Management of Vellore Institute of Technology (VIT)—Vellore campus, Tamil Nadu, India for allowing us to submit this manuscript.

Author details

Sudhagara Rajan Soundararajan[1,2], Jithin Vishnu[3], Geetha Manivasagam[1,3] and Nageswara Rao Muktinutalapati[1*]

1 School of Mechanical Engineering, Vellore Institute of Technology (VIT), Vellore, Tamilnadu, India

2 International Institute for Aerospace Engineering and Management (IIAEM), Jain University, Bangalore, Karnataka, India

3 Centre for Biomaterials, Cellular and Molecular Theranostics, Vellore Institute of Technology (VIT), Vellore, Tamilnadu, India

*Address all correspondence to: muktinutala@gmail.com

References

[1] Boyer RR. Attributes, characteristics, and applications of titanium and its alloys. Journal of Metals. 2010;**62**:21-24. DOI: 10.1007/s11837-010-0071-1

[2] Geetha M, Singh AK, Asokamani R, Gogia AK. Ti based biomaterials, the ultimate choice for orthopaedic implants—A review. Progress in Materials Science. 2009;**54**:397-425. DOI: 10.1016/j.pmatsci.2008.06.004

[3] Santhosh R, Geetha M, Nageswara Rao M. Recent developments in heat treatment of Beta titanium alloys for aerospace applications. Transactions of the Indian Institute of Metals. 2016;**70**: 1681-1688. DOI: 10.1007/s12666-016-0985-6

[4] Kolli R, Devaraj A. A review of metastable beta titanium alloys. Metals (Basel). 2018;**8**:506. DOI: 10.3390/met8070506

[5] Froes FH. Titanium - Physical Metallurgy Processing and Application. United States of America: ASM International; 2014

[6] Tang L, Hua K, Li J, Kou H, Zhang Y, Fan J, et al. Formation of slip bands and microstructure evolution of Ti-5Al-5Mo-5V-3Cr-0.5Fe alloy during warm deformation process. Journal of Alloys and Compounds. 2018;**770**:183-193. DOI: 10.1016/j.jallcom.2018.08.097

[7] Naveen M, Santhosh R, Geetha M, Nageswara Rao M. Experimental study and computer modelling of the $\beta \rightarrow \alpha + \beta$ phase transformation in Ti15-3 alloy under isothermal conditions. Journal of Alloys and Compounds. 2014;**616**: 607-613. DOI: 10.1016/j.jallcom.2014.07.163

[8] Malinov S, Sha W, Markovsky P. Experimental study and computer modelling of the $\beta \Rightarrow \alpha+\beta$ phase transformation in β21s alloy at isothermal conditions. Journal of Alloys and Compounds. 2002;**348**: 110-118. DOI: 10.1016/s0925-8388(02)00804-6

[9] Ivasishin OM, Markovsky PE, Matviychuk YV, Semiatin SL. Precipitation and recrystallization behavior of beta titanium alloys during continuous heat treatment. Metallurgical and Materials Transactions A: Physical Metallurgy and Materials Science. 2003;**34**:147-158. DOI: 10.1007/s11661-003-0216-8

[10] Devaraj A, Joshi VV, Srivastava A, Manandhar S, Moxson V, Duz VA, et al. A low-cost hierarchical nanostructured beta-titanium alloy with high strength. Nature Communications. 2016;**7**:1-8. DOI: 10.1038/ncomms11176

[11] Li C-L, Mi X-J, Ye W-J, Hui S-X, Yu Y, Wang W-Q. Effect of solution temperature on microstructures and tensile properties of high strength Ti–6Cr–5Mo–5V–4Al alloy. Materials Science and Engineering A. 2013;**578**: 103-109. DOI: 10.1016/j.msea.2013.04.063

[12] Shekhar S, Sarkar R, Kar SK, Bhattacharjee A. Effect of solution treatment and aging on microstructure and tensile properties of high strength β titanium alloy, Ti-5Al-5V-5Mo-3Cr. Materials and Design. Netherlands: Elsevier Ltd; 5 February 2015;**66**(Part B):596-610. DOI: 10.1016/j.matdes.2014.04.015

[13] Du ZX, Xiao SL, Shen YP, Liu JS, Liu J, Xu LJ, et al. Effect of hot rolling and heat treatment on microstructure and tensile properties of high strength beta titanium alloy sheets. Materials Science and Engineering A. 2015;**631**: 67-74. DOI: 10.1016/j.msea.2015.02.030

[14] Srinivasu G, Natraj Y, Bhattacharjee A, Nandy TK, Nageswara

Rao GVS. Tensile and fracture toughness of high strength β titanium alloy, Ti-10V-2Fe-3Al, as a function of rolling and solution treatment temperatures. Materials and Design. 2013;**47**:323-330. DOI: 10.1016/j.matdes.2012.11.053

[15] Zhan H, Ceguerra A, Wang G, Cairney J, Dargusch M. Precipitation of string-shaped morphologies consisting of aligned α phase in a metastable β titanium alloy. Scientific Reports. 2018; **8**:2-11. DOI: 10.1038/s41598-018-20386-1

[16] Weiss I, Semiatin SL. Thermomechanical processing of beta titanium alloys—An overview. Materials Science and Engineering. 15 March 1998;**243**(1–2):46-65. DOI: 10.1016/S0921-5093(97)00783-1

[17] Ivasishin OM, Markovsky PE, Matviychuk YV, Semiatin SL, Ward CH, Fox S. A comparative study of the mechanical properties of high-strength β-titanium alloys. Journal of Alloys and Compounds. 2008;**457**:296-309. DOI: 10.1016/j.jallcom.2007.03.070

[18] Makino T, Chikaizumi R, Nagaoka T, Furuhara T, Makino T. Microstructure development in a thermomechanically processed Ti-15V-3Cr-3Sn-3Al alloy. Materials Science and Engineering A. 1996;**213**:51-60. DOI: 10.1016/0921-5093(96)10236-7

[19] Sudhagara Rajan S, Jithin V, Geetha M, Nageswara Rao M. Processing of Beta Titanium Alloys for Aerospace and Biomedical Applications. Vol. 2. Rijeka, Croatia: Intech open; 2018. p. 64. DOI: 10.5772/32009.

[20] Hida M, Miyazawa K, Tsuruta S, Kurosawa M, Hata Y, Kawai T, et al. Effect of heat treatment conditions on the mechanical properties of Ti-6Mo-4Sn alloy for orthodontic wires. Dental Materials Journal. 2013;**32**:462-467. DOI: 10.4012/dmj.2012-118

[21] Nag S, Banerjee R, Fraser HL. Microstructural evolution and strengthening mechanisms in Ti-Nb-Zr-Ta, Ti-Mo-Zr-Fe and Ti-15Mo biocompatible alloys. Materials Science and Engineering: C. 2005;**25**:357-362. DOI: 10.1016/j.msec.2004.12.013

[22] Liu YJ, Wang HL, Li SJ, Wang SG, Wang WJ, Hou WT, et al. Compressive and fatigue behavior of beta-type titanium porous structures fabricated by electron beam melting. Acta Materialia. 2017;**126**:58-66. DOI: 10.1016/j.actamat.2016.12.052

[23] Naresh Kumar K, Muneshwar P, Singh SK, Jha AK, Pant B. Thermo mechanical working and heat treatment studies on meta-stable beta titanium alloy (Ti15V3Al3Sn3Cr) plates. Materials Science Forum. 2015;**830–831**: 151-155. DOI: 10.4028/www.scientific.net/MSF.830-831.151

[24] Cotton JD, Briggs RD, Boyer RR, Tamirisakandala S, Russo P, Shchetnikov N, et al. State of the art in beta titanium alloys for airframe applications. Journal of Metals. 2015;**67**: 1281-1303. DOI: 10.1007/s11837-015-1442-4

[25] Sauer C, Luetjering G. Thermo-mechanical processing of high strength β- titanium alloys and effects on microstructure and properties. Journal of Materials Processing Technology. 2001;**117**:311-317. DOI: 10.1016/S0924-0136(01)00788-9

[26] Ismarrubie ZN, Ali A, Satake T, Sugano M. Influence of microstructures on fatigue damage mechanisms in Ti-15-3 alloy. Materials and Design. 2011; **32**:1456-1461. DOI: 10.1016/j.matdes.2010.08.051

[27] Lütjering G, Williams JC. Titanium. 2nd ed. Berlin, Heidelberg: Springer-Verlag; 2007

[28] Chen Y, Du Z, Xiao S, Xu L, Tian J. Effect of aging heat treatment on

microstructure and tensile properties of a new β high strength titanium alloy. Journal of Alloys and Compounds. 2013; **586**:588-592. DOI: 10.1016/j.jallcom.2013.10.096

[29] Ma J, Wang Q. Aging characterization and application of Ti–15–3 alloy. Materials Science and Engineering A. 1998;**243**:150-154. DOI: 10.1016/S0921-5093(97)00793-4

[30] Sudhagara Rajan S, Swaroop S, Manivasagam G, Rao MN. Fatigue life enhancement of titanium alloy by the development of nano/micron surface layer using laser peening. Journal of Nanoscience and Nanotechnology. 2019; **19**:7064-7073. DOI: 10.1166/jnn.2019.16639

[31] Wagner L, Gregory JK. Thermomechanical surface treatment of titanium alloys. Materials Science Forum. 1994;**163–165**:159-172. DOI: 10.4028/www.scientific.net/MSF.163-165.159

[32] Schmidt P, El-Chaikh A, Christ HJ. Effect of duplex aging on the initiation and propagation of fatigue cracks in the solute-rich metastable β titanium alloy Ti 38-644. Metallurgical and Materials Transactions A, Physical Metallurgy and Materials Science. 2011;**42**:2652-2667. DOI: 10.1007/s11661-011-0662-7

[33] El Chaikh A, Schmidt P, Christ HJ. Fatigue properties of duplex-aged Ti 38-644 metastable beta titanium alloy. Procedia Engineering. 2010;**2**:1973-1982. DOI: 10.1016/j.proeng.2010.03.212

[34] Santhosh R, Geetha M, Saxena VK, Nageswara Rao M. Effect of duplex aging on microstructure and mechanical behavior of beta titanium alloy Ti-15V-3Cr-3Al-3Sn under unidirectional and cyclic loading conditions. International Journal of Fatigue. 2015;**73**:88-97. DOI: 10.1016/j.ijfatigue.2014.12.005

[35] Santhosh MNR, Geetha M, Saxena VK. Studies on single and duplex aging of metastable beta titanium alloy Ti–15V–3Cr–3Al–3Sn. Journal of Alloys and Compounds. 2014;**605**:222-229

[36] Furuhara T, Maki T, Makino T. Microstructure control by thermomechanical processing in β-Ti-15-3 alloy. Journal of Materials Processing Technology. 2001;**117**:318-323. DOI: 10.1016/S0924-0136(01)00790-7

[37] Santhosh R, Geetha M, Saxena VK, Nageswararao M. Studies on single and duplex aging of metastable beta titanium alloy Ti-15V-3Cr-3Al-3Sn. Journal of Alloys and Compounds. 2014; **605**:222-229. DOI: 10.1016/j.jallcom.2014.03.183

[38] Campanelli LC, da Silva PSCP, Bolfarini C. High cycle fatigue and fracture behavior of Ti-5Al-5Mo-5V-3Cr alloy with BASCA and double aging treatments. Materials Science and Engineering A. 2016;**658**:203-209. DOI: 10.1016/j.msea.2016.02.004

[39] Song ZY, Sun QY, Xiao L, Liu L, Sun J. Effect of prestrain and aging treatment on microstructures and tensile properties of Ti-10Mo-8V-1Fe-3.5Al alloy. Materials Science and Engineering A. 2010;**527**:691-698. DOI: 10.1016/j.msea.2009.09.046

[40] Kazanjian SM, Starke EAA Jr. Effects of microstructural modification on fatigue crack growth resistance of Ti-15V-3Al-3Sn-3Cr. International Journal of Fatigue. 1999;**21**:127-135. DOI: 10.1016/S0142-1123(99)00064-X

[41] Boyer R, Rack H, Venkatesh V. The influence of thermomechanical processing on the smooth fatigue properties of Ti–15V–3Cr–3Al–3Sn. Materials Science and Engineering A. 1998;**243**:97-102. DOI: 10.1016/S0921-5093(97)00785-5

[42] Ivasishin OM, Markovsky PE, Semiatin SL, Ward CH. Aging response of coarse- and fine-grained β titanium

alloys. Materials Science and Engineering A. 2005;**405**:296-305. DOI: 10.1016/j.msea.2005.06.027

[43] Wain N, Hao XJ, Ravi GA, Wu X. The influence of carbon on precipitation of α in Ti-5Al-5Mo-5V-3Cr. Materials Science and Engineering A. 2010;**527**: 7673-7683. DOI: 10.1016/j.msea.2010.08.032

[44] Contrepois Q, Carton M, Lecomte-Beckers J. Characterization of the β phase decomposition in Ti-5Al-5Mo-5V-3Cr at slow heating rates. Open Journal of Metal. 2011;**01**:1-11. DOI: 10.4236/ojmetal.2011.11001

[45] Wu X, del Prado J, Li Q, Huang A, Hu D, Loretto MH. Analytical electron microscopy of C-free and C-containing Ti-15-3. Acta Materialia. 2006;**54**: 5433-5448. DOI: 10.1016/j.actamat.2006.07.002

[46] Chesnutt JC, Froes FH. Effect of α-phase morphology and distribution on the tensile ductility of a metastable beta titanium alloy. Metallurgical Transactions A. 1977;**8**:1013-1017. DOI: 10.1007/BF02661592

[47] Benjamin R, Nageswara Rao M. Crack nucleation in β titanium alloys under high cycle fatigue conditions—A review. Journal of Physics Conference Series. 6th International Conference on Fracture Fatigue and Wear. Porto, Portugal: IOP Publishing Ltd; 26-27 July 2017;**843**. DOI: 10.1088/1742-6596/843/1/012048

[48] Sauer C, Lütjering G. Influence of α layers at β grain boundaries on mechanical properties of Ti-alloys. Materials Science and Engineering A. 2001;**319–321**:393-397. DOI: 10.1016/S0921-5093(01)01018-8

[49] Du Z, Xiao S, Xu L, Tian J, Kong F, Chen Y. Effect of heat treatment on microstructure and mechanical properties of a new β high strength titanium alloy. Materials and Design. 2014;**55**:183-190. DOI: 10.1016/j.matdes.2013.09.070

[50] Leyens C, Peters M. Titanium and titanium alloys: Fundamentals and applications. Weinheim: Wiley-Vch Verlag GmbH & Co. KGaA; 2003. DOI: 10.1002/3527602119. ISBN: 3-527-30534-32003

[51] Terlinde GT, Duerig TW, Williams JC. Microstructure, tensile deformation, and fracture in aged Ti 10V-2Fe-3Al. Metallurgical Transactions A. 1983;**14A**:2101-2115

[52] Dong R, Li J, Fan J, Kou H, Tang B. Precipitation of α phase and its morphological evolution during continuous heating in a near β titanium alloy Ti-7333. Materials Characterization. 2017;**132**:199-204. DOI: 10.1016/j.matchar.2017.07.032

[53] Coakley J, Vorontsov VA, Jones NG, Radecka A, Bagot PAJ, Littrell KC, et al. Precipitation processes in the beta-titanium alloy Ti-5Al-5Mo-5V-3Cr. Journal of Alloys and Compounds. 2015; **646**:946-953. DOI: 10.1016/j.jallcom.2015.05.251

[54] Gysler A, Lütjering G, Gerold V. Deformation behavior of age-hardened Ti-Mo alloys. Acta Metallurgica. 1974; **22**:901-909. DOI: 10.1016/0001-6160(74)90057-1

[55] Hagihara K, Nakano T, Todai M. Unusual dynamic precipitation softening induced by dislocation glide in biomedical beta-titanium alloys. Scientific Reports. 2017;7:1-9. DOI: 10.1038/s41598-017-08211-7

[56] Ankem S, Greene C. Recent developments in microstructure/property relationships of beta titanium alloys. Materials Science and Engineering A. 2002;**263**:127-131. DOI: 10.1016/s0921-5093(98)01170-8

[57] Shi R, Zheng Y, Banerjee R, Fraser HL, Wang Y. ω-assisted α nucleation in a metastable β titanium alloy. Scripta Materialia. 2019;**171**:62-66. DOI: 10.1016/j.scriptamat.2019.06.020

[58] Barriobero-vila P, Requena G, Warchomicka F, Stark A, Schell N, Buslaps T. Phase transformation kinetics during continuous heating of a β - quenched Ti–10V–2Fe–3Al alloy. Journal of Materials Science. 2015;**50**: 1412-1426. DOI: 10.1007/s10853-014-8701-6

[59] Karasevskaya OP, Ivasishin OM, Semiatin SL, Matviychuk YV. Deformation behavior of beta-titanium alloys. Materials Science and Engineering A. 2003;**354**:121-132. DOI: 10.1016/S0921-5093(02)00935-8

[60] Donachie M. Introduction to selection of titanium alloys. In: Titanium: A Technical Guide. Ohio: ASM International; 2000. pp. 5-11

[61] Sukumar G, Bhav Singh B, Bhattacharjee A, Siva Kumar K, Gogia AK. Ballistic impact behaviour of β-CEZ Ti alloy against 7.62 mm armour piercing projectiles. International Journal of Impact Engineering. 2013;**54**: 149-160. DOI: 10.1016/j.ijimpeng.2012.11.002

[62] Tsay L, Chang ST, Chen C. Fatigue crack growth characteristics of a Ti-15V-3Cr-3Sn-3Al alloy with variously aged conditions. Materials Transactions. 2013;**54**(3):326-331. DOI: 10.2320/matertrans.M2012349

[63] Kim HY, Ikehara Y, Kim JI, Hosoda H, Miyazaki S. Martensitic transformation, shape memory effect and superelasticity of Ti-Nb binary alloys. Acta Materialia. 2006;**54**: 2419-2429. DOI: 10.1016/j.actamat.2006.01.019

[64] Fu J, Yamamoto A, Kim HY, Hosoda H, Miyazaki S. Novel Ti-base superelastic alloys with large recovery strain and excellent biocompatibility. Acta Biomaterialia. 2015;**17**:56-67. DOI: 10.1016/j.actbio.2015.02.001

[65] Fu J, Kim HY, Miyazaki S. Effect of annealing temperature on microstructure and superelastic properties of a Ti-18Zr-4.5Nb-3Sn-2Mo alloy. Journal of the Mechanical Behavior of Biomedical Materials. 2017; **65**:716-723. DOI: 10.1016/j.jmbbm.2016.09.036

[66] Sun F, Nowak S, Gloriant T, Laheurte P, Eberhardt A, Prima F. Influence of a short thermal treatment on the superelastic properties of a titanium-based alloy. Scripta Materialia. 2010;**63**:1053-1056. DOI: 10.1016/j.scriptamat.2010.07.042

[67] Qazi JI, Rack HJ. Metastable beta titanium alloys for orthopedic applications. Advanced Engineering Materials. 2005;7:993-998. DOI: 10.1002/adem.200500060

[68] Hao YL, Niinomi M, Kuroda D, Fukunaga K, Zhou YL, Yang R, et al. Aging response of the Young's modulus and mechanical properties of Ti-29Nb-13Ta-4.6Zr for biomedical applications. Metallurgical and Materials Transactions A, Physical Metallurgy and Materials Science. 2003;**34A**:1007-1012. DOI: 10.1007/s11661-003-0230-x

[69] Bahl S, Shreyas P, Trishul MA, Suwas S, Chatterjee K. Enhancing the mechanical and biological performance of a metallic biomaterial for orthopedic applications through changes in the surface oxide layer by nanocrystalline surface modification. Nanoscale. 2015;7: 7704-7716. DOI: 10.1039/c5nr00574d

[70] Li C, Li H, van der Zwaag S. Unravelling the abrasion resistance of two novel meta-stable titanium alloys on the basis of multi-pass-dual-indenter tests. Wear. 15 December 2019;**440–441**:203094. DOI: 10.1016/j.wear.2019.203094

Chapter 11

Grain Boundary Effects on Mechanical Properties: Design Approaches in Steel

Gaurav Bhargava

Abstract

For Poly-crystalline metals, Grain boundary design plays an important role to achieve desired mechanical properties in the final product form which may be a hot rolled or cold rolled coil. The fundamental mechanical properties are yield and tensile strengths, elongation and formability where grain refinement is particularly attractive mechanism for property design. Grain boundary strengthening provides benefits both in terms of fracture toughness and mechanical behavior at lower temperatures particularly in case of hot rolled high strength structural and line pipe steels. For cold rolled steels, grain boundary effects play a crucial role in critical automotive application steels such as Bake Hardening Index in Bake-hardening Steels. Additionally, deep drawability and formability characteristics are also strongly dependent upon grain size. Mechanical properties in steels are controlled majorly by two techniques - chemical composition and Steel processing parameters. Through first method of chemical composition control, during hot rolling, the control of austenite grain size is accomplished, alongside retardation of phase transformation to lower temperature. In second method, steel manufacturing process parameters i.e. hot rolling and coil cooling parameters, cold reduction, and subsequent heat treatment parameters such as annealing play an important role.

Keywords: thermomechanical processing, recrystallization, mean flow stress, hot rolling process modeling, steel design, grain boundary strengthening, mechanical properties

1. Microstructural evolution during themomechanical processing of steel

1.1 Thermomechanical processing in hot rolling mill

Thermomechanical processing in hot rolling mill in general involves rolling of slabs or blooms to thinner sections which are classified as plate or coils by inducing reduction in thickness by rolling at controlled elevated temperatures to achieve desired mechanical properties.

As shown in **Figure 1**, typical thermomechanical hot rolling process can be divided into five basic processes.

IntechOpen

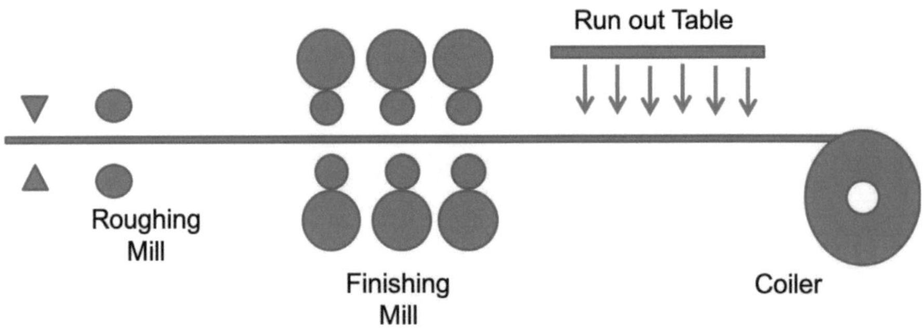

Figure 1.
Typical setup of hot strip rolling mill.

1.1.1 Reheating of slabs

The first step involves heating up of slabs to remove dendritic segregation and facilitate solutionizing of microalloying element which is intended to precipitate during hot rolling, contributing to the strength of the material.

1.1.2 Rough rolling

Before inducing final reductions at finish rolling mill, slab is first introduced to be rolled in the roughing stands where thickness of slab is reduced from 200 to 300 mm to about 50 mm in several passes, usually four or five. In the roughing process, the width increases in each pass and is controlled by vertical edge rollers. Since the temperatures are high, recrystallization takes place during this process.

1.1.3 Finish rolling

Finishing mill is generally a tandem rolling mill consisting of 5–7 rolling stands. The finishing temperature is dependent upon the rolling speed. The interpass heating, or cooling is also controlled during rolling.

1.1.4 Accelerated cooling

After finish rolling, the hot rolled coil is subjected to cooling on runout table where water is sprayed on the top and bottom of the steel at a steady flow rate to induce phase and microstructure control leading to increased strengths.

1.2 Types of hot rolling approaches

The hot rolling process can be divided approaches based upon requirement of properties. They are chiefly identified as conventional controlled rolling (CCR) and recrystallization controlled rolling (RCR) (**Figure 2**). The recrystallization rolling requires rolling at high temperatures that leads to recrystallization and control of grain size. The process is designed to have mechanisms that inhibit grain growth after recrystallization. The conventional controlled rolling approach requires rolling in no-recrystallization zone, leading to unrecrystallized grains which ultimately lead to finer sizes after phase transformation (**Figure 2**).

The conventional controlled rolling shall be presented in detail as RCR has been mainly used for higher gauge rolling, i.e., plate mills or mills with lower rolling load capacities.

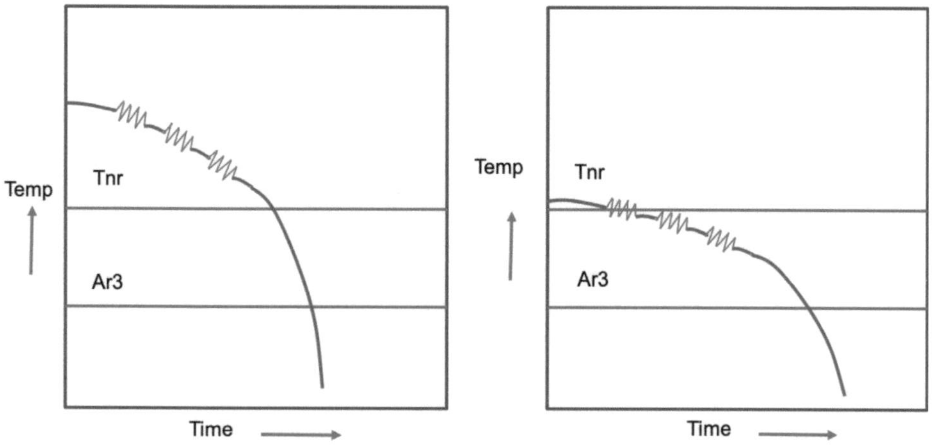

Figure 2.
Recrystallization controlled rolling and conventional controlled rolling.

In general, in hot rolling method of conventional controlled rolling (CCR), methods of achieving strength would largely be dependent upon the grain size control of austenitic phase and eventually the phase transformation which is controlled by means of addition of microalloying elements in steel chemistry alongside or singly with application thermomechanical treatment practices during rolling.

The basic underlying principle of CCR is obtaining steel which possess both high toughness and strength through grain refinement. The successive methods to achieve finer grains are carried out as explained below:

1. Repeated deformation in roughing mill at recrystallization temperature range. Here, the austenitic grains are refined due to recrystallization.

2. Finish rolling which induces the successive heavy reduction in the non-recrystallization zone, i.e., rolling just above the non-recrystallization temperature T_{NR}. During finish rolling gamma (austenite) grains are forced to elongate in rolling direction, creating annealing twin deformation bands to cause alpha (ferrite) to form with a very fine size. Thus, by inducing numerous nucleation sites for alpha grains, its size is restricted.

3. Accelerated cooling that additionally promotes refinement of the ferrite grain size and restricts formation of pearlite.

4. Addition of microalloying elements also helps in enhancing strength by restricting movement of grain boundaries by formation of precipitates and restricting transformation of gamma phase.

2. Recrystallization phenomenon during rolling

Static recrystallization is likely favored phenomenon in steels during roughing passes and for plain carbon steels it continues between finishing passes as well. The static recrystallization is favored by low alloying levels and high temperatures, strains, and strain rates.

The recovery is suppressed during finish interpasses, but as dislocation density is increasing on account of work hardening at finish rolling, dynamic recrystallization (**Figure 3**) is initiated after surpassing a critical strain value ε_C. Dynamic recrystallization is markedly identified by necklace-type grain structure. After dynamic recrystallization during rolling pass, the recrystallized nuclei continue to grow after the deformation ends, leading to a phenomenon called metadynamic recrystallization.

In controlled rolling process CCR, the addition of microalloying elements is deliberate to prevent static recrystallization. However, at low rolling temperatures, increased strain rates and lower interpass times coupled with lower precipitation, dynamic recrystallization is favored. As shown in **Figure 3**, the strain accumulates to peak strain and then decreases which differs on basis of type of steel.

There has been an established relationship [1] between the maximum peak stress σ_p and the limiting Zener-Hollomon parameter Z which is given by

$$[\sinh(\alpha\,\sigma_P)]^{n'} = AZ \tag{1}$$

typical values of $n' = 4.5$ and $A = 0.12$.

$$Z = \dot{\varepsilon}\cdot\exp\left(\frac{300000}{RT}\right) \tag{2}$$

while Sun and Hawbolt [2] have reported peak Z dependent on initial grain size d_0

$$Z_{LIM} = 5 \times 10^5\cdot\exp(-0.0155\,d_0) \tag{3}$$

The maximum peak strain ε_p [2] that can be reached for given temperature T, strain rate $\dot{\varepsilon}$, and strain ε has been established as

$$\varepsilon_p = 1.32 \times 10^{-2} d_0^{\,0.174}\varepsilon^{0.165}\exp\left(\frac{2930}{T}\right) \tag{4}$$

The dynamic recrystallization threshold strain ε_c will initiate when strain reaches 0.7 times the value of ε_p.

2.1 Determining rolling parameters for hot rolling

In order to obtain good dimensional tolerance and optimum mechanical properties after rolling, optimum rolling parameters need to be established. The usual

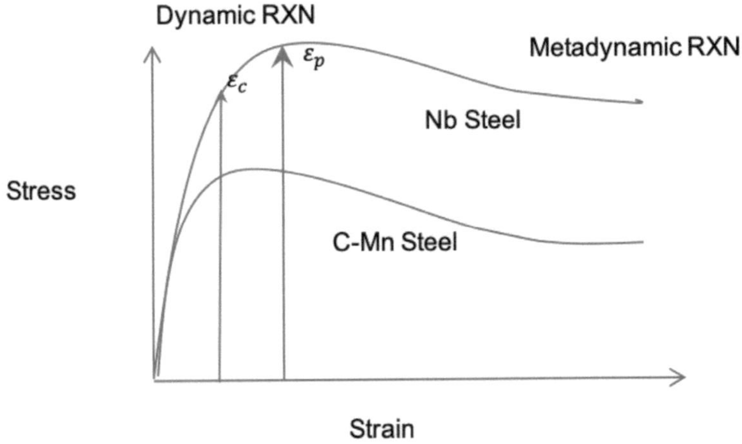

Figure 3.
Stress-strain distribution with onset of recrystallization during rolling pass.

method is to achieve this to calculate the dynamic mean flow stresses (MFS) at each rolling stand of roughing mill and more importantly at finish rolling mill. The MFS is chiefly dependent upon the alloying elements, rolling reduction, and temperature at each rolling pass and based upon these factors several models have been proposed [3].

A most widely used method, simplified rolling load versus inverse temperature, is plotted to determine the rolling conditions and has been depicted in **Figure 4**. When the mean flow stress, which is directly related to mill rolling load value, is plotted against the inverse absolute temperature, a slope kink signifying end of static recrystallization is observed. If rolling is accomplished below this temperature represented as T_{NR} (temperature of no further recrystallization), there is a sudden jump in mean flow stress which is due to additive nature of work hardening induced in each pass [4].

The onset of recrystallization during finish rolling is controlled largely by rolling interpass time. It has been reported [4] that for interpass intervals significantly longer than 1 second, static recrystallization takes place, whereas those involving interpass times of 15–100 ms, dynamic or metadynamic recrystallization is favored. Another important factor to include is consideration of strain-induced precipitation, which inhibits recrystallization phenomenon.

When high strength low-alloyed steels are finish rolled, an additive buildup of strain causes an increase in MFS consecutively after each pass. When a critical strain value is surpassed, dynamic recrystallization is initiated, and a small drop in load caused by metadynamic recrystallization is observed during end of rolling. In carbon-manganese grades, this may be associated with the austenite to ferrite transformation.

2.2 Modeling the mean flow stress to estimate the critical rolling parameters

As thermomechanical processing involves microstructural evolution in terms of static and dynamic recrystallization that takes place during the rolling process, the mean flow stress shall also depend upon these considerations also.

A model equation for mean flow stress prediction for carbon-manganese grades by Misaka [5] has been proposed as below:

$$MFS_{MISAKA} = e^{\left(0.126-1.75\ [C]+0.594\ [C]2+\frac{2851+2698[C]-1120[C]^2}{T}\right)} \tag{5}$$

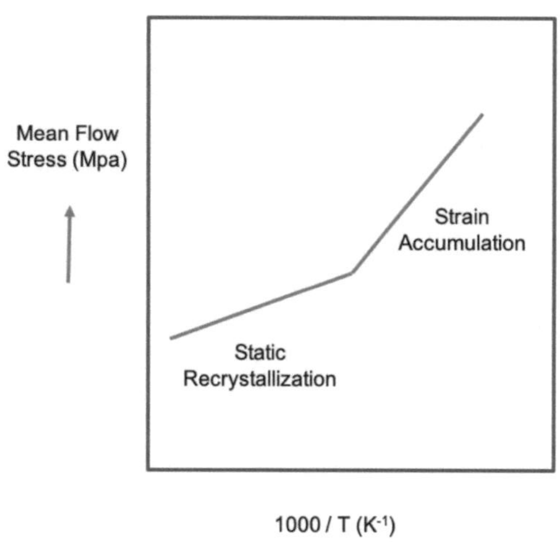

Figure 4.
Plot of mean flow stress versus inverse of absolute temperature.

Various researchers (Siciliano et al. [3], Sun and Hawbolt [2]) have worked upon refinement of the equation to include effect of recrystallization that would affect the value of MFS.

Recrystallization criteria are a function of initial grain size d_0, strain $\dot{\varepsilon}$, strain rate $\dot{\varepsilon}$, and temperature T during rolling and are initiated when the strain at a pass exceeds critical strain ε_C favored by appropriate temperature.

2.2.1 Critical strain evaluation for static recrystallization

If d_0 is the initial grain size, the critical strain to initiate dynamic crystallization is given by

$$\varepsilon_C = 5.6 \text{ x } 10^{-4} {d_0}^{0.3} Z^{0.17} \tag{6}$$

where Z is Zener-Holloman parameter defined by

$$Z = \dot{\varepsilon} \cdot \exp\left(\frac{300000}{RT}\right) \tag{7}$$

To calculate type of crystallization that occurs at a particular pass, the strain values need to be compared against critical strain value.

2.2.2 Grain size evaluation for static recrystallization

If ε_a (strain at a pass) is less than the critical strain ε_c, static recrystallization is favored, leading to grain size governed by the equation:

$$d_{SRX} = 343\, \varepsilon^{-0.5} {d_0}^{0.4} \exp\left(\frac{-45000}{RT}\right) \tag{8}$$

Time for 50% completion of recrystallization (static) is

$$t_{0.5}^{SRX} = 2.3 \text{ x } 10^{-15} \varepsilon^{-2.5} {d_0}^{2} \exp\left(\frac{230000}{RT}\right) \tag{9}$$

Grain growth during interpass time t_{ip} is governed by

$$d^2 = {d_{SRX}}^2 + 4 \text{ X } 10^7 (t_{ip} - 4.32\, t_{0.5}) \exp\left(\frac{-113000}{RT}\right) \tag{10}$$

2.2.3 Grain size evaluation for dynamic recrystallization

If ε_a (strain at a pass) is greater than the critical strain ε_c, dynamic recrystallization is favored, leading to grain size governed by equation:

$$d_{MDRX} = 2.6 \text{ x } 10^4 Z^{-0.23} \tag{11}$$

Time for 50% completion of recrystallization (dynamic) is

$$t_{0.5}^{MDRX} = 1.1 \text{ x } Z^{-0.8} \exp\left(\frac{230000}{RT}\right) \tag{12}$$

Grain growth during interpass time t_{ip} is governed by

$$d^2 = d_{MDRX}{}^2 + 1.2 \, X \, 10^7 \left(t_{ip} - 2.65 \, t_{0.5}\right) \exp\left(\frac{-113000}{RT}\right) \quad (13)$$

2.2.4 *Evaluation for work hardening and fractional softening*

Accumulated strain ε_a at each pass is governed by the following equation:

$$\varepsilon_a = \varepsilon_n + (1 - X)\varepsilon_{n-1} \quad (14)$$

where fractional softening X is governed by recrystallization:

$$X = 1 - \exp\left(0.693\left[\frac{t}{t_{0.5}}\right]^q\right) \quad (15)$$

The value of q shall depend upon the type of recrystallization.
For static recrystallization (SRX), q = 1.0.
For metadynamic recrystallization (MDRX), q = 1.5.

2.2.5 *Evaluation for predicting mean flow stress (MFS) in each rolling pass*

The modified mean flow stress modeling equation incorporating effect of manganese addition shall be

$$MFS = (0.78 + 0.137[Mn]) X \, MFS_{MISAKA} X \, 9.8 \, X \, (1 - X_{DYN}) + K\sigma_{SS} X_{DYN} \quad (16)$$

where X_{DYN} is

$$X_{DYN} = 1 - \exp\left(0.693\left[\frac{\varepsilon - \varepsilon_c}{\varepsilon_{0.5}}\right]^2\right), \quad (17)$$

where

$$\varepsilon_{0.5} = 1.44 \, x \, 10^{-3} \dot{\varepsilon}^{0.05} {d_0}^{0.25} \exp\left(\frac{6420}{T}\right), \quad (18)$$

and where σ_{SS} is defined as steady state of stress after peak stress is achieved:

$$\sigma_{SS} = 7.2 \, \dot{\varepsilon} \, \exp\left(\frac{300000}{RT}\right)^{0.09} \quad (19)$$

3. Effect of chemical composition

High-strength steel requires tensile properties as main requirement, whereas the requirements such as weldability and ductility are also of chief importance. Therefore, carbon which is the chief source of strength should not exceed very high values, and hence high-strength steel requires addition of alloying elements.

The addition of microalloying elements can be divided into two categories [6]:

1. Microalloying elements: niobium, vanadium, titanium, aluminum, and boron.

2. Substitutional elements: silicon, manganese, molybdenum, copper, nickel, and chromium.

3.1 Addition of niobium, titanium, and vanadium

Microalloying elements such as niobium, titanium, and vanadium are principally carbide-forming elements. Although the addition of these elements in steel raises its Ar_3 temperature, they retard austenite transformation to ferrite by restricting carbon diffusion. Strengthening by addition of one or all of niobium, vanadium, or titanium has shown a remarkable increase in strength of steel. The strengthening phenomenon is caused by fine precipitation of nitrides, carbides, or carbonitrides which are coherent with ferrite matrix but induce strengthening by impeding dislocation movement.

One of the most significant effect of adding individually or simultaneous addition of V, Nb, and Ti is to decrease recrystallization temperature. Which contributes in generating a finer size of gamma (austenite) grains during finish rolling.

The two principal mechanisms that inhibit recrystallization and eventually grain growth are particle pinning and solute drag.

The grain boundary movement can be accounted on strain-induced precipitation of micro carbides on gamma grain boundaries that limit the gamma grain size. Addition of titanium or niobium helps in suppressing gamma grain growth by means of nitride or carbonitride precipitates which are majorly present at grain boundaries and inhibit their movement.

In the case of vanadium addition, addition of nitrogen can be helpful in increasing the strength and toughness. The vanadium nitride precipitates are useful in imparting strength to the steel. The addition of nitrogen however attributes to poor weldability. Likewise, the strengthening may be achieved by adding niobium, but a higher niobium content is bound to give poor weldability. Hence the conventional methods require simultaneous addition of V and Nb.

3.2 Manganese-based strengthening

The improvement of toughness can be achieved through addition of manganese that leads to decrease in Ar_3 temperature. Due to decrease in Ar_3 coupled with low coiling temperatures, the alpha (ferrite) grains are refined, thus increasing the strength. Additionally, the fine precipitate size is contributed by niobium carbonitrides and vanadium nitrides.

4. Effect of controlled hot rolling parameters

4.1 Reheating temperatures at reheating furnace

In general, austenitic grains starts to recrystallize at temperatures above 1050°C. Since an initial finer gamma grain size is helpful in creating a final finer size of alpha, lower reheating temperatures are effective. Also, the microalloying elements also add to refinement of the austenitic grain size by means of undissolved carbides and nitrides that restrict initial austenitic grain size.

Eventually, a lower slab reheating temperature by contributing fine austenitic grain size and lower temperature rolling at roughing mill will induce even finer grain size by reduction at lower temperatures.

4.2 Repeated recrystallization in roughing mill

Due to repeated reduction in roughing mill, both recrystallization and precipitation are competing phenomena. However, at higher temperatures recrystallization will be the first phenomenon. As a basic principle of controlled rolling demands that precipitation should occur during finish rolling, it has been recommended to have higher roughing temperatures along with short rougher interpass intervals.

The gamma grain is refined by repeated static recrystallization caused during roughing mill action. Increasing rolling strain has marked effects on facilitating static recrystallization caused by higher dislocation density and increased nucleation sites caused by fine size of austenite which in turn leads to softening of material. However, there is a limiting value of grain refinement.

4.3 Finish rolling at no-recrystallization temperatures

As has been elucidated above, no-recrystallization temperature (T_{NR}) is important in design of controlled rolling process. This temperature determines where strain is multiplied for austenite grains, leading to strain-induced precipitation of carbonitrides as well as enhanced sites for a fine-grain size ferrite to be nucleated at the sites. Hot rolling being a dynamic process, no-recrystallization temperature depends upon deformation parameters. The influencing factors for T_{NR} are composition of the steel, strain values applied in each pass, the strain rate, and the rolling interpass time [7, 8].

During finish rolling the value of T_{NR} tends to dynamically lower down as the strain value or the reduction increases. This phenomenon is attributable on account of static recrystallization caused by increased recrystallization sites owing to finer grains and higher dislocation density induced during each rolling pass.

The strain rate value is also a determining factor for the onset of dynamic recovery and facilitates static recrystallization which eventually decreases the T_{NR}.

During controlled rolling, the interpass time during each rolling reduction also plays an important role as the prime requirement is to roll below T_{NR} temperatures. The precipitation kinetics are accelerated due to strains induced when rolling below T_{NR}. A lower interpass time is preferable as higher interpass will lead to coarsening of precipitate sizes as well as increased tendency of recrystallization detrimental to final strength value of steel.

4.4 Accelerated cooling

The runout table and the coiler in general act like post heat treatment unit which makes possible to achieve phase transformation through control of cooling to generate coils with varied properties and microstructures.

Accelerated cooling after hot rolling leads to further refinement of grains and phase control, leading to enhancement of properties. The phenomenon for strengthening of microstructure is phase transformations in terms of microstructures avoiding pearlitic transformations, precipitation strengthening through carbides, and nitride precipitates which along with controlled cooling rates lead to the refinement of grain size in the resulting microstructure. The accelerated cooling may be classified into two techniques—continuous accelerated cooling and interrupted accelerated cooling.

The final mechanical properties after accelerated cooling are majorly influenced by the alloying content and hot rolling parameters.

Author details

Gaurav Bhargava
Product Development and Quality Control Department, JSW Steel Limited, Karnataka, India

*Address all correspondence to: gaurav.iitr.mse@gmail.com

References

[1] Bowden JW, Samuel FH, Jonas JJ. Effect of interpass time on austenite grain refinement by means of dynamic recrystallization of austenite. Metallurgical Transactions A. 1991;**22** (12):2947-2957

[2] Sun WP, Hawbolt EP. Comparison between static and metadynamic recrystallization-an application to the hot rolling of steels. ISIJ International. 2007;**37**(10):1000-1009

[3] Siciliano F, Jonas JJ. Mathematical modeling of the hot strip rolling of microalloyed Nb, multiply-alloyed Cr-Mo, and plain C-Mn steels. Metallurgical and Materials Transactions. 2000;**31**(2): 511-530

[4] Jonas JJ. Dynamic recrystallization—scientific curiosity or industrial tool? Materials Science and Engineering A. 1994;**184**(2):155-165

[5] Misaka Y, Yoshimoto T. Formulation of mean resistance of deformation of plain carbon steel at elevated temperature. The Japan Society for Technology Plasticity. 1967–1968;**8**(79): 414-422

[6] Tamura I, Sekine H, Tanaka T. Thermomechanical Processing of High-Strength Low-Alloy Steels. Butterworth & Co. Ltd., 1988

[7] Bai DQ, Yue S, Sun WP, Jonas J. Effect of deformation parameters on the no-recrystallization temperature in nb-bearing steels. Metallurgical Transactions A. 1993;**24 A**:1993-2153

[8] Vega MI, Medina SF, Chapa M, Quispe A. Determination of critical temperatures (Tnr, Ar3, Ar1) in hot rolling of structural steels with different Ti and N contents. ISIJ International. 1999;**39**(12):1304-1310

Chapter 12

Welding Residual Stresses to the Electric Arc

Lino Alberto Soares Rodrigues, Ednelson da Silva Costa, Tárcio dos Santos Cabral and Eduardo Magalhães Braga

Abstract

Arc welding processes are widely used in the industrial sector, mainly for productivity and continuity. However, these processes have several undesirable results, such as distortions and residual stresses (RS). When compared to other welding processes, the RS level can make the welded joint unfeasible. Many studies on these arc welding discontinuities have been carried out in experimental and numerical areas about their measurement, analysis, and control, however, not yet clearly enlightened, since it is a complex topic, both for industry and academia, needing to be deepened. This study aims to present a contextualized approach to destructive and non-destructive techniques used to measure RS generated by arc welding, as well as the influence of these distortions and stresses on the welded structures and, finally, to present possible control techniques. Finally, this study highlights the use of CW-GMAW welding, which achieved a reduction in stress and distortion levels, due to the introduction of a non-energized wire in the arc of the GMAW process, as evidenced by the results of RS measured by X-ray diffraction (XRD) and acoustic birefringence (AB). Thus, in this context, the approach to RS in arc welding presented here is extremely relevant for researchers involved with the topic.

Keywords: arc welding, welding residual stresses, acoustic birefringence, CW-GMAW, NDT

1. Introduction

The search for improvements in the control of residual stresses resulting from arc welding processes has been intense due to the production requirements imposed by the industrial sector, referring to the fact that stresses affect the mechanical properties of materials, such as strength, plasticity and surface integrity, therefore it is extremely important to measure and evaluate the levels of these stresses in welded joints. The RS exist in structures, parts, components from different manufacturing processes, resulting from interactions of temperature, stresses, and microstructure, which makes its evaluation quite complex. Among the numerous manufacturing processes, the arc welding process generates high levels of RS, formed by the thermal changes imposed. The control of the RS continues from the moment of the conception of the project, passing through the choice of the construction procedures until its completion. There are several ways to measure RS, which can be computational or experimental. During the research of this work it was observed that there is still no standardized system for the distribution of residual stresses, each author uses what suits him, there is no right or wrong way to observe the RS. But a main point can be considered, being the consensus of many authors is the magnitude of the RS, whether it is compressive or tensile, each type of stress has a specific attribute, which may be beneficial or not for the evaluated component. A methodology worth mentioning for the evaluation of RS is the ultrasonic technique of acoustic birefringence (AB), with this technique it is possible to evaluate a metallic component in a non-destructive and entirety. This line of research, presents an interesting approach, but not much explored. The use of different RS measurement and control technologies, widen the options for existing assessment, with the combination of these technologies it is possible to achieve a reduction in RS levels or even almost its complete elimination. Countless welding processes have been used to control RS. Recently a process has been introduced, the CW-GMAW process (Cold Wire—Gas Metal Arc Welding), showing promising results. The premise of the process is to reduce the temperature of the fusion arc/weld pool. In this context, this work addresses a review of the concepts of residual stresses generated by arc welding, as well as their magnitude and implications for welded structures. However, ways of measuring them are also explored, so that more efficient control methodologies, welding processes are developed, which together with the welding procedures, promote significant results in reducing the values of residual stresses and deformations generated.

2. Mechanisms for generating residual stresses in electric arc welding

The importance of this study is based on the principle that the most widespread concept of residual stresses (RS) refers to stresses that remain in the component even though the external forces applied on the body are removed [1]. Otherwise, residual stresses are those that are not necessary to maintain the balance between the body and its environment [2].

The state of stresses causes a residual deformation that is self-balancing and, therefore, the resulting forces and moments that tend to zero [Eqs. (1) and (2)]. These equations describe the state of residual stresses considering a generic volume of the material and the moment of the forces acting on the material, respectively. Here, dV is the volume and dM is the resulting moment.

Mainly regarding metallic materials, residual stresses are a consequence of the interactions between time/temperature, stress/strain, and microstructure, that is, residual stresses arise from misfits *(eigentrains)* between different regions or different phases within the material, or even in different layers of atomic arrangements [3].

The material or characteristics related to this that influence the development of residual stresses include thermal conductivity, calorific capacity, thermal expansivity, modulus of elasticity and Poisson coefficient, thermodynamics and kinetics of transformations, transformation mechanisms and transformation plasticity [3].

$$\int \sigma . dV = 0 \tag{1}$$

$$\int dM = 0 \tag{2}$$

The RS are present in all materials that go through manufacturing processes, and in general, in metallic materials they are more evident and extremely studied, a fact that is due to the great use of these materials in the industrial one. Thus, in metallic materials, residual stresses are a consequence of the interactions between temperature, heating time, stress-strain, and microstructure, that is, residual stresses arise from mismatches between different regions or different phases within the material, or still, in different layers of atomic arrangements [3]. Regarding the intrinsic characteristics of these materials, which influence the suggestion of RS, there are thermal conductivity, heat capacity, thermal expansiveness, elasticity modulus and Poisson's coefficient, in addition to the thermodynamics and kinetics of transformations, of the mechanisms of transformations and transformation plasticity [3]. Hence, therefore, the importance of studying residual stresses in metallic materials, due to their great generation complexity.

Another important problem surrounding the RS is its classification, and this is not yet well established, however, some of these classifications will be presented below, according to the point of view of some areas [4]:

a. The most common occurs according to the scale to which they self-balance:

- Type I: These are stresses that act on macroscopic scales, involving several adjacent grains of the material, having an almost homogeneous character. Each interference in the balance of forces and moments of a volume containing this type of stresses may change its dimensions.

- Type II: They correspond to the average stresses within each phase, are almost homogeneous in all microscopic areas, of a grain or parts of it in a material and are balanced through enough grains. This type of stress is also known as pseudo-macrostresses [5].

- Type III: They are heterogeneous in the submicroscopic areas of a material, can be caused by accumulations of displacements (variations in interatomic distances) within a grain or elastic stresses around precipitates and are balanced through small parts of a grain.

b. According to their origins, that is, by the causes as they arose [2];

c. Or according to the effect on the behavior of the welded structure [4]

However, it is known that metal components need to go through at least one manufacturing process, even considering current processes, such as additive or more traditional manufacturing such as lamination, casting, forging, machining, thermal sprinkling or welding, each generates its pattern of residual stresses in the product, which is intended to produce [1]. However, few influences or generate high levels of residual stresses as do electric arc welding processes.

Thus, the joining of metallic parts using the electric arc as a heat source brings together many welding processes used on a large scale in the industry. These arc welding processes cause abrupt thermal changes through the local heat source at different scales giving rise to the known residual welding stresses, that is, the consequent mismatch between different parts (base metal—BM, zone affected by heat—HAZ and weld metal—WM), different phases (microstructures) or different regions of the same part (different grains in the HAZ) contributing to their formation [2]. However, the phenomenology of this process is complex and the following phenomena occur almost simultaneously:

1. In the heating occurs the formation of the weld pool (until the weld pool forms the piece heats up a lot, varying for each material) that expands and generates compressive flow from the neighborhood and when the weld pool cool causes

the formation of contraction forces [6]; more particularly, the mass of the heated volume with the restrictive combinations and the contraction of the weld metal originate the thermal stresses.

2. The heat transfer and the flow of the liquid metal generate thermal gradients in the welded joint, besides actively acting in the shape and size of the melting pool, that is, in the volume of liquid metal. The measurement of heat transfer through the cooling rate helps in the prediction of Thermal Stresses (TS) that will affect the level of residual stresses formed in the welded joint.

3. The volume of the liquid metal depends on the interaction of the welding parameters and the displacement of the electric arc that subjects the welded joint to various thermal cycles, can be differentiated when: (a) if the welding process uses only the electric arc to fuse the parts to be joined, i.e. when only the base metal is melted, autogenous welding (e.g. GTAW) or (b) if transfer of molten metal from the consumable occurs, mixing with the molten base metal simultaneously with the movement of the arc, the weld is deposition (SMAW, GMAW, FCAW, SAW, etc.).

4. Thermal stresses are stresses formed during the thermal cycle, both in heating and cooling, not existing in liquid metal; however, these stresses until they reach room temperature become the forms of residual stresses, distortions and/or defects [7].

5. During the solidification of the weld pool, phase transformations occur influencing or not the formation of residual deformations, depending on the chemical composition of the metal alloy, there is less or greater influence on the formation of residual welding stresses [8].

One way to study the mechanism of generation of residual stresses in welded structures beyond experimentation and measurement is based on thermomechanical processes associated with computational mathematics and simulation with specific software. It is possible to obtain satisfactory results that help in the prediction, control and relief of the real state of residual stresses caused in the structure after welding in a qualified manner, that is, to perceive the effect that each factor or parameter alone has on the magnitude of such stresses. The starting principle is based on the models of [9, 10] referring to the effects of the temperature distribution of the heat source during welding and through these models the possible effects on residual stresses originated as proposed by [11, 12].

However, the great predominance of studies today has been consolidated with residual stresses being influenced by four groups of interrelated welding technology parameters.

- ***Welded structure design parameters***: the thickness of the plate or pipes (thin, medium or coarse), joint geometry (butt weld, fillet, double fillet, etc.) and type of chamfer (V, X, K, U, etc.), pass numbers, chemical composition of the base metal and consumable (wire or rod).

- ***Available choice parameters***: Welding processes (autogenous process), such as GTAW or with metal deposition such as GMAW, FCAW, SAW, SMAW, PAW; the mode used for welding (manual, mechanized, automated), etc.

- ***Application parameters of operational techniques***: welding sequence, welding passes (continuous, intermittent, alternating, reverse, tensioning), pre and post heating, structure with or without restrictions.
- ***Primary operating parameters:*** such as I (current), U (voltage), welding speed (mm/s). These are, however, the most used, because they constitute the energy involved and used for metal fusion, its formula is described according to Eq. (3). Known as heat input (Hp), generic term of welding energy, usually in KJ/mm. Which also interfere in the shape and volume of the weld bead.

$$Hp = \eta . \frac{U.I}{\upsilon_s} \tag{3}$$

First, due to the greater constraints of safety criteria, the design of a conscious welded structure currently involves the skills of welding engineers and designers, considering not only the structure itself, but also the prior knowledge of complex interactions between metallurgical phenomena, aspects related to the mechanisms of connections and the mechanical behavior of all materials involved [13]. Where residual stresses, distortions and failures are perceived to be costly and complex control phenomena. Thus, considering that some areas of industrial production, for example, shipbuilding do not all projects quantify the initial welding imperfections explicitly, but these imperfections reduce the strength of the structure, besides being accumulative [14] and dangerous in the case of vessels when in operation.

In a way, the control of welding residual stresses starts from the choice of the type of material to be welded, that is, carbon steel, because it presents a composition basically formed by Fe and C and few alloying elements that interfere in phase transformations, it is considered in practice that this mechanism has negligible participation in the generation of RS in the weld metal of this type of material. Unlike medium and high alloy steels, stainless steel, Monel, Stellite, etc. these will have their inherent stress levels affected by chemical composition [15]. Thus, welding processes were developed through the formulation and manufacture of special consumables produced with this objective based on the principle of phase transformations at low temperatures (LTTW—low temperature transformation welding), where compressive residual stress formations and distortion reduction are induced [16, 17]. Other characteristics intrinsic to the material such as physicochemical properties (thermal conductivity, thermal expansion, density, specific heat, etc.), may favor for the generation or need preheating to relieve RS, such as copper and aluminum and their alloys, because they are good heat conductors, dissipate it quickly, most often requiring more intense localized sources, due to the difficulty for local fusion of the weld.

Other factors, very relevant that are correlated to the formation or increase of RS in welded components, refer to the parameters of joint geometry such as thickness (plate, pipes, flanges, etc.), type and angle of bevel. The temperature gradient is what differentiates the origin of residual stresses into a thin and coarse component. Where, in thin thicknesses, the weld pool can be considered 2D and in thick thicknesses, 3D, changing the process of heat transfer of the part and, consequently, the distribution of temperature that directly influences the process of RS formation [18]. The geometry of the joint in the form of butt weld, fillet weld, superimposed, among others also affects this distribution, by the way the heat is distributed, being the T-joint is the one with the highest heat removal coefficient [19].

Using models through the application of finite element methods (FEM) applied to the butt weld, with the decrease in thickness the RS increase [20]. A possible argument to describe the fact is that the absorption energy per unit of volume in thin plates is higher than in the thick ones, causing the inverse relationship between the thickness of the plate and the residual stresses. However, in T-joint welds, with the increase in plate thickness, the non-uniformity of the temperature alters the thermal expansion and the contraction during cooling, consequently, increases the RS. In this same type of joint, the increase in flange thickness strengthens the internal restriction by increasing these stresses in T joints [21]. The number of passes also influences the distribution of residual stresses, in fillet weld in T joint, a single pass on one side generates more stresses than when welding both sides [22]. The measurement of stresses at each weld pass was simulated in a joint containing six passes in plates of 16 mm thick hardened steel, where it was observed that the first 3 passes tend to generate more compressive stresses and the following passes 4, 5 and 6 showed the tendency to tensile stresses.

The choice of process also affects the magnitude of RS of electric arc welding, one of the studies that most involves processes with this type of local source, used four of these processes (SAW, DC GMAW, GMAW pulsed and CMT Fronius) and two more laser (one autogenous and one hybrid) applying in naval carbon steel (ASTM A131), where it was observed that the processes showed very similar peaks (near to 400 MPa) proportional to temperature peaks, differing by the width of the peaks, where the SAW has wider peaks [10]. On the other hand, the process and RS can present unexpected results, using the SMAW, FCAW and GTAW processes in ballistic steel butt welds, it was observed that the HAZ when subjected to projectile penetration testing, the SMAW process with higher residual stresses, withstood the test, and the others even with lower stress levels were penetrated [23].

Regarding the application parameters of the techniques, the two most researched are the constraints of the welded structure and the welding sequence. The degree of restriction refers to the resistance of the welded joint to the contraction and thermal expansion free of the heated material [24]. The constraint of the welded joint has a strong influence on the level of residual stresses, so much so that it can be considered that there are the inherent residual stresses produced naturally by the internal misfit and auto equilibrium and the reaction stresses that are a consequence of welded parts usually trapped by mechanical mechanisms. However, the basic principle for good welding practices reveals that longitudinal constraints more efficiently decrease residual welding stresses [25]. Otherwise, the deposition sequence of the welds also has a direct impact on the distribution of RS and distortions in the most varied forms and welded geometries. In studies involving the simulation of butt welds in plates and circumferential welds in pipes, this influence became notorious [26]. Simulating the J bevel deposition sequence in austenitic stainless steel tube-block joints (SUS304), the results indicated that this sequence has not only significant influence on the gradient of RS, but the last pass has a more significant gradient at the end [27].

Finally, no class of factors is further studied than the primary parameters of electric arc welding (U, I and ν_s), the interaction between these parameters generates the energy required for arc opening, producing thermal cycles and temperature peaks that generate thermal stresses and are the driving force of phase transformations during weld pool cooling. From this arc energy only part of it participates effectively in the fusion that generates the weld metal. It is important to mention that heat input has been erroneously referenced in many articles as synonymous with welding energy, which are not equal. However, its quantification is difficult due to numerous experimental difficulties [7, 28]. The basic relationship of the heat input with RS refers to its origin from the conversion of thermal stresses, as already mentioned.

Thus, any change in the primary parameters will change the formation settings of the residual stresses. That is, when the welding speed growth, the heat input decreases and residual stresses increase by producing a strait isotherm [12, 29]. In addition, the increase in welding energy is directly proportional to the magnification in RS present in the welded joint, also causing an enlargement the peak of tensile stresses [9, 24]. Otherwise, welding energy is so important by controlling the transformation temperature of weld pool phases that it is possible to combine the microstructure with acceptable resistance and toughness with low tensile or even compressive residual stresses. These transformations are governed almost entirely by austenitic transformation in the case of ferrous alloys.

3. Evaluation and measurement of residual stresses in welded structures

3.1 Directions and magnitudes

The evaluation of RS states is often uncertain and the reason for this is related to several specific aspects that should be considered in the measurement of residual stress, since analyses are sometimes problematic and dubious and that residual stress notation is not always used adequately by some authors [30]. Because there is no standardization system that convinces and guides the distribution of residual stresses, each author uses what suits him. However, this point is paramount for proper understanding and knowing the possible effects that can cause on a welded part or structure. Thus, considering that there is no right or wrong way, the criterion used is based on the one most referenced by the specialized academic community, according to the model in **Figure 1**. Generally, the analyses are performed two-dimensionally, considering the longitudinal (σ_y) and transverse (σ_x) directions studied, always having as reference the weld bead. The stresses of the normal plane (σ_z) are less measured, but should not be discarded in any hypothesis, even though the thickness of the plate is conditioned. Finally, it is important mentioning that, in practice, the most significant component is composed of longitudinal stresses, and generally equals, on average, three times the transverse stresses to the weld bead, where welds of a single pass are considered and that there are no temperature gradients on the z axis [25, 31].

Still, within the study of residual stress generated by arc welding two parameters are essential for understanding it: (i) the behavior of this RS (in MPa), that is, if it

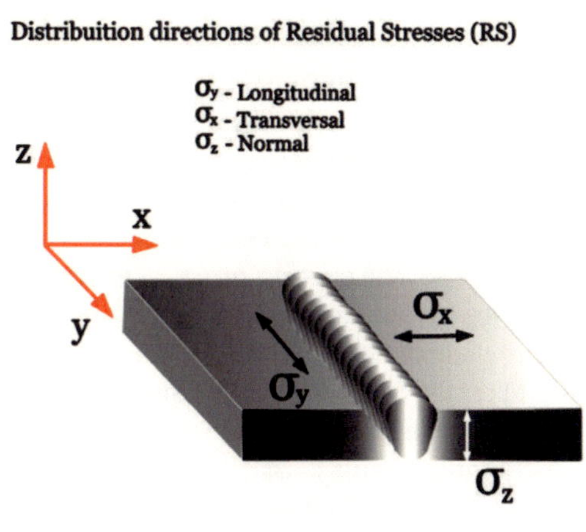

Figure 1.
Schematic representation of the distribution of RS directions on a three-dimensional plate.

is tensile to positive (+) or compressive values for negative (–) values; (ii) and its effect on the welded component, i.e. defects or failures related to residual stresses. Therefore, it is worth mentioning that the relationship between tensile and compressive residual stresses lies in the fact that the first has a degradative effect on the welded part, while the other can contribute in a beneficial way in many cases [3, 32]. However, this statement is often justified, due to the absence of knowledge of the existing residual stress state and its consequences, which is used to explain unexpected failures most often.

Based on the various works already published, it is possible to have a predictability of RS profiles in any of the directions of the plane, considering the metal alloy and the geometry of the welded joint, with this, represents a pattern of longitudinal residual stresses for certain alloys considering plates with butt welds [11].

When it comes to the effect of residual welding stresses on the behavior of materials, depending on their magnitude these stresses affect not only the welded joint, but also, on some occasions, the entire structure. Generally, tensile stresses are attributed to decreased fatigue life and corrosion resistance (cracks by stress corrosion), in addition to hydrogen embrittlement, etc. It is often suggested that the maximum magnitude of the tensile residual stresses should not approach the yield limit of the material, which would lead this or the welded joint to premature failures, especially near the weld. On the other hand, compressive residual stresses when at high magnitudes are usually attributed to decreased buckling resistance of the base metal, provided that the component is subject to compressive loading. Finally, in this way, it is expected that any welded structure, the simplest, will have residual stresses at any magnitude at controlled levels, or that approach the neutral line, since the peaks, even unprovoked, only increase the risk of unwanted and un predictable failures.

3.2 Measurement techniques

There are different measurement techniques to evaluate the residual welding stresses in a welded component. Some are based on the measure of relieved deformation, due to localized removal of material, called destructive techniques. Others are based on the interaction between the residual stress field and the physical properties of the material, called non-destructive (NDT). However, numerous authors also classify some techniques as semi-destructive, since their use does not compromise the physical structure of the material.

3.2.1 Destructive techniques

Among the various techniques for measuring residual stresses considered destructive, there is the sectioning technique that is based on the measurement of deformation due to the release of residual stress after removal of the material from the sample. The sectioning method consists of making a cut with appropriate instrument in the sample in order to release the residual stresses that are present in the cutting line. For this purpose, the cutting process used must not introduce plasticity or heat into the sample, so that the original residual stress can be measured without the influence of the effects of plasticity on the surface of the cutting planes [33]. In [32, 34], a cut-off sequence used by the technique to measure residual stress is shown schematically in literature.

Another technique belonging to this classification is the contour technique, which is based on solid mechanics, which determines RS through an experiment that involves carefully cutting a sample into two parts, measuring the resulting deformation due to the redistribution of residual stresses. The measured strain data is used to calculate residual stresses through an analysis that involves a finite element model that considers stiffness and geometry as just one parameter in the analysis, providing only one result [32].

There is also the technique of removal of layers, whose principle of action consists in the removal of material that contains residual stress, this removal of material causes unbalance in the workpiece that may result in deformation of the same. Removal of the tensioned material is performed continuously and the measurement of the curvature (deflection) of the sample is also done at the same time of removal, the residual stresses originally present can be deduced. Therefore, this technique provides a quick determination of residual stress as a function of the depth below the surface [35]. According to [34], the variation of deflection (curvature) after the removal of a layer of material, from a thickness e', can be related to the stress σ_e, which acted on that layer.

Other destructive techniques are the Hole-Drilling Method and the Ring Core Method. The hole-drilling technique is the most used general technique for measuring residual stresses in materials. It uses standardized procedures and has good accuracy and reliability. The test procedure involves some damage to the sample, but this is often tolerable or repairable. For this reason, the technique is sometimes called semi-destructive.

Otherwise, the hole-drilling technique is the most used due to its greater ease of use and to cause less damage to the sample. Finally, there is the Deep Hole Drilling Method, which is considered a variant of the hole-drilling and ring core techniques, with the difference of performing an analysis on thicker materials [33]. According to [36] the basic procedure involves the machining of a reference hole through the sample and the subsequent removal of a column of material, centered on the reference hole, using a trepanning technique. The diameter of the reference hole is accurately measured along its length before the material column is removed. Because, when the material column is removed, the stresses are relaxed, the dimensions of the hole diameter of the reference column are changed. In this context, the dimensions of the column and the reference hole are measured again and the residual stresses are calculated from the dimensional changes caused by the removal of the material from the greatest deformation of the sample in the analyzed area.

3.2.2 Non-destructive techniques (NDT)

Within this classification of techniques, there are the magnetic techniques that are based on the relationship between magnetization and the elastic deformation existing in ferromagnetic materials, as experiments demonstrate that a piece of steel wire, once magnetized, will undergo elongation in the direction of magnetization, while once pulled it will magnetize in the direction of the pull. Two techniques have been extensively explored in the literature, in addition to being applicable in industry: the Barkhausen noise technique and the magneto-striction technique. The first is based on the change in the magnetic microstructure caused by the presence of stresses, while the second is based on measurements of the permeability and magnetic induction of the material [34].

The most common magnetic technique is the Barkhausen noise magnetic technique. Ferromagnetic materials present magnetically ordered microscopic regions, called domains, where each domain is magnetized based on the crystallographic directions preferred to the magnetization. Furthermore, a domain does not coincide with a grain, since within a grain there are several domains, which are separated by walls, in which the direction of magnetization generally changes by 90° or 180°. In [34] also states that when a magnetic field or mechanical stress is applied to a ferromagnetic material, changes occur in the structure of the domains caused by the sudden movement of the walls. These changes cause variations in the average magnetization of the component, as well as in its dimensions. Thus, if a conductive coil is placed close to the sample while the domain wall is moving, the resulting change in magnetization will induce electrical pulses in the coil. When these electrical pulses are produced by the movement of all domains, a signal is generated, called "Barkhausen noise." The extent of movement of the domain walls, that is, the intensity of Barkhausen noise, depends on the stresses present and the material's microstructure. The measurement depth for practical applications of this technique on steel varies between 0.01 and 1 mm. The authors [2] present results of stresses measured by this technique.

Another non-destructive technique is Neutron Diffraction, which, similarly to the X-ray diffraction technique, measures the crystallographic spacing between the crystalline planes. This spacing is affected by RS or applied stress [37]. This technique can measure the elastic deformations induced by residual stresses in the entire volume of the relatively thick steel components with a spatial resolution as small as 1 mm^3 [2]. The authors [3, 36] claim that the greatest advantage of neutron diffraction over x-ray diffraction is the great depth of penetration that neutrons can obtain, which makes it capable of measuring a greater depth, reaching 25 mm in aluminum and 25 cm in steel. According to [36], due to the high spatial resolution, neutron diffraction can provide complete three-dimensional deformation maps in an engineering component, that is, for each measurement point, the deformation can be measured in three orthogonal directions along the axis Sample. Practical applications of the technique and the theoretical background can be seen in [34, 38, 39].

There is also the x-ray diffraction technique (XRD). This XRD technique was first proposed by Lester and Aborn in 1925 [34]. However, the technique is restricted, its main restriction being the depth of analysis of the samples, since the beam of x-rays can only penetrate the distance of some atomic planes, about 1–50 μm [36]. For [40], the penetration is around 25 μm and for [41] it ranges from 5 to 20 μm, that is, the XRD makes a subsurface assessment of the stresses. To overcome this restriction, [37] states that for measurements at a greater depth, that is, greater than 0.013 mm, destructive techniques such as the layer removal technique should be used. In the evaluation by XRD, the residual stress is calculated from the measurement of the deformation in the crystal of the polycrystalline aggregate, compared to the network parameters of the crystal of this same material without suffering deformation. When a beam of x-rays is directed towards the surface of a body, a part of these rays is absorbed by the atoms while another part is sent back in all directions of the irradiated area. This technique basically measures the maximum diffracted ray intensity for a given scanning angle. From this angle it is possible to obtain the interplanar spacing of the diffraction planes determined by Bragg's Law [5, 42]. For the measurement of residual stresses using XRD, there are three basic techniques [3, 42]. Techniques, double exposure, single exposure, and multiple exposures or sen2ψ. The amount of exposure refers to the amount of exposure angles, such as angles between the normal at the surface of the part and the plane formed by the incident X-ray beam and the diffracted beam. The sen2ψ technique is one of the most classic [43]. This method is capable of measuring stresses with an accuracy of ±20 MPa, with a penetration depth in the order of microns under the sample surface [2, 4]. This technique and its history of development are described in [5, 43, 44].

Finally, there are the ultrasonic techniques, which are based on the acustoelastic effect, which is the influence of the state of stress on changes in the speed of propagation of the ultrasonic wave as it travels through the material, developed by [45], using the theory of finite strain and third order terms of [46] elastic strain. The techniques that use Critically Refracted Longitudinal Waves (Lcr) and shear waves [32] stand out. The ultrasonic technique that uses Lcr waves, according to [47], is a special case, as these to be generated, must be introduced into the material with an angle of incidence slightly greater than the critically refracted angle (first critical angle), based in Snell's Law. This wave propagates parallel to the surface of the material to be analyzed, as can be seen in the works of [47–49]. In addition to these, numerous studies are available in the literature, such as [41, 50–52].

The other ultrasonic technique is acoustic birefringence (AB), which relates the relative difference between the velocities or the time of two shear ultrasonic waves with polarization directions orthogonal to each other, indicating the degree of anisotropy of the material, where birefringence is determined by (Eq. (4)) [53–59]. In this equation, V_l is the velocity of the shear ultrasonic wave with the polarization direction aligned with that of the lamination, V_t is the velocity of the shear ultrasonic wave with the transverse to lamination polarization direction, t_l is the travel time of the ultrasonic wave with the polarization direction aligned with the material lamination direction and tt the travel time of the ultrasonic wave with the polarization direction perpendicular to the lamination direction.

$$B = \frac{V_l - V_t}{\frac{V_l + V_t}{2}} = \frac{-(t_l - t_t)}{\frac{t_l + t_t}{2}} \tag{4}$$

The acoustic birefringence B depends, therefore, on the initial anisotropy B_0 and on the main difference in tension (σ_1-σ_2), as well as the anisotropic speed of the wave not being directly linked to the effect of stress due to the presence of this initial anisotropy of the material [60]. When the directions of the principal stresses coincide with the axis of the initial anisotropy, the relationship between the difference of the principal stresses with the birefringence is established according to Eq. (5) [61].

$$B = B_0 + k(\sigma_1 - \sigma_2) \tag{5}$$

In this equation, B_0 is the birefringence for the material in the stress-free state and k is the acustoelastic constant that relates the stress variations with the birefringence. This technique has been used a lot lately in practical measurements of residual stresses. In Brazil, since the end of the 1990s by [62] in projects with Petrobras. More recently, by [63], where the technique was used to measure residual stresses generated by GMAW and CW-GMAW welds. **Figure 2** shows the assembly of the RS measurement equipment by acoustic birefringence, identical to that used by researchers.

A comparison between non-destructive methods can be seen in **Figure 3**, where x-ray diffraction techniques (the mean between σ_y and σ_x) and acoustic birefringence (the difference between σ_y and σ_x in MPa), in order to verify the main similarities and differences regarding the values obtained and the magnitudes found. The measurement was made on naval steel plate (ASTM A131 grade AH32) with a thickness of 9.5 mm and dimensions of 1200 mm by 800 mm, considering 3 equidistant lines separated by 300 mm and 100 mm from the edge in the longest direction. Each line has 9 points separated by 120 mm from each other on the same line.

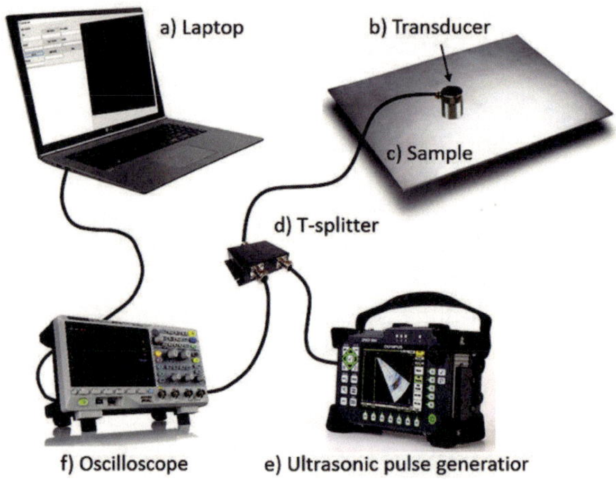

Figure 2.
Ultrasonic system used to measure RS by acoustic birefringence.

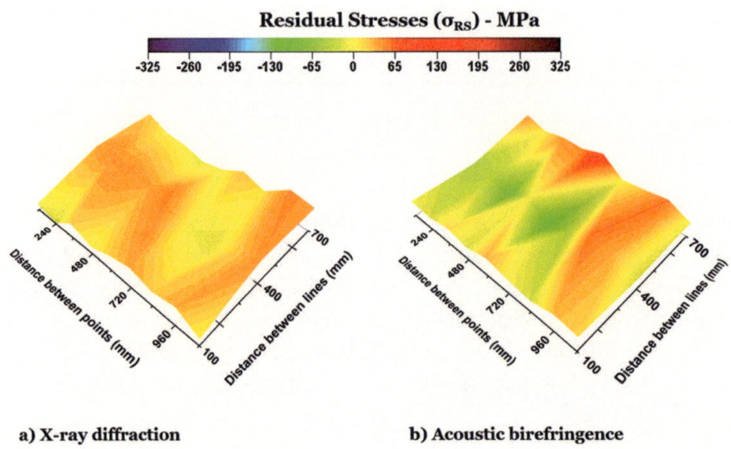

Figure 3.
Comparative measurement in a naval steel sheet (ASTM A131 grade AH32) between the methods: (a) X-ray diffraction (XRD) and (b) acoustic birefringence (AB).

Through **Figure 3**, there is a reasonable difference between the data obtained, where the plate measured by XRD is predominated by yellow and some peaks in more orange, however the range of residual stresses varied from −17 MPa to 50 MPa, while measuring with AB, it showed the central line with more compressive points, in green and other red peaks indicating tensile stresses, with values ranging from −112 MPa to 103 MPa. However, this disagreement in the results is mainly due to two factors, first, due to the difference in the depths analyzed between the measurement techniques, where the XRD method is more superficial and the AB method analyzes the entire plate thickness. Second, due to the specific methodology of each technique, that is, the XRD shows the punctual and unidirectional RS, while the AB shows the average difference of the main RS, but this result was already predicted. Although both techniques were able to clearly show the presence of RS in the component welded by CW-GMAW.

4. Techniques for controlling or reducing residual stresses and distortions

The relief of residual stress in welding has as premise the production of a rearrangement of atoms or molecules from their position of momentary equilibrium, from where the material leaves from a larger state to another of lower tension (lower potential energy), a more stable position. The analysis of residual stress and welding distortion, seen from a historical perspective, developed largely independently of each other, although, from the physical point of view, they are closely related [11].

To design and manufacture a structure with the least number of defects it is essential to have: an appropriate design; an appropriate selection of materials; suitable welding equipment and procedures; good manpower; and strict quality control [1]. The authors [64] showed that the procedures for reducing RS's are directly linked to the reduction of deformations generated during the joining process. There are several examples of methodologies that can be used to perform the relief of RS's, classified into three major categories: Thermal, Mechanical and Chemical, which can be performed before, during and after welding. There are many stress relief techniques that can be classified as conventional, due to their extensive use and the most varied applications in welded components, such as: hammering, heat treatment (pre- and post-heating), shot peening, mechanical tensioning, among others. However, non-conventional alternative techniques will be approached.

4.1 Procedures for the adequacy of the quantity of material deposited in the weldment

Since the RS in welding are the result of non-uniform deformations caused by the thermal gradient of the process, which can be attenuated by reducing the volume of weld metal deposited and adapting the chamfer design, which tends to decrease the heat transferred to the part and consequently, it causes a decrease in the RS levels and the degree of distortion of the weldment. The adequacy of the quantity of material must be chosen at the stage of development of the project and welding procedures. **Figure 4a** shows the representation of angular deformations in butt welds with various thicknesses, demonstrating that the angular distortion increases as a thicker plate is used, due to the greater amount of deposited material, resulting in a greater contraction during the solidification process. For small thicknesses, the angular deformation is not significant due to the high homogeneity of the temperature field through the thickness of the sheet. Plates with

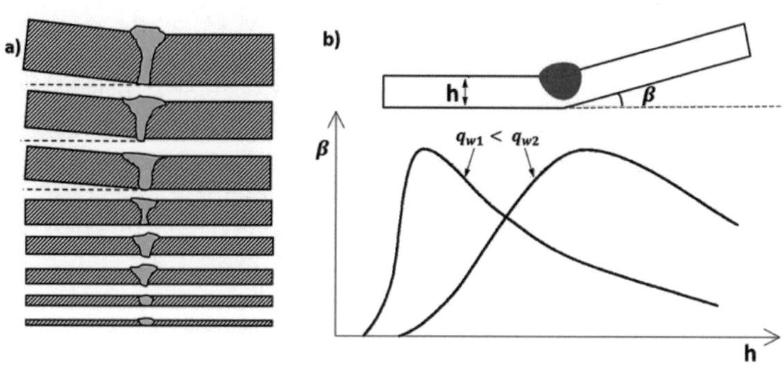

Figure 4.
Representation of angular distortions. (a) plate distortions in butt joints with various thicknesses and (b) angular distortion depending on the thickness of the plate and the heat input q_w.

intermediate thickness correspond to the highest angular distortion value [64–66]. However, for plates with great thickness the value of angular deformation became low, due to its greater rigidity. **Figure 4b** shows, in a qualitative way, the angular deformation as a function of the thickness of the plate and the heat input, q_w [67]. The increase in heat delivered to the plate can be represented using different welding parameters, such as increasing the feed value, in the case of the GMAW process, using the welding torch manipulation, weaving, and reducing the welding speed. The variation of these parameters generates an increase in material deposited per unit of length, consequently greater heat input, which shifts the curve to the right (**Figure 4b**). The authors [68] observed in their study that by applying a variant of the GMAW process to the CW-GMAW process, a reduction in the width of HAZ was obtained. This trend would be consistent with faster cooling rates produced using the CW-GMAW process, which suppresses ferrite nucleation at the grain boundaries. This reduction in the cooling rate resulted in less misalignment of the welded joint in the CW-GMAW process, compared to welding performed by the GMAW process.

4.2 Processes of construction and arrangement of the weldment

The authors [64] highlight the need to establish in advance the meanings of the welding sequence and the deposition sequence used in the construction of a structure. The welding sequence is closely linked to the assembly sequence, which must be established during construction planning, observing the manufacturing feasibility in order to minimize residual stresses and distortions. The manufacture of a structure formed by several unions, must follow some basic rules. In summary, the welding process must be performed symmetrically in relation to the neutral axis of the structural set, in order to counterbalance the forces arising from the contraction of the weld beads. However, it is not always possible to follow welding procedures determined in the project, in practice, these procedures will be decided when the construction inconveniences are observed in the field.

The deposition sequence refers to the progression of the formation of the beads during the execution of the welding, being able to use several types of progression that can be combined in different ways. As examples of sequences and progressions one can mention: continuous sequence (continuous pass); symmetric sequence; sequence with guided passes; progression by continuous passes and sequence of backwards passes.

The researchers [69, 70] developed studies on welding sequences, in order to promote the reduction of deformations in stiffened panels. The authors [69] used the robotic FCAW welding process in T joints of ASTM 131 grade AH32 stiffened steel panels. **Figure 5** shows examples of welding sequences used. After welding, they observed that sequence 3 (S3) presented the lowest global distortions value. In addition to the deformation analysis, he quantified the RS values using a portable X-ray diffraction equipment. **Figure 6** presents the 3D representation of the mean RS values in the longitudinal σ_y and transversal σ_x directions along the X axis. As the RS is closely linked to the behavior of the distortions, the panel that presented the lowest RS values was the sequence S3, with 59 and 86% for longitudinal and transverse compression stresses, respectively.

The influence of direction, welding sequence and reverse pass welding were analyzed by [70] on the distortions levels. The author uses in his work the GMAW process in short circuit mode, in T-joints of low carbon steel stiffened panels, where five welding sequences were used. Six welds of the same length were deposited in each panel to fix the reinforcements, obeying the order and sequence of welding.

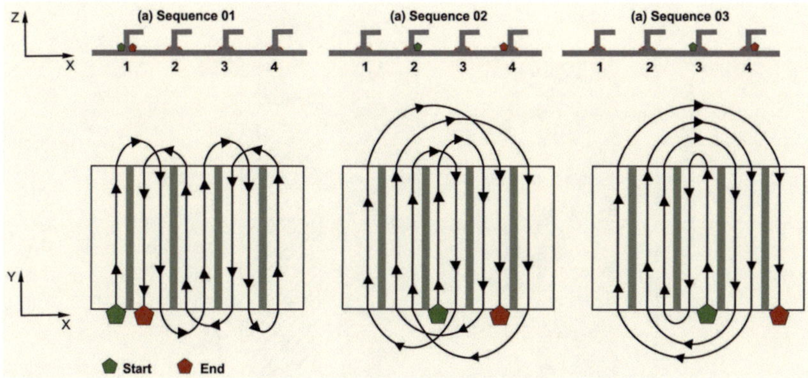

Figure 5.
Examples of welding sequences and directions in the union of hardened panels.

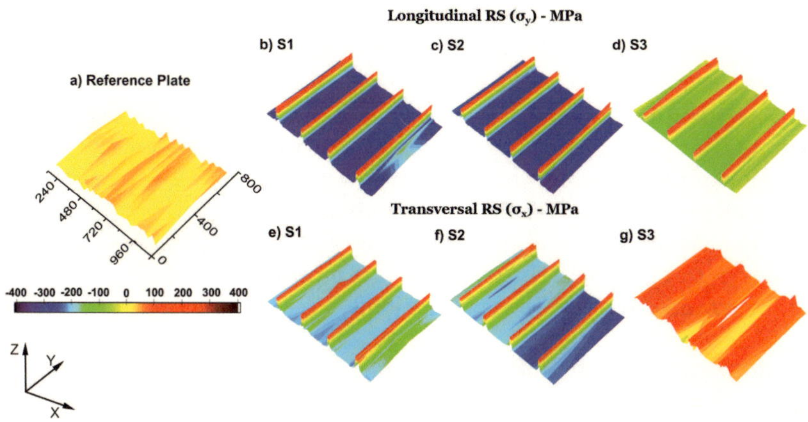

Figure 6.
TR's values in the 3D perspective for (a) reference plate; (b) sequence 1; (c) sequence 2; (d) sequence 3 in the longitudinal direction; (e) sequence 1; (f) sequence 2; and (g) sequence 3 in the transverse direction.

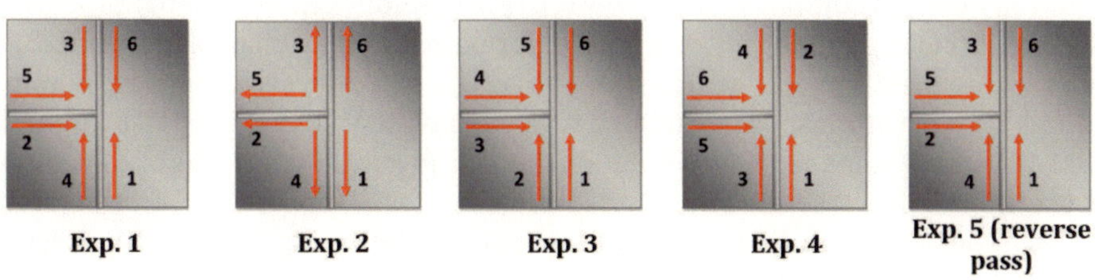

Figure 7.
Schematics of welding sequences and directions for the stiffened panels.

The welding sequences and directions for fixing the stiffeners can be seen in **Figure 7**. In the fifth sequence, the reverse pass was used, since the weld was divided into three equal segments. The arrow indicates the welding direction.

The taking of the values of distortions of the panels was carried out in the four corners of each panel, adding the values of the distortions to obtain the global distortion of the panel. The average global distortion values measured, **Figure 8**, show that the lowest distortion values are obtained from the pass in the central direction to the ends (Exp. 2) and to the reverse pass (Exp. 5). For Exp. 2, the sequence used

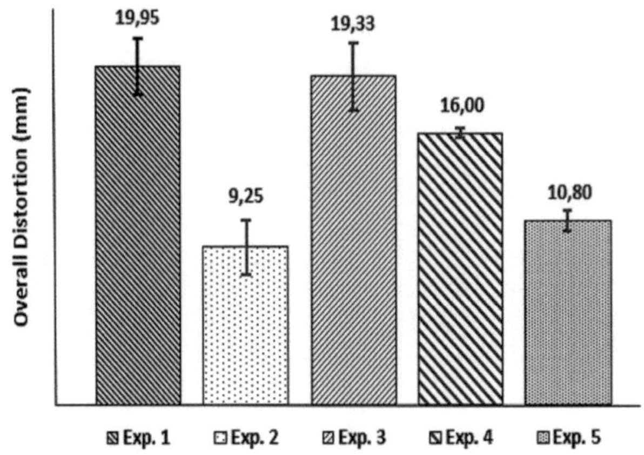

Figure 8.
Overall distortion of the experiments.

generated values consistent with those found in numerical studies [71–73]. Similarly, the results are corroborated by the work of [69], where it was observed that the welding sequence that provided the lowest distortion value must be performed from a more rigid point (central part) to one of less rigidity (extremities), resulting in less flexion of the panel and consequently lower values of RS. Exp. 5 obtained values very close to Exp. 2, due to the greater control of the temperature differential applied to the welded sheet and to a more uniform distribution of residual stresses [71]. However, this gain in the distortion values must be well considered, since a longer time was used to make the stiffener joint.

4.3 Vibratory stress relief (VSR)

The VSR process has achieved great prominence in the relief of RS's induced by thermal processes, such as welding, casting, but not those induced by cold work, being applied in several materials, low and medium carbon steels, stainless steels and aluminum alloys, not having an expected effect on copper alloys. VSR offers several advantages compared to the PWHT (Post Weld Heat Treatment) process: low time and energy spent, low thermal deformation and no change in the mechanical or metallurgical properties of the material [11, 65]. However, there are numerous conditions that must be considered when using the VSR and PWHT processes. Within the conditions employed by the authors, a lower RS value was obtained for the use of both processes.

The basic premise of this method is the relief of the workpiece RS with a region where the natural stress has been changed. When the part is subjected to vibrations below its new frequency, the metal absorbs energy, gradually redistributing the stresses and the resonant frequency returns to the point corresponding to a residual, or almost free, state [20]. The search for greater productivity for the arc welding process has generated efforts by researchers to develop the VSR process, to act during the welding process, that is, Vibration assisted welding (VAW), which can reduce most expenses related to post-weld vibrations or heat treatments [74].

4.4 Optimized welding processes

Aiming at reducing the values of deformations and RS's, the market has been promoting the improvement of welding processes, such as, for example, better stability in metal transfer, methods of controlling the waveform of sources, feedback of parameters during the process, reduction of heat input on the material to be welded, among these processes the CW-GMAW (CW)® stands out. It is a process derived from the GMAW process. The CW-GMAW process uses the feeding procedures to those of the GTAW process with automated feeding, it presents itself as an alternative to increase productivity without increasing the heat input in the melting arc/puddle system. The respective process uses the introduction of an additional wire, at room temperature, in the atmosphere of the arc, generated by the main wire, through an independent feeder and an injector connected to the welding torch [70, 75–78]. In detail, **Figure 9** illustrates the entry of the non-energized wire into the arc atmosphere in the CW-GMAW process.

It is possible to highlight several advantages of the CW-GMAW process over conventional processes, among which we have the one presented in the work of [70, 75]. The authors pointed out that the introduction of an additional wire improves the melting rate, tending to decrease the heat input to the workpiece, thus, there is a decrease in the values of distortions and consequently of RS. Distortions levels were compared using the GMAW, Surface Tension Transfer (STT) and CW-GMAW processes [70]. With the global deformation values, **Figure 10**, it was possible to observe that the CW-GMAW process obtained the lowest global distortion value, in comparison with the other investigated processes.

In another study, RS values were compared using two measurement techniques, X-ray diffraction and Acoustic Birefringence (BA), using the GMAW and CW-GMAW processes [63]. The analyzes were performed in simple deposition welds, on ASTM 131 grade AH32 naval steel plates, rigidly attached to a support to ensure a condition close to that found in real welding, simulating the dimensional restriction levels of a welded structure. After the deposition of the welds, the values of AB and XRD were measured at previously established points, in the regions of the base metal, heat-affected zone (HAZ) and weld metal (WM). Through **Figure 11** the measurements obtained by XRD showed that the use of the CW-GMAW process decreases the longitudinal stresses in the region near and in the weld bead, a difference was not observed for the transverse stresses. BA measurements showed that the difference between longitudinal and transverse residual stresses tends to decrease when using the CW-GMAW process, compared to the GMAW. These results suggest that the addition of an extra wire to the conventional process reduces the amount of heat supplied to the welded joint and, consequently, prevents the generation of residual welding stresses.

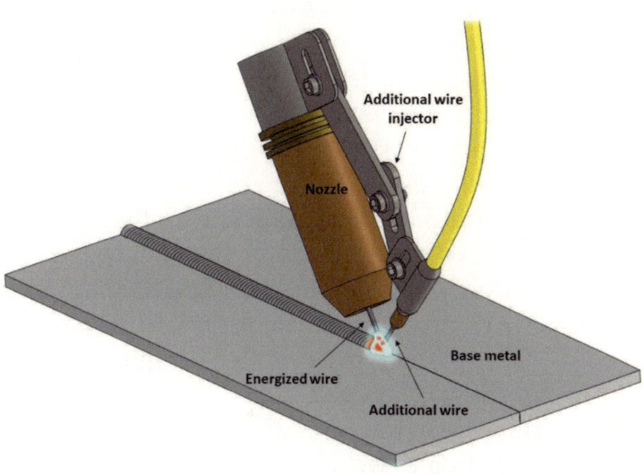

Figure 9.
CW-GMAW process scheme.

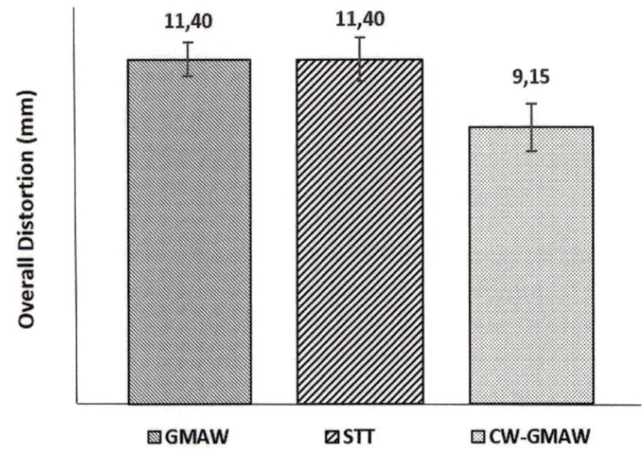

Figure 10.
Overall distortion of the experiments for each process.

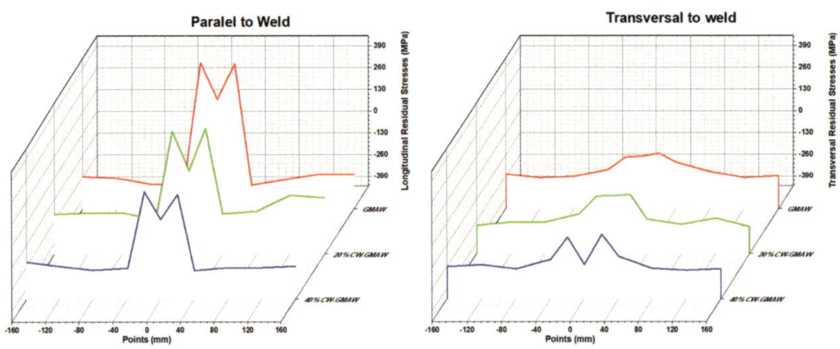

Figure 11.
Comparative between longitudinal and transverse RS with GMAW and CW-GMAW measuring by XRD.

The authors [68] presented evidence that the use of the CW-GMAW process reduces the residual stress and increases the resistance to fatigue, when compared to samples welded by the conventional GMAW process. The weld bead was made on plates with V joints of ASTM 131 grade A steel. Through a micrographic analysis it was shown that welding by the CW-GMAW process promoted a decrease in the amount of intergranular ferrite and an increase in hardness in the HAZ. SN fatigue resistance curves can be seen in **Figure 12**. Analysis of the results revealed that, for lower levels of reliability, joints manufactured using the GMAW process have a fatigue life at high voltage amplitude levels and greater fatigue life at lower stress amplitudes. However, when a higher level of confidence is considered, weldments made using the CW-GMAW process showed greater resistance to fatigue at both high and low amplitude stress levels.

The versatility and low cost of application of the CW-GMAW process is presented by [76], the authors developed a study on the viability of narrow bevel welding (Narrow Gap Welding-NGW). The tests showed an improvement in stability with the CW-GMAW process instead of the GMAW, and an increase in the melting rate, from 4.9 to 9.7 kg/h, promoting a complete filler weld with just 3 passes (root, filler and finishing), **Figure 13**.

Through high speed filming, in the GMAW process the arc attaches to the side wall, causing erosion and leading to welding discontinuities shown in **Figure 13e**, while for CW-GMAW welding this is not observed, **Figure 13a**. For a better understanding of the metal transfer mechanism [76], welds were deposited on a metal plate with the addition of increasing amounts of extra wire, shown in **Figure 14**.

DOI: http://dx.doi.org/10.5772/intechopen.93533

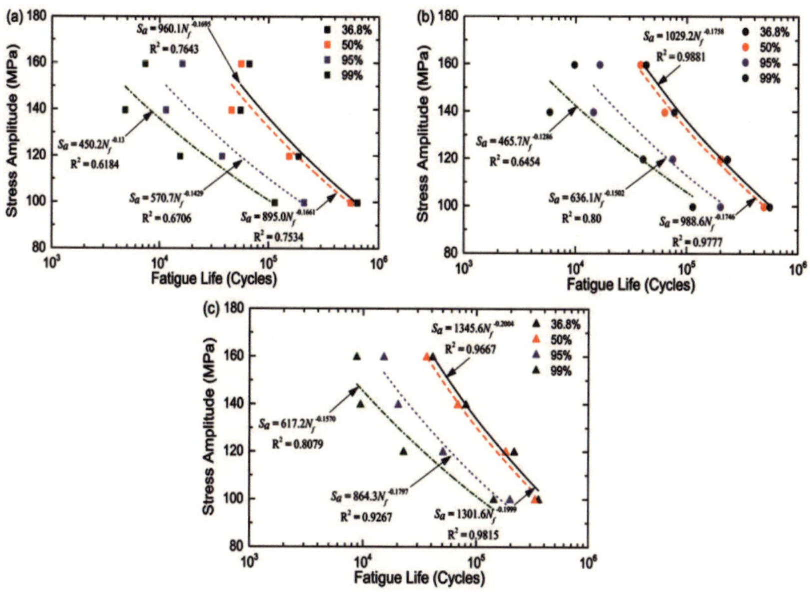

Figure 12.
Fatigue life for different levels of reliability: (a) GMAW, (b) CW-GMAW additional wire of 0.8 mm, and (c) CW-GMAW additional wire of 1.0 mm.

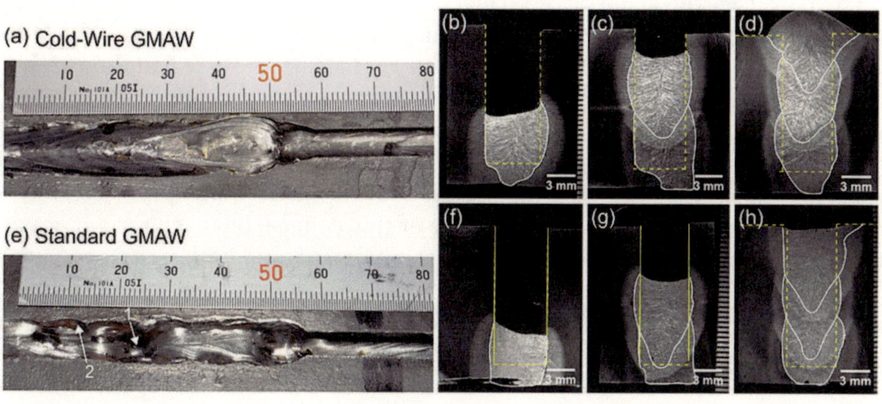

Figure 13.
Appearance and cross section of the bead for welds: (a-d) CW-GMAW; (e-h) GMAW.

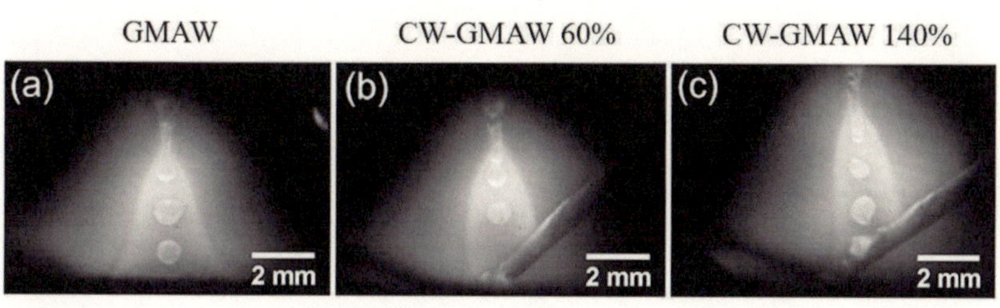

Figure 14.
Influence of the addition of additional wire on the cathode point of the arch: (a) represents the GMAW process, and (b) and (c) represent CW-GMAW welding with the addition of 60 and 140% additional wire, respectively.

It was found that as the amount of additional wire increases, the position fixing the arc changes from the weld pool to the additional wire, promoting greater stability during the welding process. When larger amounts of additional wire, 140% mass more is inserted into the system, the cathode point moves to the additional wire, which is being feed, shown in **Figure 14c**. This change in the cathode point for the additional wire allows for a more stable welding, thus there is no erosive effect on the wall base metal, the smallest variation in the heat input value for the CW-GMAW points to the stability of the process and its ability to increase the deposition rate without changing the welding heat input.

5. Conclusion

It is perceived that identifying, predicting and measuring residual stresses is not an easy labor, the notion of the behavior of residual stresses in a welded component, often implies not only in theoretical terms of their formation and generation, but also to discern which technique best fits to certain types of analysis, although, it is known that all techniques have their advantages and disadvantages. Thus, to help the authors involved in this area to adopt the technique that best suits their research, a broad approach on destructive and non-destructive techniques was explored. Therefore, reaching the conclusion that in addition to the behavior of residual stresses in the welded structure, other parameters must be considered when choosing the technique, for example, the thickness of the joint. Thus, the use of destructive and non-destructive techniques depends not only on having them for use, but also on the need for analysis. In general, non-destructive techniques are more widely used and acoustic birefringence (AB) appears as a promising technique with excellent results compared to more consolidated techniques such as X-ray diffraction. Despite some limitations, the depth of analysis, the ease handling and the low cost of the equipment are very attractive advantages when compared to other methods of the same class or not. Finally, it is clear that, while having knowledge of the mechanisms for generating residual stress by arc welding, its behavior in the weld structure, as well as the techniques used for its measurement, it is certainly more viable to weld without the generation of this voltage, but as this is not possible, a way to control or reduce this voltage is important, and within this perspective there are countless ways to do this, and one result that drew attention within this study was the welding process CW-GMAW, which in addition to reducing the levels of tensile stresses, generates less deformation in the structure. Therefore, in general, a study was successful in its objectives, which was to provide a range of relevant information on arc welding stresses to the community studying the topic.

Acknowledgements

We would like to thank CAPES for the scholarships that each one had during our study and FINEP for the financing of projects that helped us in the acquisition of the equipment and assembly of the laboratory. And we also thank Proderna (UFPA) for the opportunity to be part of an excellent research program.

Conflict of interest

The authors declare no conflict of interest.

DOI: http://dx.doi.org/10.5772/intechopen.93533

Author details

Lino Alberto Soares Rodrigues*, Ednelson da Silva Costa, Tárcio dos Santos Cabral and Eduardo Magalhães Braga
Postgraduate Program in Natural Resources Engineering in the Amazon (PRODERNA) - Federal University of Pará (UFPA), Belém-PA, Brazil

*Address all correspondence to: lino@ufpa.br

References

[1] Masubuchi K. Analysis of Welded Structures: Residual Stresses, Distortion, and their Consequences. Vol. 33. Elsevier; 2013. p. 650

[2] Withers PJ, Bhadeshia HKDH. Residual stress. Part 1-Measurement techniques. Materials Science and Technology. 2001;**17**(4):355-365. DOI: 10.1179/026708301101509980

[3] Totten GE. Handbook of Residual Stress and Deformation of Steel. Ohio, USA: ASM international; 2002. p. 499

[4] Withers PJ et al. Recent advances in residual stress measurement. International Journal of Pressure Vessels and Piping. 2008;**85**(3):118-127. DOI: 10.1016/j.ijpvp.2007.10.007

[5] Noyan IC, Cohen JB. Residual Stress: Measurement by Diffraction and Interpretation. New York, USA: Springer; 2013. p. 286

[6] Colegrove P et al. Welding process impact on residual stress and distortion. Science and Technology of Welding and Joining. 2009;**14**(8):717-725. DOI: 10.1179/136217109X406938

[7] Scotti A. Modelos de cinco barras e de uma barra para geração de tensões térmicas na ZF. ZAC e MB durante soldagem a arco. Soldagem & Inspeção. 2014;**19**(1):82-90

[8] Nitschke-Pagel T, Dilger K. Sources and consequences of residual stresses due to welding. Materials Science Forum. 2014;**783-786**:2777-2785

[9] Goldak J, Chakravarti A, Bibby M. A new finite element model for welding heat sources. Metallurgical Transactions B. 1984;**15**(2):299-305. DOI: 10.1007/BF02667333

[10] Dong P. Residual stresses and distortions in welded structures: A perspective for engineering applications. Science and Technology of Welding and Joining. 2005;**10**(4):389-398. DOI: 10.1179/174329305X29465

[11] Radaj D. Heat Effects of Welding: Temperature Field, Residual Stress, Distortion. Springer Science & Business Media; 1993. p. 367

[12] Gery D, Long H, Maropoulos P. Effects of welding speed, energy input and heat source distribution on temperature variations in butt joint welding. Journal of Materials Processing Technology. 2005;**167**:393-401. DOI: 10.1016/j.jmatprotec.2005.06.018

[13] Machado IG. Novos paradigmas para especificação de juntas soldadas. Soldagem & Inspeção. 2012;**17**(3):278-288. DOI: 10.1590/S0104-92242012000300012

[14] Khedmati MR, Ghavami K, Rastani M. A comparative study on three different construction methods of stiffened plates-strength behaviour and ductility characteristics. Rem: Revista Escola de Minas. 2007;**60**(2):365-379. DOI: 10.1590/S0370-44672007000200019

[15] Leggatt RH. Residual stresses in welded structures. International Journal of Pressure Vessels and Piping. 2008;**85**(3):144-151. DOI: 10.1016/j.ijpvp.2007.10.004

[16] Çam G, Özdemir O, Koçak M. Progress in low transformation temperature (LTT) filler wires. In: Proceedings of the 63rd Annual Assembly & International Conference of the International Institute of Welding. İskenderun, Hatay and Pendik, İstanbul, Turkey; 2010. pp. 759-765

[17] Thomas SH, Liu S. Analysis of low transformation temperature welding

(LTTW) consumables—Distortion control and evolution of stresses. Science and Technology of Welding and Joining. 2014;**19**(5):392-401. DOI: 10.1179/1362171814Y.0000000199

[18] Khurram A et al. Parametric study of welding temperature distribution in t-joint fillet weld using fem. In: Advanced Materials Research. Harbin, China: Trans Tech Publications Ltd; 2011. pp. 492-496. DOI: 10.4028/www.scientific.net/AMR.328-330.492

[19] Tusek J et al. Influence of type of welded joint on welding efficiency. Science and Technology of Welding and Joining. 2003;**8**(3):157-164

[20] Teng T-L, Lin C-C. Effect of welding conditions on residual stresses due to butt welds. International Journal of Pressure Vessels and Piping. 1998;**75**(12):857-864. DOI: 10.1016/S0308-0161(98)00084-2

[21] Teng T-L et al. Analysis of residual stresses and distortions in T-joint fillet welds. International Journal of Pressure Vessels and Piping. 2001;**78**(8):523-538. DOI: 10.1016/S0308-0161(01)00074-6

[22] Fu G et al. Influence of the welding sequence on residual stress and distortion of fillet welded structures. Marine Structures. 2016;**46**:30-55. DOI: 10.1016/j.marstruc.2015.12.001

[23] Reddy GM, Mohandas T. Influence of welding process and residual stress on ballistic performance. Journal of Materials Science Letters. 1996;**15**(18):1633-1635. DOI: 10.1007/BF00278111

[24] Schroepfer D, Kromm A, Kannengiesser T. Engineering approach to assess residual stresses in welded components. Welding in the World. 2017;**61**(1):91-106. DOI: 10.1007/s40194-016-0394-9

[25] Gurova T, Quaranta F, Estefen S. Monitoramento do estado das tensões residuais durante fabricação de navios. In 21º CONGRESSO NACIONAL DE TRANSPORTE AQUAVIÁRIO, CONSTRUÇÃO NAVAL E OFFSHORE. Anais... Rio de janeiro. 2006

[26] Teng T-L, Chang P-H, Tseng W-C. Effect of welding sequences on residual stresses. Computers & Structures. 2003;**81**(5):273-286. DOI: 10.1016/S0045-7949(02)00447-9

[27] Deng D. Influence of deposition sequence on welding residual stress and deformation in an austenitic stainless steel J-groove welded joint. Materials & Design. 2013;**49**:1022-1033. DOI: 10.1016/j.matdes.2013.02.065

[28] Mishchenko A, Scotti A. Tensões residuais em soldagem a arco: uma visão holística. Soldagem & Inspeção. 2018;**23**(1):93-112. DOI: 10.1590/0104-9224/si2301.10

[29] Francis JA, Turski M, Withers PJ. Measured residual stress distributions for low and high heat input single weld beads deposited on to SA508 steel. Materials Science and Technology. 2009;**25**(3):325-334. DOI: 10.1179/174328408X372074

[30] Hauk V. Structural and Residual Stress Analysis by Nondestructive Methods: Evaluation-Application-Assessment. Aachen, Germany: Institut für Werstoffkunde, Elsevier; 1997. p. 640

[31] Thomas N-P, Helmut W. Residual stresses in welded joints–sources and consequences. In: Materials Science Forum. Braunschweig, Germany: Trans Tech Publications Ltd; 2002. pp. 215-226. DOI: 10.4028/www.scientific.net/MSF.404-407.215

[32] Schajer GS. Practical Residual Stress Measurement Methods. Vancouver, Canada: University of British Columbia, John Wiley & Sons; 2013. p. 321

[33] Rossini NS et al. Methods of measuring residual stresses in

components. Materials & Design. Bari, Italy, Ireland: Politecnico di Bari, School of Mechanical & Dublin City University; 2012;**35**:572-588. DOI: 10.1016/j.matdes.2011.08.022

[34] Lu J. Handbook of Measurement of Residual Stresses. Fairmont Press; 1996. p. 245

[35] Kruth J-P, Bleys P. Measuring residual stress caused by wire EDM of tool steel. International Journal of Electrical Machining. 2000;**5**:23-28

[36] Kandil FA, Lord JD, Fry AT, Gran PV. A Review of Residual Stress Measurement Methods—A Guide to Technique Selection. Teddington, Middlesex, United Kingdom; 2001. Available from: http://eprintspublications.npl.co.uk/1873/

[37] Ruud CO. A review of selected non-destructive methods for residual stress measurement. NDT International. 1982;**15**(1):15-23. DOI: 10.1016/0308-9126(82)90083-9

[38] Webster GA, Wimpory RW. Polycrystalline Materials—Determinations of Residual Stresses by Neutron Diffraction. ISO/TTA3 Technology Trends Assessment. Vol. 20. Geneva: Imperial College - London; 2001

[39] Dann JA et al. A comparison between Engin and Engin-X, a new diffractometer optimized for stress measurement. Physica B: Condensed Matter. 2004;**350**(1-3):E511-E514. DOI: 10.1016/j.physb.2004.03.139

[40] Hilly ME. Residual Stress Measurement by X-Ray Diffraction. SAE Information Report. Vol. 784. Warrendale, Pennsylvania, USA; 1971

[41] Song W et al. Nondestructive testing and characterization of residual stress field using an ultrasonic method. Chinese Journal of Mechanical Engineering. 2016;**29**(2):365-371. DOI: 10.3901/CJME.2015.1023.126

[42] Prevey PS. X-ray diffraction residual stress techniques. In: ASM Handbook. Vol. 10. Metals Park. Ohio: ASM International; 1986. pp. 380-392

[43] Cullity BD. Elements of X-Ray Diffraction. Indiana, USA: University of Notre Dame, Addison-Wesley Publishing; 1956

[44] Raj B et al. X-ray diffraction based residual stress measurements for assessment of fatigue damage and rejuvenation process for undercarriages of aircrafts. Journal of Nondestructive Evaluation. 2009;**28**(3-4):157

[45] Hughes DS, Kelly JL. Second-order elastic deformation of solids. Physical Review. 1953;**92**(5):1145. DOI: 10.1103/PhysRev.92.1145

[46] Murnaghan FD. Finite Deformation of an Elastic Solid. São Paulo, Brazil: Instituto Tecnológico de Aeronáutica, Wiley; 1951

[47] Andrino MH. Avaliação de Tensões Residuais em Soldas de Dutos Utilizando o Efeito Acustoelástico. São Paulo, Brazil: Universidade de Campinas (UNICAMP); 2003

[48] Ricardo dos Santos F. Avaliação da Profundidade de Penetração de Ondas Longitudinais Criticamente Refratadas. São Paulo, Brazil: Universidade de Campinas (UNICAMP); 2007

[49] Santos CS et al. Avaliação da Influência de Variáveis envolvidas no Comportamento da Velocidade de Propagação e Tempo de Percurso das Ondas Lcr utilizando Planejamento Experimental. In: Conferência sobre Tecnologia de Equipamentos (COTEQ). Porto de Galinhas: Anais... Recife; 2011

[50] Egle DM, Bray DE. Measurement of acoustoelastic and third-order

elastic constants for rail steel. The Journal of the Acoustical Society of America. 1976;**60**(3):741-744. DOI: 10.1121/1.381146

[51] Tanala E et al. Determination of near surface residual stresses on welded joints using ultrasonic methods. NDT&E International. 1995;**28**(2):83-88. DOI: 10.1016/0963-8695(94)00013-A

[52] Xu C et al. Nondestructive testing residual stress using ultrasonic critical refracted longitudinal wave. Physics Procedia. 2015;**70**:594-598. DOI: 10.1016/j.phpro.2015.08.030

[53] Hsu NN. Acoustical birefringence and the use of ultrasonic waves for experimental stress analysis. Experimental Mechanics. 1974;**14**(5):169-176. DOI: 10.1007/BF02323061

[54] Okada K. Stress-acoustic relations for stress measurement by ultrasonic technique. Journal of the Acoustical Society of Japan (E). 1980;**1**(3):193-200. DOI: 10.1250/ast.1.193

[55] Allen DR, Sayers CM. The measurement of residual stress in textured steel using an ultrasonic velocity combinations technique. Ultrasonics. 1984;**22**(4):179-188. DOI: 10.1016/0041-624X(84)90034-9

[56] Clark AV Jr. On the use of acoustic birefringence to determine components of plane stress. Ultrasonics. 1985;**23**(1):21-30. DOI: 10.1016/0041-624X(85)90007-1

[57] Hirao M, Ogi H, Fukuoka H. Advanced ultrasonic method for measuring rail axial stresses with electromagnetic acoustic transducer. Research in Nondestructive Evaluation. 1994;**5**(3):211-223. DOI: 10.1007/BF01606409

[58] Schneider E. Ultrasonic birefringence effect—Its application for materials characterisations. Optics and Lasers in Engineering. 1995;**22**(4-5):305-323. DOI: 10.1016/0143-8166(94)00032-6

[59] Ortega LPC et al. Introdução à avaliação de tensões por ultrassom. Rio de Janeiro, RJ: Editora Virtual Científica; 2011

[60] Hirao M, Ogi H. Electromagnetic Acoustic Transducers. Tokyo, Japan: Springer; 2017

[61] Iwashimizu Y, Kubomura K. Stress-induced rotation of polarization directions of elastic waves in slightly anisotropic materials. International Journal of Solids and Structures. 1973;**9**(1):99-114. DOI: 10.1016/0020-7683(73)90035-8

[62] Marcelo de Siqueira Queiroz B. Desenvolvimento de Um Sistema de Medida de Tempo Decorrido da Onda Ultra-Sônica e Análise do Estado de Tensões em Materiais Metálicos Pela Técnica da Birrefiingência Acústica. Rio de Janeiro, Brazil: Instituto de Engenharia Nuclear; 2000

[63] Costa ES et al. Residual stresses in cold-wire gas metal arc welding. Science and Technology of Welding and Joining. 2017;**22**(8):706-713. DOI: 10.1080/13621718.2017.1306014

[64] Okumura T, Taniguchi C. Engenharia de soldagem e aplicações. Japan, Brazil: University of Tokio, University of São Paulo; 1982. p. 493

[65] Michaleris P. Minimization of Welding Distortion and Buckling: Modelling and Implementation. 1st ed. Philadelphia, PA, USA: Woodhead Publishing Limited; 2011. p. 316

[66] Kou S. Welding metallurgy. New Jersey, USA: John Wiley & Sons; 2003. p. 446

[67] Pilipenko A. Computer simulation of residual stress and distortion of thick

plates in multielectrode submerged arc welding: Their mitigation techniques [thesis]. Trondheim: Norwegian University of Science and Technology; 2001

[68] Marques LFN et al. Fatigue life assessment of weld joints manufactured by GMAW and CW-GMAW processes. Science and Technology of Welding and Joining. 2017;**22**(2):87-96. DOI: 10.1080/13621718.2016.1194735

[69] Rodrigues LAS et al. Welding procedures influence analysis on the residual stress distribution and distortion of stiffened panels welded via robotized FCAW. Thin-Walled Structures. 2019;**141**:175-183. DOI: 10.1016/j.tws.2019.03.055

[70] Cabral T d S, Braga EM, Mendonça EAM, Scott A. Influence of procedures and transfer modes in MAG welding in the reduction of deformations on marine structure panels. Welding International. 2015;**29**:928-936. DOI: 10.1080/09507116.2014.932993

[71] Feng Z. Processes and Mechanisms of Welding Residual Stress and Distortion. Ohio, USA: The Ohio State University, Elsevier; 2005. p. 354

[72] Tsai CL, Park SC, Cheng WT. Welding distortion of a thin-plate panel structure. Welding Journal-New York. 1999;**78**:156-s

[73] Chen BQ, Guedes Soares C. Effect of welding sequence on temperature distribution, distortions, and residual stress on stiffened plates. International Journal of Advanced Manufacturing Technology. 2016;**86**:3145-3156. DOI: 10.1007/s00170-016-8448-0

[74] Jose MJ, Kumar SS, Sharma A. Vibration assisted welding processes and their influence on quality of welds. Science and Technology of Welding and Joining. 2016;**21**:243-258. DOI: 10.1179/1362171815Y.0000000088

[75] Ribeiro RA, Santos E, Assunção P, Maciel R, Braga E. Predicting weld bead geometry in the novel CW GMAW process. Welding Journal. 2015;**94**:301s-311s

[76] Assunção P, Ribeiro RA, Dos Santos EBF, Gerlich AP, de Magalhães Braga E. Feasibility of narrow gap welding using the cold-wire gas metal arc welding (CW-GMAW) process. Welding in the World. 2017;**61**:659-666. DOI: 10.1007/s40194-017-0466-5

[77] Assunção P, Ribeiro RA, Moreira PMGP, Braga EM, Gerlich AP. A preliminary study on the double cold wire gas metal arc welding process. International Journal of Advanced Manufacturing Technology. Ohio, USA: The Ohio State University; 2020;**106**:5393-5405. DOI: 10.1007/s00170-020-05005-6

[78] Ribeiro RA, Dos Santos EBF, Assunção P, Braga EM, Gerlich AP. Cold wire gas metal arc welding: Droplet transfer and geometry. Welding Journal. 2019;**98**:135-149. DOI: 10.29391/2019.98.011